TRACTOR

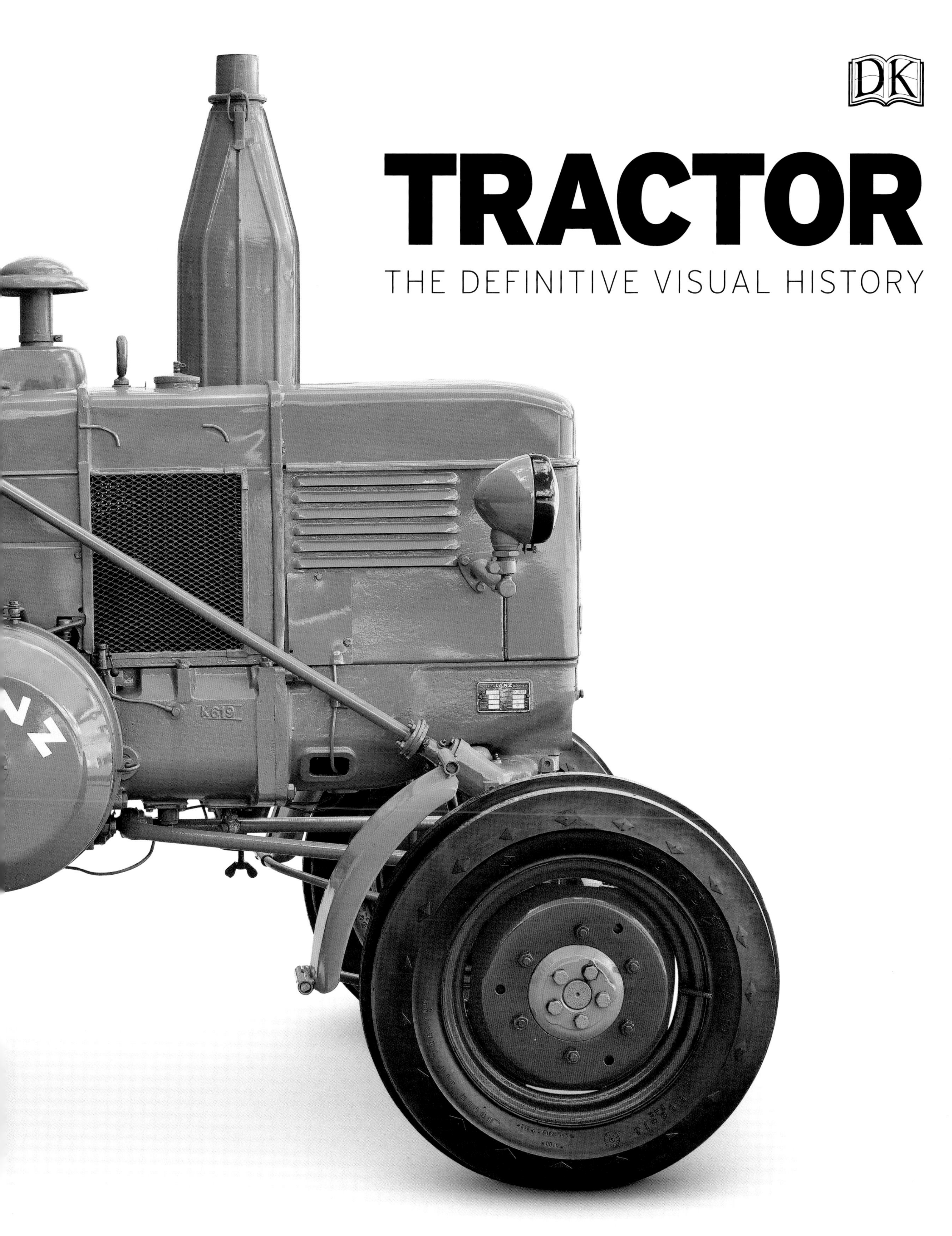

DK

TRACTOR

THE DEFINITIVE VISUAL HISTORY

DORLING KINDERSLEY
Senior Editor Jemima Dunne
Senior Art Editor Sharon Spencer
Editors Sam Kennedy, Miezan van Zyl
US Editors Christine Heilman, Margaret Parrish
Design Assistance Renata Latipova
Photographer Gary Ombler
Picture Research Nic Dean
DK Picture Library Claire Bowers, Claire Cordier, Romaine Werblow
Jacket Designer Mark Cavanagh
Jacket Assistant Claire Gell
Jacket Design Development Manager Sophia MTT
Producer, Pre-Production Adam Stoneham
Producer Linda Dare
Managing Editor Esther Ripley
Managing Art Editor Karen Self
Art Director Phil Ormerod
Associate Publishing Director Liz Wheeler
Publishing Director Jonathan Metcalf

DK INDIA
Senior Editor Sreshtha Bhattacharya
Senior Art Editor Anjana Nair
Project Editor Suparna Sengupta
Art Editors Namita, Pallavi Narain
Managing Editor Pakshalika Jayaprakash
Managing Art Editor Arunesh Talapatra
Production Manager Pankaj Sharma
Pre-production Manager Balwant Singh
Senior DTP Designer Sachin Singh
DTP Designers Vishal Bhatia, Nand Kishor Acharya, Mohammad Usman
Picture Researcher Aditya Katyal
Picture Research Manager Taiyaba Khatoon

Contributors Stuart Gibbard, Simon Henley, Peter Longfoot, Steve Mitchell, Michael Williams, David Williams, Martin Rickatson

First American Edition, 2015

Published in the United States by
DK Publishing, 345 Hudson Street, New York, New York 10014

15 16 17 18 19 10 9 8 7 6 5 4 3 2 1
001—274736—May 2015

Published in Great Britain by Dorling Kindersley Limited.

A catalog record for this book is available from the Library of Congress.

ISBN: 978-1-4654-3599-6

DK books are available at special discounts when purchased in bulk for sales promotions, premiums, fund-raising, or educational use. For details, contact: DK Publishing Special Markets, 345 Hudson Street, New York, New York, 10014 or SpecialSales@DK.com.

Printed and bound in China by Leo Paper Products Limited

A WORLD OF IDEAS:
SEE ALL THERE IS TO KNOW

www.dk.com

Contents

1900–1920: THE EARLY YEARS

Pioneers began developing agricultural motors in the late nineteenth century. Later, vast prairie giants ruled the North American plains while smaller "agrimotors" began to replace the horse in Europe. World War I convinced the initially skeptical farming community of the tractor's utility.

1921-1938: COMING OF AGE

Despite worldwide economic depression, this was an era of invention, in which Henry Ford pioneered new methods of construction and Harry Ferguson's three-point linkage and hydraulic lift revolutionized tractor design.

1939-1951: WAR AND PEACE

World War II sparked a massive demand for tractors to perform a military and not just agricultural role. Production rates soared but there was little evolution in design over the course of the war. The food shortages of the postwar years stimulated a boom in the industry, and new ranges of styled tractors began to appear in the US.

1952-1964: GOLDEN AGE

A parade of new models with improved features was introduced as new manufacturers entered the market and farms became increasingly mechanized. With postwar

shortages resolved, farmers demanded more and more from their machines and their dealerships. The race for higher power and greater productivity had begun.

1965–1980: THE NEW GENERATION

Tractor manufacturers began to aim at markets around the globe. A greater focus on operator comfort saw the introduction of safety and quiet cabs. Demand for power meant larger and more reliable machines but as efficiency rose, fewer tractors were needed and sales started to fall.

1981–2000: THE NEW TECHNOLOGY

High-speed tractors, powershift transmissions, and computerized controls defined a period of technological advancement. In the face of difficult economic realities, manufacturers consolidated into global corporations.

AFTER 2000: 21ST CENTURY

Stringent emission standards and the demands of modern agriculture have led to a greater focus on fuel efficiency. Precision farming, aided by satellite technology, ensures that tractors operate at the very highest levels.

HOW TRACTORS WORK: TRACTOR TECHNOLOGY

Tractors are exceptionally complex machines, set apart by their extraordinary hydraulics and the power of their engines. This chapter explains the basics of tractor engineering and provides an overview of the most important historical evolutions and improvements.

The Tractor Revolution

Gold! Gold! Gold! These three words, shouted through the streets of San Francisco after the discovery of gold in the American River, sparked the California Gold Rush. Fortune seekers from around the country and the world flocked to the then sleepy town of San Francisco, which swelled in population from 1,000 in 1847 to more than 25,000 by 1850, with far-reaching consequences.

To begin the story of the tractor in North America with the discovery of a gold nugget on January 24, 1848, might seem like a strange place to start. Yet it was the mass migration of people during the Gold Rush—who came to California with their skills, ideas, and needs—that created the ideal conditions for the advent of modern farming.

Gold drew the masses west, but few struck it rich, and many were forced to return to their trades or to farming to earn a living. And from necessity came invention. Instead of finding gold in the form of metal, some cultivated another form of gold—wheat—which was planted and harvested in the fertile areas around San Francisco to feed the expanding population.

With this sudden influx of people also came an influx of farming technology: advances were made in the manufacture of plows, wagons, threshers, reapers, harvesters, and tractors. More than 60 tractor companies, among them Caterpillar, got their start in California. Not all of these companies survived, but they all helped to make California one of the world's largest suppliers of food and agricultural commodities.

This burst of ingenuity would not have been possible had it not been for the improvements made in metallurgy during the Industrial Revolution of the previous century. Innovators had laid considerable groundwork: John Deere introduced the steel plow in 1836; Silas Nye developed the Nye plow in 1855; and Scottish inventor James Watt, father of the steam engine, had, by 1782, already coined horsepower as the unit of measurement of power by which the tractor is measured.

What can be dubbed the country's "Black Gold Rush" also happened in the latter half of the 19th century. During this period, oil, improved fuels and internal combustion engines began to take on a greater importance, first in automobiles, and later in tractors. Abraham Cesner, a Canadian geologist, produced an illuminating oil he called kerosene in 1845, while in 1859 Edwin Laurencine Drake, founder of the petroleum industry, drilled the first oil well at Titusville, Pennsylvania. The four-stroke combustion engine developed by Nicolaus August Otto in Germany in 1876 was soon known worldwide, and by 1877 the "Otto Silent" was widely used throughout the United States.

It was John Froelich, of Waterloo, Iowa, who in 1892 produced the first gas traction engine that combined all the essential elements: internal combustion engine, power train, clutch, an engaging forward and reverse gear, manual steering, and a drawbar, while Hart-Parr of Iowa was the first American company to manufacture gas traction engines on a commercial scale.

"The **farm tractor**... is a mighty **industry** in the **making.**"

BARTON W. CURRIE, US JOURNALIST AND INDUSTRY COMMENTATOR, *THE COUNTRY GENTLEMAN*, 1916

The mass movement of people is almost always a catalyst for change. The Homestead Act of 1862 is considered one of the most important pieces of legislation in the United States. In signing it, President Abraham Lincoln turned over vast tracts of public land to private citizens. Ten percent of the land in the United States—270 million acres—was claimed and settled by homesteaders, primarily in the open prairies of the Midwest. These great expanses of farmland—today's breadbasket—were ideally suited for mechanized farming.

The outbreak of war in Europe in 1914 further fueled tractor innovation, as horses were drafted into the war effort. Industrialist Henry Ford introduced a highly popular, small, and inexpensive tractor called the Fordson during World War I. The war also called for specialized tractor adaptations: Pliny Holt made gun mounts for tractors, and R. G. LeTourneau developed earthmoving equipment for road- and dam-building.

In 1918 the number of US tractor makers was at its peak. With the Great Depression that followed the 1929 stock-market crash, many small companies went under, with others swallowed up by larger makers. Seven manufacturers, among them International Harvester and John Deere, emerged as the dominant players.

With the outbreak of World War II, the nation again faced labor shortages, creating tremendous pressure for farm mechanization and rapid technological advances. The peak year of tractor production was 1951, with the market approaching saturation shortly thereafter. American manufacturers responded by producing ever-larger tractors, as farms were also growing in size. By 1963 John Deere had overtaken International Harvester in a declining market, and it is still the largest presence in agricultural equipment today.

The story of the tractor is ongoing. Today's tractor is a sophisticated beast, incorporating the most up-to-date electronics, computers, data communications, and satellite guidance systems. The engines are finely engineered to deliver maximum output on minimum fuel, while meeting the latest emissions regulations. A transmission is no longer a simple box of cogs, but can be semi-automatic or constantly variable. Powerful hydraulic systems are optimized for efficiency, while modern cabs are not just utilitarian workplaces, but comfortable environments with ergonomic seating and controls. The machines are a far cry from the early, simple iterations of a tractor.

The tractor is recognized as one of the three greatest influences on farming in the 20th century—along with pesticides and plant breeding—and the primary tool that has made farming on such a massive scale possible. As the world's population continues to grow, the evolution of the tractor will continue as well. Its main purpose, however, will remain the same: harvesting food that feeds the world.

LORRY DUNNING

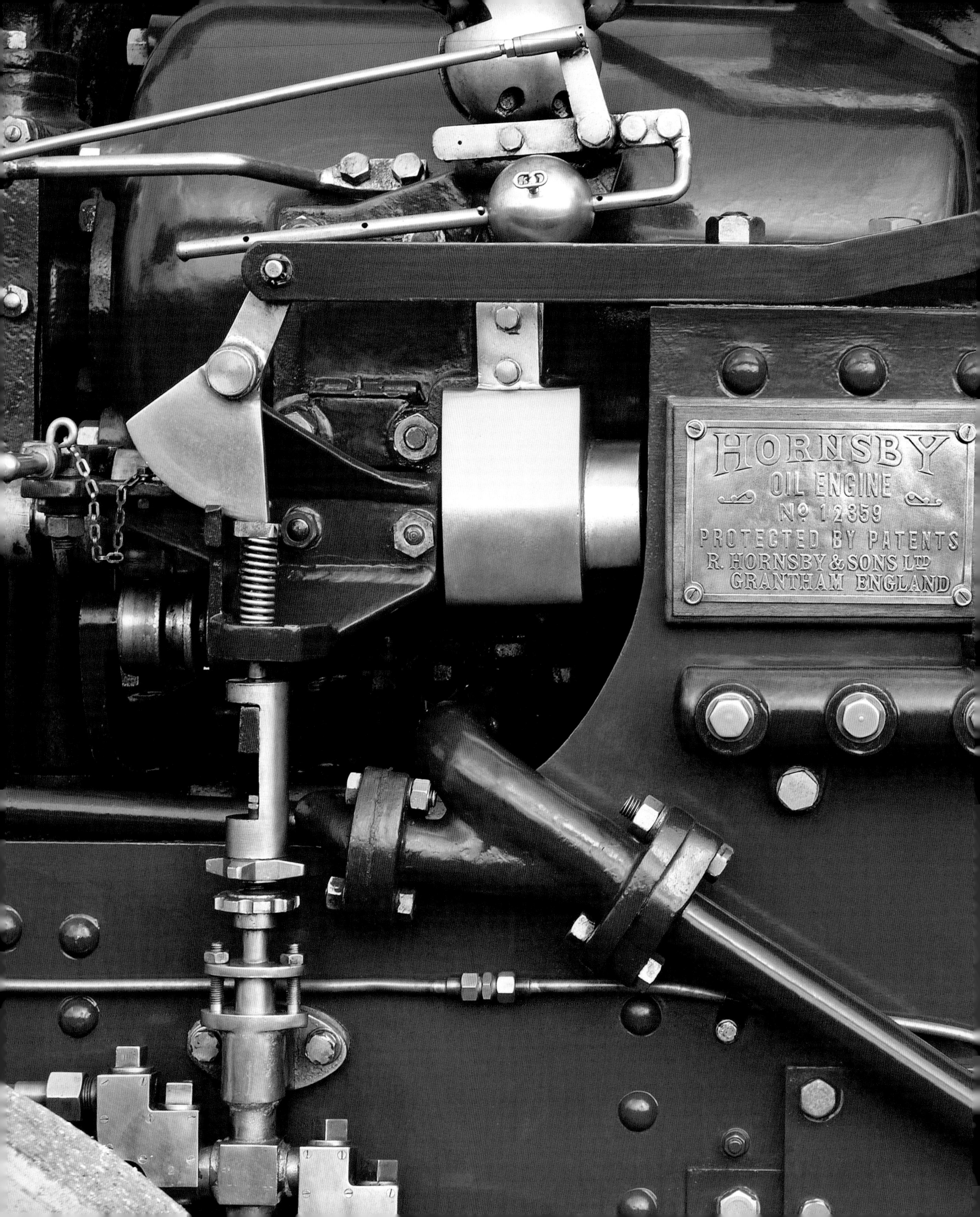
HORNSBY
OIL ENGINE
Nº 12359
PROTECTED BY PATENTS
R. HORNSBY & SONS LTD
GRANTHAM ENGLAND

1900–1920

THE EARLY YEARS

TWIN CITY TRACTORS

THE EARLY YEARS

By the end of the 19th century, several pioneers had explored the concept of what they called an "agricultural motor," but none had gone beyond the prototype stage. In 1901, Hart-Parr of Iowa developed an "agricultural gasoline traction engine"—16 of which were sold over the next couple of years, allowing their manufacturer to justifiably claim to be founders of the tractor industry. Hart-Parr's "No. 1" model was not the first tractor built, but it was the first to enjoy commercial success.

△ **Simms agricultural motor**
Probably the earliest British gasoline tractor, this "agrimotor" was developed by motor industry pioneer Frederick R. Simms in 1902.

Once the tractor industry got off the ground, US manufacturers blazed a different trail than those in the UK and Europe. In North America, the priority was to produce big gasoline tractors for breaking the prairies, similar to the steam engines that had gone before them. Across the Atlantic, the fledgling tractor manufacturers built lightweight "agrimotors" for farmers who wanted to replace their horses with something more progressive.

Eventually, the prairie tractor boom went bust due to oversupply and a declining market as the great ranches were subdivided into smaller holdings. By 1910 the US manufacturers had also turned their attention to lightweight machines, but the tractor was far from being an instant success. Most farmers viewed these mechanical contrivances with trepidation, regarding them as expensive novelties or even potentially dangerous contraptions. It would take a cataclysmic event, World War I, to convince them of the contribution that motorized farming could make to safeguarding food supplies. The conflict hastened tractor development, and the arrival of Henry Ford's Fordson Model F in 1917 set the benchmark that all future designs would follow.

"To lift **farm drudgery** off flesh and blood and lay it on **steel and motors** has been my most **constant ambition.**"

HENRY FORD (1863-1947)

◁ **A catalog issued in 1919** by Minneapolis Steel & Machinery Co. for its big Twin City tractors.

Key events

▷ **1889** The Charter Gas Engine Co. of Chicago mounts an engine on a Rumely steam engine chassis; it works, but is not a practical farm tractor.

▷ **1892** John Froelich of Waterloo, IA, builds an experimental tractor.

▷ **1896** Richard Hornsby & Sons of Grantham constructs UK's first tractor with a compression-ignition engine.

▷ **1901** Hart-Parr introduces its "No. 1" gasoline traction engine.

▷ **1902** Dan Albone forms Ivel Agricultural Motors Ltd. to produce farm tractors.

▷ **1905** Henry Ford builds experimental tractor using automobile components.

▷ **1907** German firm Deutz introduces its first tractor, the *Pfluglokomotive*.

▷ **1908** The first Winnipeg tractor trials are organized in Canada. The event was held annually until 1913.

▷ **1914** The Bull tractor, priced at $395, outsells all machines on the US market.

▷ **1916** German U-boat attacks on Allied shipping convince the UK that it needs tractors to increase food production. The first consignment of Fordson tractors is shipped from Dearborn, MI, to the UK the following year.

▷ **1920** The Lincoln tractor trials, held in England, attract machines from the UK, the US, Italy, and Switzerland.

△ **Plowing the prairies**
Photographed in Asquith, Saskatchewan, Canada, in 1912, this "five-bottom" Parlin & Orendorff plow is hitched to an International Titan Type D tractor.

Early Power

From the middle of the 19th century, the steam engine in its various forms slowly began to augment the horse as a source of power on farms around the world. The first farm task to make use of it was threshing; then attention was turned to cultivation by steam. On the dry soils of the Americas, certain parts of Europe, and Russia, direct traction was the favored method. On the moist soils of the UK, Germany, and Eastern Europe, the heavy steam engine could not be used without damage to the soil structure, and steam cable engines were employed until they were replaced by the crawler tractor.

△ **Marshall traction engine**
Date 1908 **Origin** UK
Engine Marshall
Cylinders Single
Horsepower 7 nhp (nominal horsepower)

Marshall had a very loyal customer base for its engines and threshing machines. Traction engine production continued into WWII, although numbers produced were small.

◁ **Port Huron**
Date 1915 **Origin** UK
Engine Port Huron
Cylinders Tandem compound
Horsepower 65 hp

The layout of the Port Huron traction engines was similar to that adopted by most US makers. The cylinder was mounted to one side of the boiler, driving on to a disc crank with the flywheel at the other end. Tanks on the driver's footplate carried water for the boiler.

▽ **Avery Under-type**
Date 1911 **Origin** US
Engine Avery
Cylinders 2
Horsepower 40 hp

Though unconventional for US steam engine builders, this tractor was built using the best materials and engineering practices. While expensive, it gained a very good reputation during the brief reign of the steam traction engine on the prairies.

▽ Wallis & Steevens

Date 1916 **Origin** UK
Engine Wallis & Steevens
Cylinders Single
Horsepower 7 nhp

Wallis & Steevens introduced the expansion engine, which had a type of valve gear that was unusual for traction engines. When in good condition, it did give some economy of operation, but when worn out, the effect was negligible, outweighing the extra cost and complication.

△ Clayton & Shuttleworth

Date 1919 **Origin** UK
Engine Clayton & Shuttleworth
Cylinders Single
Horsepower 7 nhp

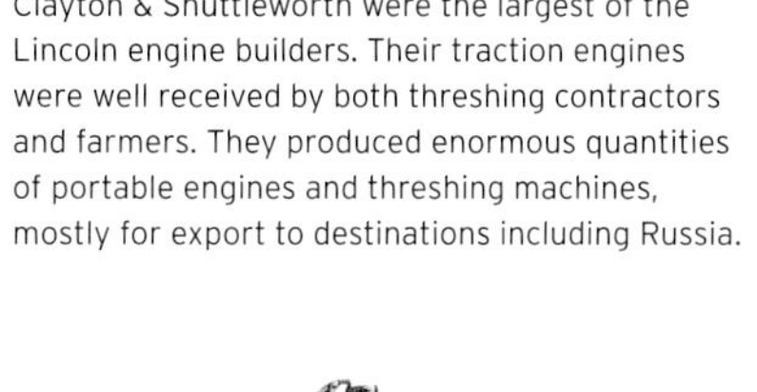

Clayton & Shuttleworth were the largest of the Lincoln engine builders. Their traction engines were well received by both threshing contractors and farmers. They produced enormous quantities of portable engines and threshing machines, mostly for export to destinations including Russia.

△ Fowler K7

Date 1919 **Origin** UK
Engine Fowler
Cylinders Compound
Horsepower 12 nhp

The K7 was the smallest Fowler cable plowing engine in general use. Since it required the same number of men to operate a set of small engines as it did for the largest set, smaller engines were not generally economical. Early examples of K7s were usually rated at 10 nhp, while the later ones rated at 12 nhp.

◁ Fowler BB1

Date 1920 **Origin** UK
Engine Fowler
Cylinders Compound
Horsepower 16 nhp

The BB1 was one of the most successful cable plowing engines made in the UK. Originally designed by Fowler in 1917 to Ministry of Munitions specifications, 46 sets were delivered to help with food production in 1918. BB1s were built until the mid-1920s.

Plowing with an Ivel tractor c.1906

Great Manufacturers
Ivel Agricultural Motors

British farmers had a sworn aversion to mechanical complexities, and the internal combustion engine was regarded as a fearful contraption. Despite this background of negativity, Dan Albone's Ivel tractor found its supporters and went on to become an early export success.

SEVERAL NAMES ARE associated with early tractor development in the UK, including Professor John Scott, Frederick Simms, Drake & Fletcher of Kent, Sharps of York, and Herbert Saunderson of Elstow in Bedfordshire. However, it was another Bedfordshire pioneer, Dan Albone, who first broke through the conservatism of the farming community and its allegiance to the horse.

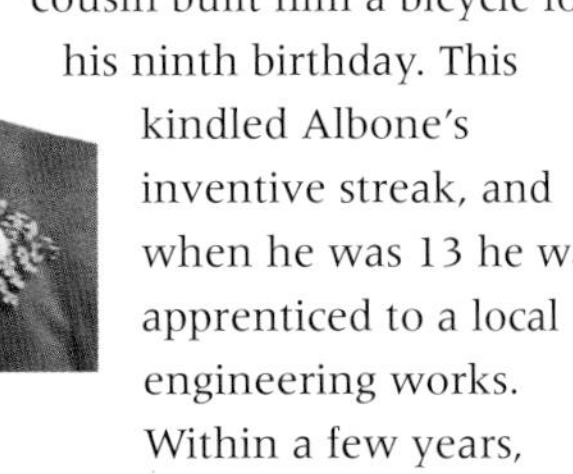
Dan Albone (1860–1906)

Albone's success with his Ivel tractor was undoubtedly down to the man himself. He realized that the design had to be simple and compact to succeed, and he seized every opportunity to underline the tractor's capabilities, organizing demonstrations and creating sales where others had failed.

Born on September 12, 1860, Albone was the youngest of a family of eight. His father was a farmer, innkeeper, and carpenter. The Albone family lived in Biggleswade at the Ongley Arms in Shortmead Street, which was part of the Great North Road.

Albone's affinity with things mechanical began when his cousin built him a bicycle for his ninth birthday. This kindled Albone's inventive streak, and when he was 13 he was apprenticed to a local engineering works. Within a few years, Albone was constructing his own bicycles and had become a well-known racing cyclist.

In 1880, Albone acquired the premises next door to the Ongley Arms, turning the building into the Ivel Hotel, named after the Ivel River, which flowed nearby, as a meeting center for cyclists. The yard attached to the property became the site of the Ivel Cycle Works, where Albone manufactured bicycles and ball bearings before progressing to motorized vehicles.

After a period experimenting with motorcycles, powered tricycles, and even a car, Albone began developing a farm tractor. It was a simple, three-wheel design with a single-speed (forward or reverse) transmission. The engine, a two-cylinder horizontal unit of 24 hp, was supplied by a fellow cyclist, Walter Payne of Coventry.

Export sales
An Ivel tractor being crated at the Ivel Works in Biggleswade for shipment to Russia via the port of Riga. By 1908, the company was exporting to 22 countries worldwide.

Sales promotion
This lavish brochure for the Ivel was issued in 1913 after sales had begun to fall.

The first tractor was completed in 1902, and Ivel Agricultural Motors Ltd. was formed the same year. Albone had a gift for promoting his products, and he demonstrated his Ivel tractor widely. He also had influential friends, including the motoring pioneer Selwyn Francis Edge and the racing driver Charles Jarrott, who joined the company as directors.

Within a few years, the Ivel was being exported worldwide. Its success looked assured, but a twist of fate would alter the course of events. On October 30, 1906, Dan Albone died from a seizure brought on by being struck by lightning while demonstrating his tractor at night in a thunderstorm. His death at the age of 46 left the company in limbo. Without his inventive streak and inspired leadership, many planned projects, including a larger Ivel tractor, a two-speed transmission, and a "new pattern" engine, were shelved.

> "The Ivel agricultural motor is **light** and **handy** ... a new field of activity [is] **opening.**"
>
> *DAILY TELEGRAPH*, SEPTEMBER 5, 1902

The Ivel tractor was awarded numerous prizes at agricultural shows, and was widely feted by the local and national press. Each demonstration was well attended by a large crowd eager to see this latest mechanical contrivance in action. Orders eventually began to flow in from home and abroad.

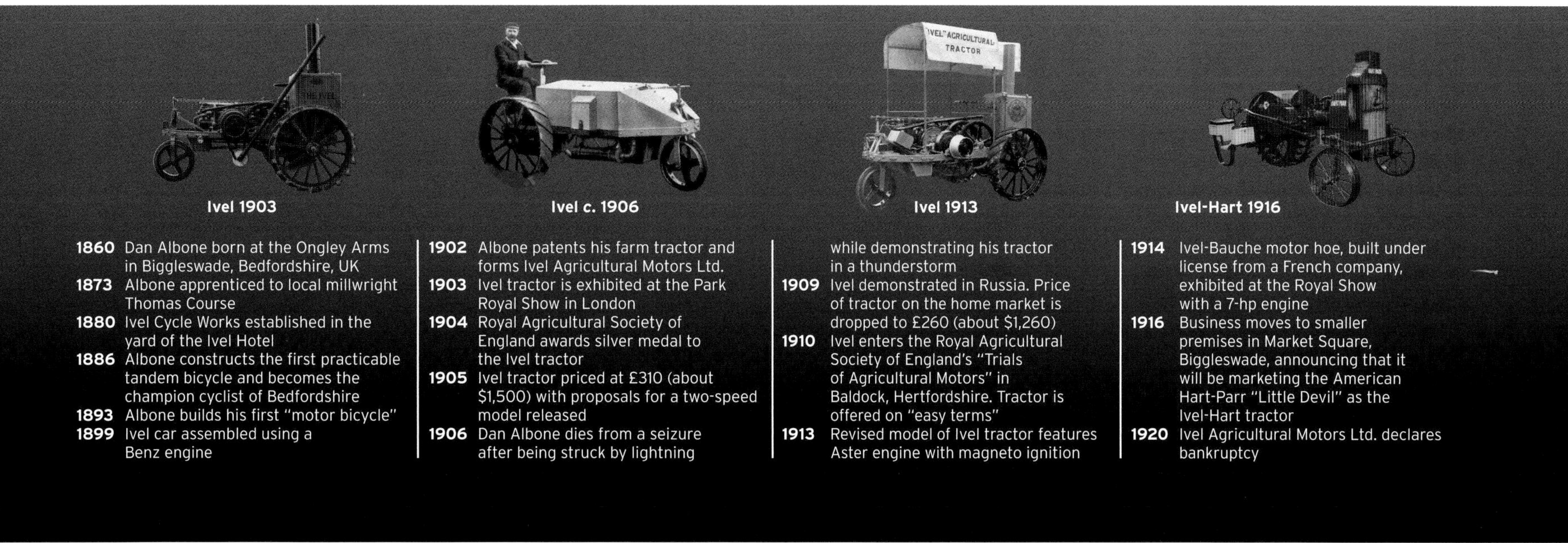

Ivel 1903

Ivel *c.* 1906

Ivel 1913

Ivel-Hart 1916

1860 Dan Albone born at the Ongley Arms in Biggleswade, Bedfordshire, UK
1873 Albone apprenticed to local millwright Thomas Course
1880 Ivel Cycle Works established in the yard of the Ivel Hotel
1886 Albone constructs the first practicable tandem bicycle and becomes the champion cyclist of Bedfordshire
1893 Albone builds his first "motor bicycle"
1899 Ivel car assembled using a Benz engine
1902 Albone patents his farm tractor and forms Ivel Agricultural Motors Ltd.
1903 Ivel tractor is exhibited at the Park Royal Show in London
1904 Royal Agricultural Society of England awards silver medal to the Ivel tractor
1905 Ivel tractor priced at £310 (about $1,500) with proposals for a two-speed model released
1906 Dan Albone dies from a seizure after being struck by lightning while demonstrating his tractor in a thunderstorm
1909 Ivel demonstrated in Russia. Price of tractor on the home market is dropped to £260 (about $1,260)
1910 Ivel enters the Royal Agricultural Society of England's "Trials of Agricultural Motors" in Baldock, Hertfordshire. Tractor is offered on "easy terms"
1913 Revised model of Ivel tractor features Aster engine with magneto ignition
1914 Ivel-Bauche motor hoe, built under license from a French company, exhibited at the Royal Show with a 7-hp engine
1916 Business moves to smaller premises in Market Square, Biggleswade, announcing that it will be marketing the American Hart-Parr "Little Devil" as the Ivel-Hart tractor
1920 Ivel Agricultural Motors Ltd. declares bankruptcy

Ivel Agricultural Motors Ltd. soldiered on, making minor tweaks to the design, including changing the engine supplier from Payne to Aster, a French concern. Ivel was one of just four manufacturers, including two steam engine builders, to enter the Royal Agricultural Society of England's 1910 trials in Baldock, Hertfordshire. The Biggleswade manufacturer acquitted itself well, but the gold medal was awarded to a McLaren steam tractor.

The other motor tractor producer at the trials was Herbert Saunderson, whose designs were beginning to overtake those of Ivel, now in danger of becoming outdated. After the latest US machines were imported in large numbers during World War I, Ivel Agricultural Motors Ltd. simply could not compete on performance or on price.

Sales of the Ivel tractor began to dwindle, and the company moved into smaller premises. Sidelines in building French Bauche and US Hart-Parr machines under license came to nothing as the business ran into debt. In 1920 Ivel Agricultural Motors Ltd. declared bankruptcy.

Threshing demonstration
An Ivel tractor is shown driving a threshing machine for a local contractor at a public demonstration staged by Dan Albone on a farm near Biggleswade in 1904.

Pioneer Machines

The early years of tractor development were a time for new ideas. Some designers chose three wheels, others preferred four, and tracklayers added another option. There were heavyweight models to compete with steam traction engines, while other tractor pioneers, particularly in the UK, developed lighter, more versatile designs to replace horses. Kerosene and gasoline engines dominated the market, and there was little enthusiasm initially for the first semi-diesel-powered tractors. Most manufacturers gave little thought to driver comfort.

△ Froelich Gasoline Traction Engine

Date 1892 **Origin** US
Engine Van Duzen single-cylinder gasoline
Horsepower 20 hp
Transmission 1 forward, 1 reverse

The first Froelich was built in 1892, making it one of the earliest tractors ever built. The Waterloo Gasoline Traction Co., which built Froelich tractors, was later bought by John Deere to start their tractor business. This is a replica of the original tractor.

△ Ivel

Date 1903 **Origin** UK
Engine Payne twin-cylinder horizontally opposed
Horsepower 14 hp
Transmission 1 forward, 1 reverse

Dan Albone's Ivel was probably the first successful tractor designed to replace horses. As well as the standard model, Albone also built a special orchard version—probably the first-ever fruit tractor—and a bullet-proof Ivel designed as a battlefield ambulance.

▽ Sharp's Auto-Mower

Date 1904 **Origin** UK
Engine Humber 4-cylinder gasoline
Horsepower N/A
Transmission 1 forward, 1 reverse

The tractor built by William Sharp in 1904 was not a commercial success, but the prototype worked on his Lancashire farm for almost 50 years. The original power unit was a Daimler engine, but when this was damaged by frost it was replaced by a Humber engine.

△ Hornsby-Akroyd Patent Oil Tractor

Date 1897 **Origin** UK
Engine Hornsby-Akroyd single-cylinder hot-bulb with airless injection
Horsepower 20 hp
Transmission 3 forward, 1 reverse

Richard Hornsby designed a tractor to work as a steam traction engine. The first tractor sold in the UK was a Hornsby, and a tracklaying version underwent trials with the UK's War Office in 1909.

▷ Hart-Parr 20-40

Date 1912 **Origin** US
Engine Hart-Parr 2-cylinder kerosene
Horsepower 40 hp
Transmission 2 forward, 1 reverse

Almost all of Hart-Parr's tractors built before 1918 featured a heavyweight design, a kerosene-burning two-cylinder engine, and a prominent front tower for the oil-based cooling system. The 20-40 included all of these features plus a single front wheel.

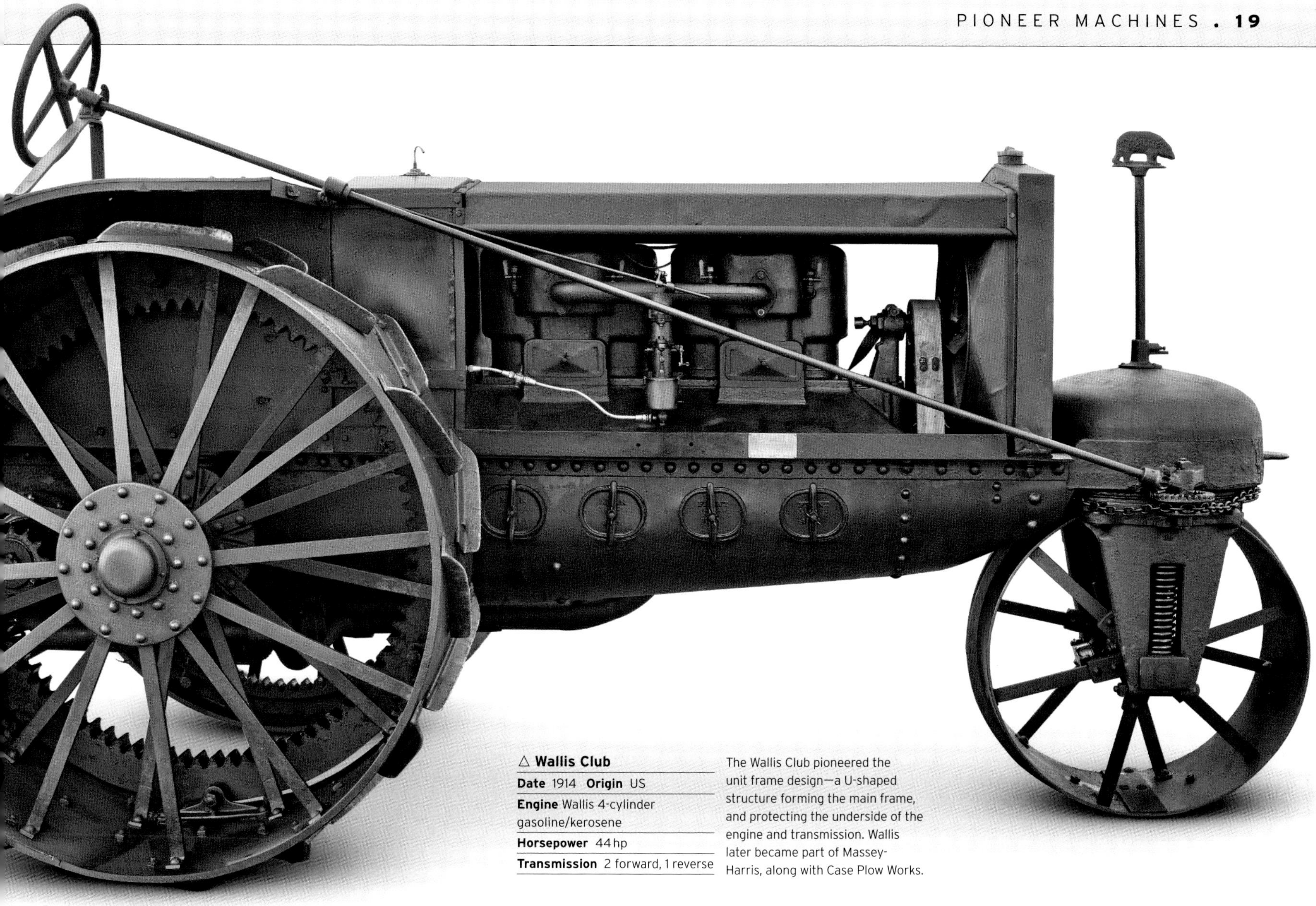

△ **Wallis Club**

Date 1914 **Origin** US
Engine Wallis 4-cylinder gasoline/kerosene
Horsepower 44 hp
Transmission 2 forward, 1 reverse

The Wallis Club pioneered the unit frame design—a U-shaped structure forming the main frame, and protecting the underside of the engine and transmission. Wallis later became part of Massey-Harris, along with Case Plow Works.

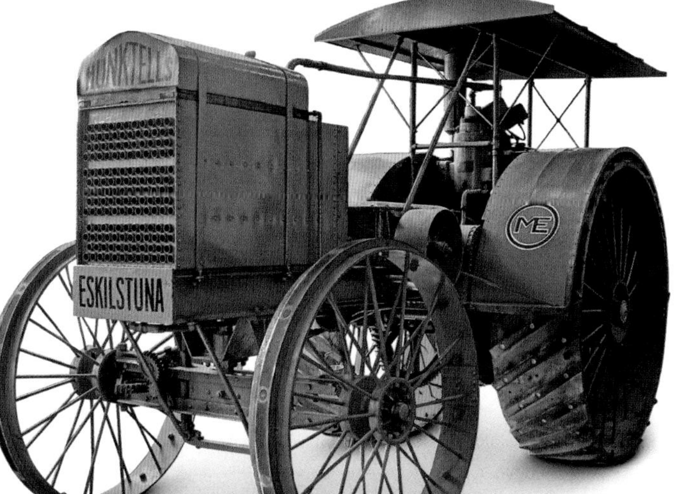

▷ **Allis-Chalmers 10-18**

Date 1914 **Origin** US
Engine Allis-Chalmers twin-cylinder horizontally opposed gasoline/kerosene
Horsepower 18 hp
Transmission 1 forward, 1 reverse

The first Allis-Chalmers production tractor was the 10-18 with an unconventional design that placed the single front wheel in line with the right hand rear wheel. The final drive used a large-diameter ring gear unprotected from mud or stone damage.

△ **Munktells 30-40**

Date 1913 **Origin** Sweden
Engine Munktells 2-cylinder 2-stroke semi-diesel
Horsepower 40 hp maximum
Transmission 3 forward, 1 reverse

This was the first and the biggest tractor built by Munktells, which is now part of the Volvo car and truck company. About 30 of the heavyweight tractors were built, equipped with 7-ft- (2-m-) diameter drive wheels and a 14.4-liter semi-diesel engine.

◁ **Little Bull**

Date 1914 **Origin** US
Engine Gile 2-cylinder horizontally opposed gasoline
Horsepower 12 hp
Transmission 1 forward, 1 reverse

The small size and unusual features of the Little Bull attracted interest, and it was the bestselling tractor in the US in 1914. It lacked power, with only 5 hp available at the drawbar, and with just one drive wheel to provide traction.

Hornsby-Akroyd

Unveiled in 1896, the Hornsby-Akroyd tractor was manufactured by Richard Hornsby & Sons of Grantham to the designs of inventor Herbert Akroyd Stuart. It was the first British tractor, and also the world's first vehicle with a compression-ignition engine, Akroyd Stuart's experiments predating those of German engineer Rudolph Diesel. Its single-cylinder engine, rated at 20 hp, was capable of running on a variety of fuels.

HERBERT AKROYD STUART got the idea for a compression-ignition engine after witnessing a vapor explosion at his father's tinplate works in Buckinghamshire. After patenting various oil engine designs, he sold the manufacturing rights to Hornsby in 1891. The Lincolnshire firm perfected the design and built several sizes of oil engine before their factory manager, David Roberts, suggested developing an "agricultural locomotive."

The first Hornsby-Akroyd Patent Oil Tractor was completed in 1896, and the firm began to assemble another three the following year. The hot-bulb engine was started by compressed air from an air receiver, and exhausted through a chimney, the blast inducing a draft of air for the cooling system. The first tractor was sold to the philanthropist H.J. Locke-King, but the other three remained unfinished until they were exported to Australia several years later.

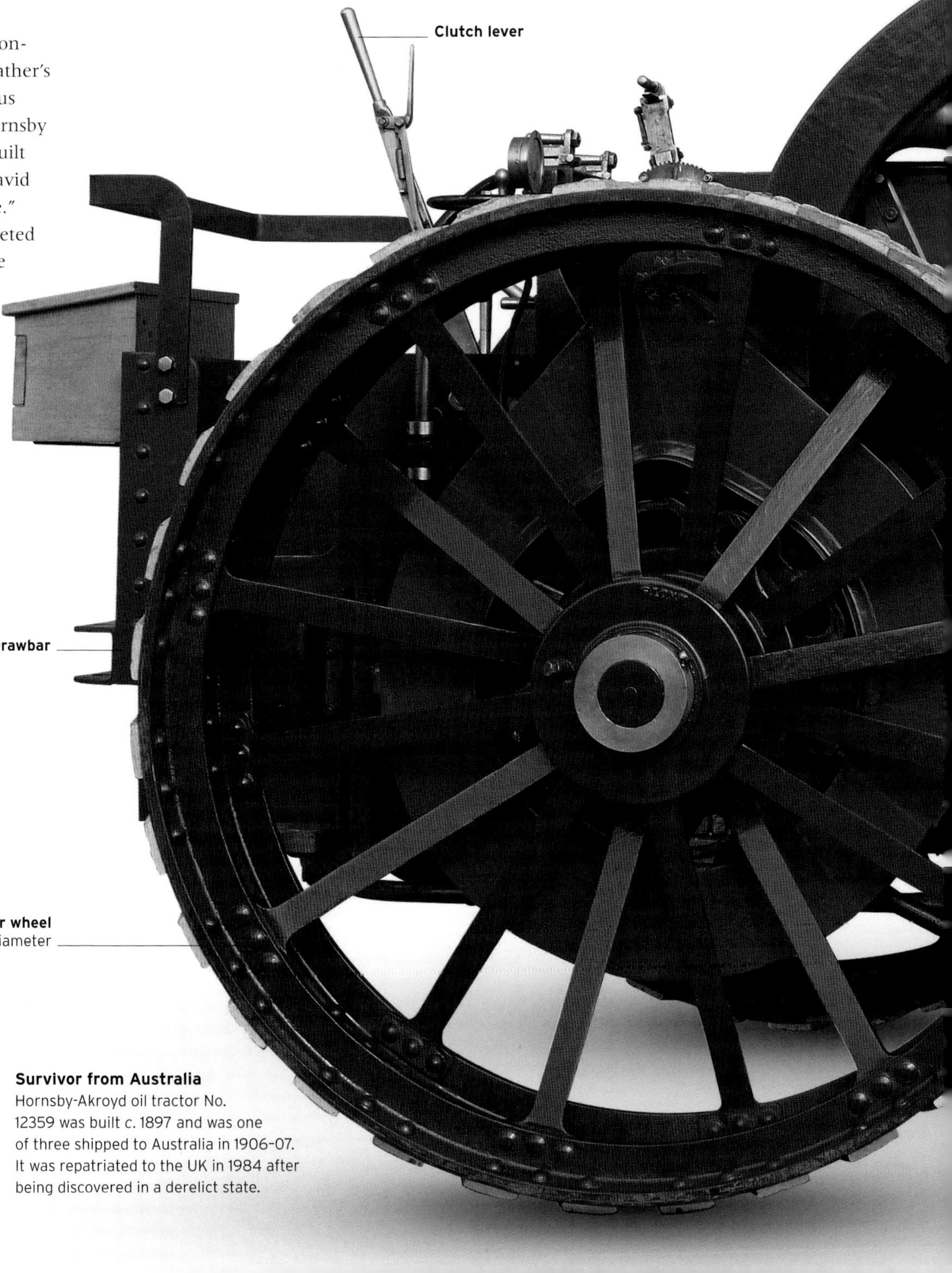

FRONT VIEW

REAR VIEW

Clutch lever

Drawbar

Rear wheel
6 ft (1.8 m) in diameter

Survivor from Australia
Hornsby-Akroyd oil tractor No. 12359 was built *c.* 1897 and was one of three shipped to Australia in 1906–07. It was repatriated to the UK in 1984 after being discovered in a derelict state.

SPECIFICATIONS	
Model	Hornsby-Akroyd Patent Oil Tractor
Built	1897
Origin	UK
Production	4
Engine	20 hp Hornsby-Akroyd single-cylinder hot-bulb with airless injection
Capacity	2,290 cu in (37,530 cc)
Transmission	3 forward, 1 reverse
Top speed	Not known
Length	15 ft 10 in (4.5 m)
Weight	8.5 tons (7.7 metric tons)

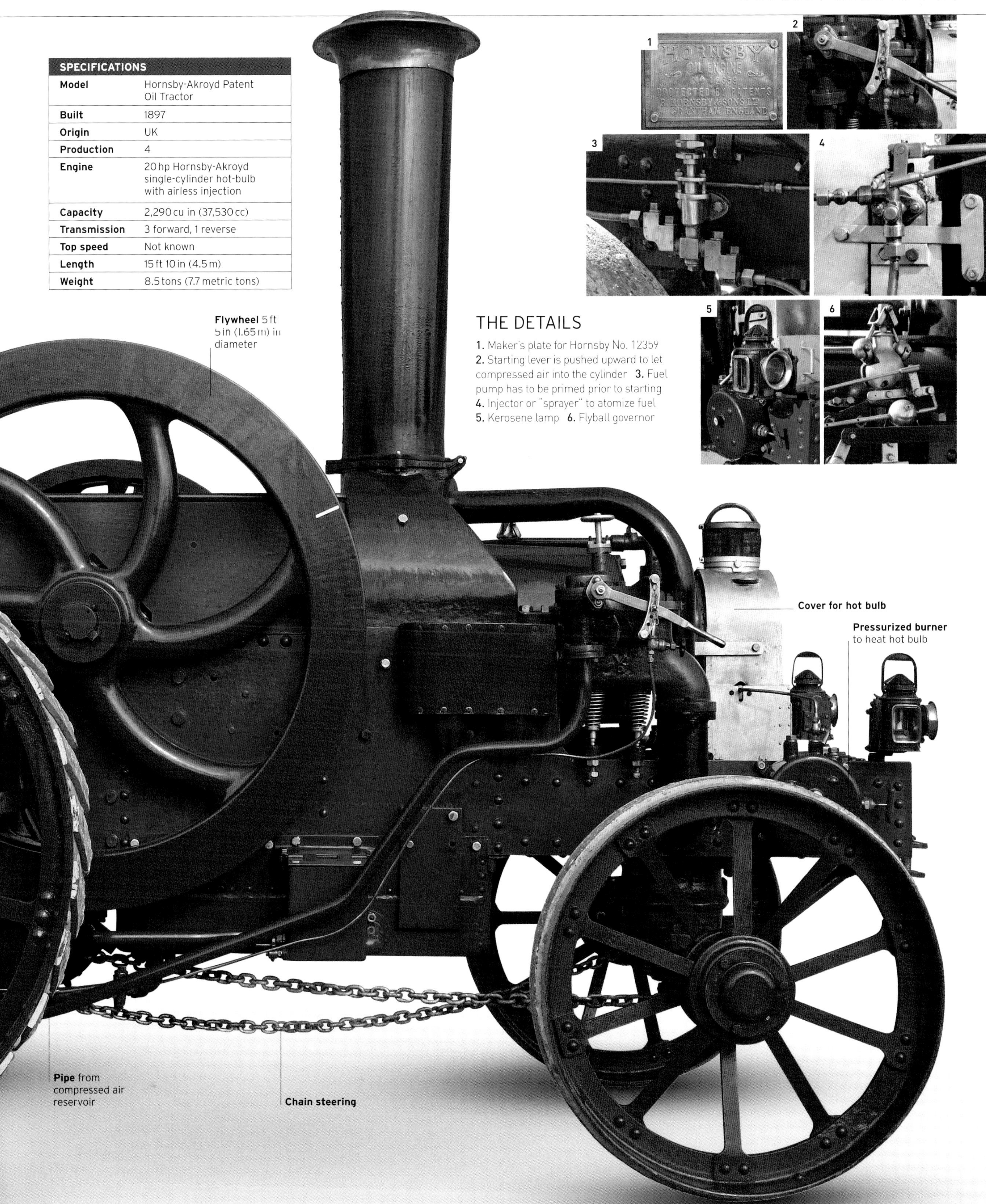

THE DETAILS

1. Maker's plate for Hornsby No. 12359 **2.** Starting lever is pushed upward to let compressed air into the cylinder **3.** Fuel pump has to be primed prior to starting **4.** Injector or "sprayer" to atomize fuel **5.** Kerosene lamp **6.** Flyball governor

Prairie Heavyweights

Building big tractors for big farms was the recipe for success for many leading companies during the early years of tractor history in the US and Canada. The prairie farms in Canada, and large-scale crop production in Australia and Africa, also provided export sales for big tractors built by some British companies. The tractors that were in demand were sturdy and, by contemporary standards, many of them offered plenty of power. However, customers were also demanding simple, no-frills reliability, and the temperamental ignition and fuel systems used during the first 25 or so years of tractor history ensured that reliability remained a major issue.

△ Hart-Parr 30-60 "Old Reliable"

Date 1910 **Origin** US
Engine Hart-Parr 2-cylinder horizontal gasoline/kerosene
Horsepower 60 hp
Transmission 1 forward, 1 reverse

It is not clear if the 30-60 actually earned the nickname "Old Reliable" or whether the name was just a clever marketing idea. It was a popular tractor in spite of a starting process involving 19 steps, which included manually turning the 1,000 lb (450 kg) flywheel to spin the engine.

△ International Mogul Junior 25

Date 1912 **Origin** US
Engine International single-cylinder kerosene
Horsepower 25 hp
Transmission 1 forward, 1 reverse

In 1911, International opened its new Tractor Works in Chicago, and this was one of the first new models to emerge. Production began at six machines a day, but increased to 12, then 14 as the tractor boom peaked in 1912. It remained in production until early 1913; 812 were built before it reappeared as the largely similar Mogul 15-30.

△ International Titan Type D

Date 1910 **Origin** US
Engine International single-cylinder horizontal kerosene
Horsepower 25 hp
Transmission 1 forward, 1 reverse

International's Titan tractor range was sold through the company's Deering dealer network in the US and Canada, while Mogul models were offered by McCormick outlets. The Titan Type D was built in 20 hp and 25 hp versions plus a small number of 45 hp models.

△ Marshall Colonial Class A

Date 1908 **Origin** UK
Engine Marshall 2-cylinder gasoline/kerosene
Horsepower 30 hp
Transmission 3 forward, 1 reverse

Marshall's venture into motorized tractors began in 1904 when it started building tractors for broad-acre farming. Distributed in South America, Australia, Africa, India, Russia, and Canada, these huge machines ranged from 16 hp to 32 hp. By 1914, 300 had been sold; production ceased during WWI.

▷ Case 20-40

Date 1913 **Origin** US
Engine J.I. Case 2-cylinder gasoline/kerosene
Horsepower 40 hp
Transmission 2 forward, 1 reverse

Case built its first tractor in 1892 but it was not a success. It returned to the market in 1911 with a prototype machine that led to full production from 1912 in time for the WWI sales boom. The 20-40 was a bestselling model and a gold medal winner at the Canadian tractor trials.

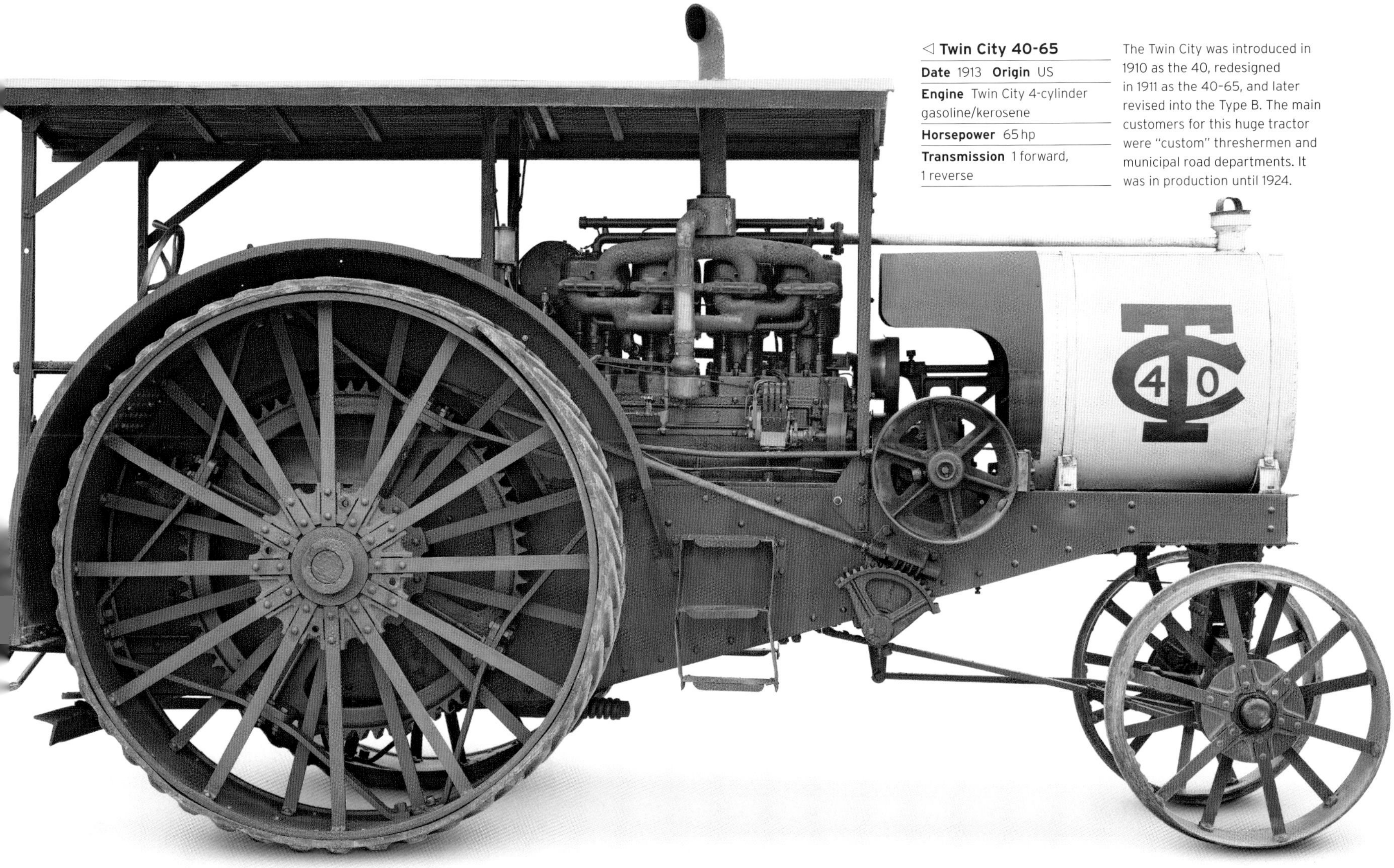

◁ Twin City 40-65

Date 1913 **Origin** US

Engine Twin City 4-cylinder gasoline/kerosene

Horsepower 65 hp

Transmission 1 forward, 1 reverse

The Twin City was introduced in 1910 as the 40, redesigned in 1911 as the 40-65, and later revised into the Type B. The main customers for this huge tractor were "custom" threshermen and municipal road departments. It was in production until 1924.

◁ Fairbanks Morse 15-30

Date 1913 **Origin** Canada

Engine Fairbanks Morse single-cylinder kerosene

Horsepower 30 hp

Transmission 1 forward, 1 reverse

Fairbanks-Morse tractor production was centered in Beloit, WI, and Toronto, Canada. A Model N 15-25 found a ready market in western Canada, and it was said that the sharp report of its engine could be heard for miles. In 1912 the 15-25 was superseded by the 15-30.

△ Avery 18-36

Date 1916 **Origin** US

Engine Avery 4-cylinder horizontally opposed gasoline/kerosene

Horsepower 36 hp

Transmission 2 forward, 1 reverse

Avery was a leading steam traction engine company before joining the tractor industry. A distinctive feature of most of their early heavyweight tractors was the radiator tower where water was cooled while circulating through copper pipes in an airflow induced by exhaust gases.

▷ Rumely OilPull Type G 20-40

Date 1919 **Origin** US

Engine Advance-Rumely 2-cylinder gasoline/kerosene

Horsepower 46 hp maximum

Transmission 2 forward, 1 reverse

The OilPull name emphasized the Rumely engine's ability to use cheaper, low-grade kerosene, which needed oil instead of water for cooling. The Model G arrived in 1919, one of the last new OilPulls in the traditional heavyweight series.

Prairie Tractor Boom and Bust

The early North American gasoline-powered tractors were large, ponderous beasts, designed with just one job in mind: breaking new land across the prairies. Their success led to a boom in the North American tractor industry, which lasted for only a few years—from about 1907 until the bubble burst around 1912, after the big wheat bonanza slowed and the great ranches were subdivided into smaller holdings.

RUMELY OILPULL

The Advance-Rumely Thresher Company of La Porte, IN, was one of the manufacturers fueling the big tractor boom. The firm built its first model in 1909, coining the OilPull name for its tractor line, which employed oil cooling instead of water to keep the operating temperature of the engine high enough to vaporize the kerosene fuel. By 1912 with a turnover of $16 million, Rumely was the third-largest tractor manufacturer in the US, with 2,000 staff building 2,500 tractors per year.

The prairie monsters were usually bought on credit. Crop failures led farmers to default on payments and go out of business. The market, already oversupplied, collapsed almost overnight, forcing Rumely to adopt more lightweight designs.

Contrasting motor power with horsepower: a single-cylinder Rumely OilPull 15-30 tractor breaking new ground in Maynard, Saskatchewan, in 1912.

The British Pioneers

The biggest problem for UK tractor makers before World War I was lack of customers. Horses and people continued to provide power on British farms, with steam limited mainly to specialty contractor services, and there was little interest in tractors. When the war started and brought an urgent need for tractor power, most of the demand was met by imports from the US. None of the British companies that started tractor production before or soon after the war ended achieved long-term success.

◁ **Saunderson Universal**

Date 1916 **Origin** UK

Engine Saunderson 2-cylinder gasoline/kerosene

Horsepower 23 hp

Transmission 3 forward, 1 reverse

This forward-control tractor was based on the previous Universals, but with the driving position moved to the front. Saunderson had used this layout on some earlier models, and on the Universal version it provided good forward visibility, but a restricted rear view.

△ **Alldays General Purpose**

Date 1917–18 **Origin** UK

Engine Alldays & Onions 4-cylinder gasoline/kerosene

Horsepower 25–30 hp

Transmission 3 forward, 1 reverse

The Alldays General Purpose from Alldays & Onions made a brief appearance. Price was a problem, caused by the high specification including front and rear suspension, a canopy protecting the driver, plus both transmission and rear-axle brakes.

▷ **Saunderson Universal Model G**

Date 1916 **Origin** UK

Engine Saunderson 2-cylinder gasoline/kerosene

Horsepower 23 hp

Transmission 3 forward, 1 reverse

Saunderson's success peaked during WWI when its Universal Model G tractors were needed to help the wartime food production campaign. One of the Model G orders was from the royal farms in Norfolk—the king's first tractor.

Motor Plows

The motor plow fashion was brief and patchy. It lasted through World War I and the early 1920s, and then disappeared. Demand was strong in the US and some European countries, but with much less interest elsewhere. Motor plows often cost less than tractors, but this advantage vanished when the Fordson arrived. Disadvantages included poor versatility and awkwardness of use. Attempts to build costly, high-output motor plows, such as the 65-hp Stock from Germany, attracted few customers.

△ **Crawley Agrimotor**

Date 1920 **Origin** UK

Engine Buda 4-cylinder gasoline/kerosene

Horsepower 30 hp

Transmission 2 forward, 1 reverse

A prototype version of this tractor was designed and built on the Crawley brothers' farm in Essex c. 1912–13. A small number were built by Garretts of Leiston before production was taken over by the family-owned Crawley Agrimotor Co.

△ Glasgow

Date 1919 **Origin** UK

Engine Waukesha 4-cylinder gasoline or gasoline/kerosene

Horsepower 27 hp

Transmission 2 forward, 1 reverse, with 3-wheel drive

Glasgow tractors were thoroughly Scottish, designed by a Scot and built near Glasgow. Unusual features included three-wheel drive, plus a tilt facility to keep the engine upright. Production fell far short of the 5,000 per year target, and ended in 1924.

△ Austin Model R

Date 1919 **Origin** UK

Engine Austin 4-cylinder gasoline/kerosene car engine

Horsepower 25 hp

Transmission 2 forward, 1 reverse

The Model R achieved some success in the UK with output briefly reaching 60 per week in 1922. It was also popular in France, where production started in about 1926, the British market was later supplied with an updated, French-built version.

▷ Weeks-Dungey New Simplex

Date 1919 **Origin** UK

Engine Waukesha 4-cylinder gasoline/kerosene

Horsepower 22.5 hp

Transmission 3 forward, 1 reverse

Hugh Dungey designed the tractor he wanted for his farm in Kent, and William Weeks, a local engineering company, built it for him c. 1914. It attracted orders from other farmers and an improved version, the New Simplex, was produced in 1918.

◁ Garner All-purpose

Date 1919 **Origin** UK

Engine Garner 4-cylinder gasoline/kerosene

Horsepower 30 hp

Transmission 3 forward, 1 reverse

The Garner was a British tractor built in the US by the William Galloway Co. of Waterloo, IA. A selling point claimed by Henry Garner Ltd., the British importer and distributor, was that it was suitable for women drivers recruited for the wartime food-production campaign.

◁ Praga Model X

Date 1918-20

Origin Czechoslovakia

Engine Praga single-cylinder gasoline

Horsepower 10 hp

Transmission Not known

The Praga Co. was building cars and trucks before they started making motor plows around 1912, and tractors followed later. The first Praga motor plows were the 40-hp K5 and 32-hp X designs, followed by the 10-hp (shown) and 20-hp models for smaller farms.

△ Fowler Motor Plough

Date 1920 **Origin** UK

Engine Fowler 2-cylinder gasoline or gas/kerosene

Horsepower 14 hp

Transmission 2 forward, 1 reverse

Falling demand for steam power caused problems for the Fowler company, and the motor plow offered an opportunity to diversify. In spite of the prestigious Fowler name, sales were disappointing and the search for other products continued.

The Age of Experiment

World War I provided a huge boost for the tractor industry, especially in North America. As well as being a period of rapid growth, it was also an age of experiment, with four-wheel drive, three-wheelers, various combinations of wheels and tracks, and even motor plows all making an appearance. Improvements introduced in the decade before 1920 included the first tractor with an electric starter, while International announced the first tractor with a true power takeoff (PTO).

△ **Ford Model B**

Date c. 1916 **Origin** US
Engine Gile 2-cylinder horizontally opposed gasoline/kerosene
Horsepower 16 hp
Transmission 1 forward

The Ford Model B arrived after news broke that Henry Ford was developing a tractor. The company that built it was not connected with Henry Ford; it was started by businessmen hoping to attract investors and customers who were confused by the name.

△ **Samson Sieve-Grip 6-12**

Date 1917 **Origin** US
Engine Samson single-cylinder horizontal gasoline/kerosene
Horsepower 12 hp
Transmission 1 forward, 1 reverse

The Sieve-Grip name refers to the tractor's steel wheels. They were designed to grip the ground efficiently, and the idea seems to have worked—the Sieve-Grip tractor was so popular that the Samson company was bought by General Motors in 1917.

△ **Moline Universal Model D**

Date 1918 **Origin** US
Engine Moline 4-cylinder gasoline
Horsepower 27 hp
Transmission 1 forward, 1 reverse

For many farmers, motor plows were the first step into power farming, and the Moline Plow Co.'s Universal was one of the best-selling models. Increasing sales allowed Moline to build their own engine, and in 1918 they added electric starting and lights to the specification.

△ **Gray 18-36**

Date 1918 **Origin** US
Engine Waukesha 4-cylinder gasoline
Horsepower 36 hp
Transmission 2 forward, 1 reverse

The Gray tractor had a corrugated metal sheet to keep the engine dry. The drive at the rear was a single wide drum to reduce soil compaction, which Gray claimed increased the tractor's pulling power, but it slipped badly when tested in Nebraska.

▷ **Parrett 12-25**

Date 1919 **Origin** US
Engine Buda 4-cylinder gasoline/kerosene
Horsepower 25 hp
Transmission 2 forward, 1 reverse

Parrett tractors were called Clydesdales in the UK, and this one probably took part in the 1920 trials at Lincoln. Big front wheels were said to need less power over rough ground and bearing wear was reduced because the hubs were farther from mud and dust.

△ **Nelson 20-28**

Date 1919 **Origin** US

Engine Wisconsin Type A 4-cylinder gasoline/kerosene

Horsepower 28 hp

Transmission 3 forward, 1 reverse

US experiments with four-wheel drive (4WD) date back to about 1910, but benefits of improved traction were outweighed by higher prices and poor maneuverability. The 4WD models made by the Nelson Blower & Furnace Co. were not successful.

△ **Rumely OilPull Model H 16-30**

Date 1919 **Origin** US

Engine Advance-Rumely 2-cylinder horizontal gasoline/kerosene

Horsepower 30.5 hp

Transmission 2 forward, 1 reverse

Rumely tractors burned kerosene with water injection to improve fuel consumption and increase the power when running under load. To keep the operating temperature high enough, oil cooling was employed instead of water—hence the OilPull name.

△ **Waterloo Boy Model N**

Date 1920 **Origin** US

Engine Waterloo twin-cylinder horizontal gasoline/kerosene

Horsepower 25 hp

Transmission 2 forward, 1 reverse

The Waterloo Boy was the first tractor to complete a Nebraska test. Selling the Model N through his garage prompted Harry Ferguson's interest in tractor design. John Deere entered the tractor industry when it bought the Waterloo Boy Co.

Alternative Power Sources

There were no limits in the design of power units for use in general haulage and agriculture. Everything was tried—light and heavy steam tractors, self-propelled cultivators, oil-fueled cable-cultivating engines, both as original designs and as conversions. Even into the 1920s, novel designs were still being introduced to the market; most had little or no success.

▷ **Cooper Steam Digger**

Date 1900 **Origin** UK

Engine Cooper double crank compound

Cylinders 2

Horsepower 25 hp

The Cooper digger was an attempt to replace the plow for cultivating land. The digging attachment could be removed and the engine used for general farm haulage. Cooper even offered a winding drum attachment to convert the engine to cable cultivation.

△ **Foden Colonial**

Date 1913 **Origin** UK

Engine Foden

Cylinders Compound

Horsepower 8 nhp

The products of Foden's Ltd. were engineered to a very high standard, but not many were built. These machines enjoyed a good reputation among both drivers and owners for being economical and lively. The Foden engines were noted for the larger-than-normal rear wheels.

▽ **Huber**

Date 1915 **Origin** US

Engine Huber

Cylinder 1

Horsepower 15 hp

Huber traction engines differed from most US types. They incorporated a return flue boiler and the cylinder was placed over the firebox with the crankshaft toward the front. Boilers of this type were economical as well as being short and able to use a variety of fuels.

▽ **Fowler Diesel Conversion**

Date 1918 **Origin** UK

Engine Mercedes-Benz 6-cylinder diesel

Transmission 2 forward, 2 reverse

Horsepower 100 hp

Several Fowler steam plowing engines were converted to diesel in the 1950s. The conversion of this particular engine was done by Beeby Bros. of Rempstone, Leicestershire, for use in their dredging business. With diesel it was no longer necessary to cart coal and water to the engines.

△ Mann

Date 1920 **Origin** UK	
Engine Mann	
Cylinders Compound	
Horsepower 25 hp	

The Mann "low-cost steam tractor" was yet another attempt by a steam-engine builder to produce a machine to compete with the highly successful internal-combustion-engined tractors. Single or compound cylinder versions were offered; both failed to achieve worthwhile sales.

TECHNOLOGY

Best's Engines

Daniel Best started out building grain cleaners in 1869, and he formed the Daniel Best Agricultural Works in San Leandro, California in 1885. He went on to build combine harvesters, and in 1893 his Best Manufacturing Co. produced these and steam and gas traction engines. In 1908 he sold out to The Holt Manufacturing Co. of Stockton.

For hills The Best's 110 hp engine had an oil-fired, upright boiler, so it could be used in hilly areas where the incline affects the water level.

▽ Walsh & Clark Oil Plowing Engine

Date 1919 **Origin** UK	
Engine Walsh & Clark 2-cylinder horizontally opposed gasoline/kerosene	
Transmission 2 forward, 1 reverse	
Horsepower 35 hp	

Walsh & Clark designed these engines to look like steam engines in an attempt to persuade steam-engine users to buy them. In the end, they proved unsuccessful and few were built. The early internal combustion engines did not have the power or versatility to compete with the steam engine.

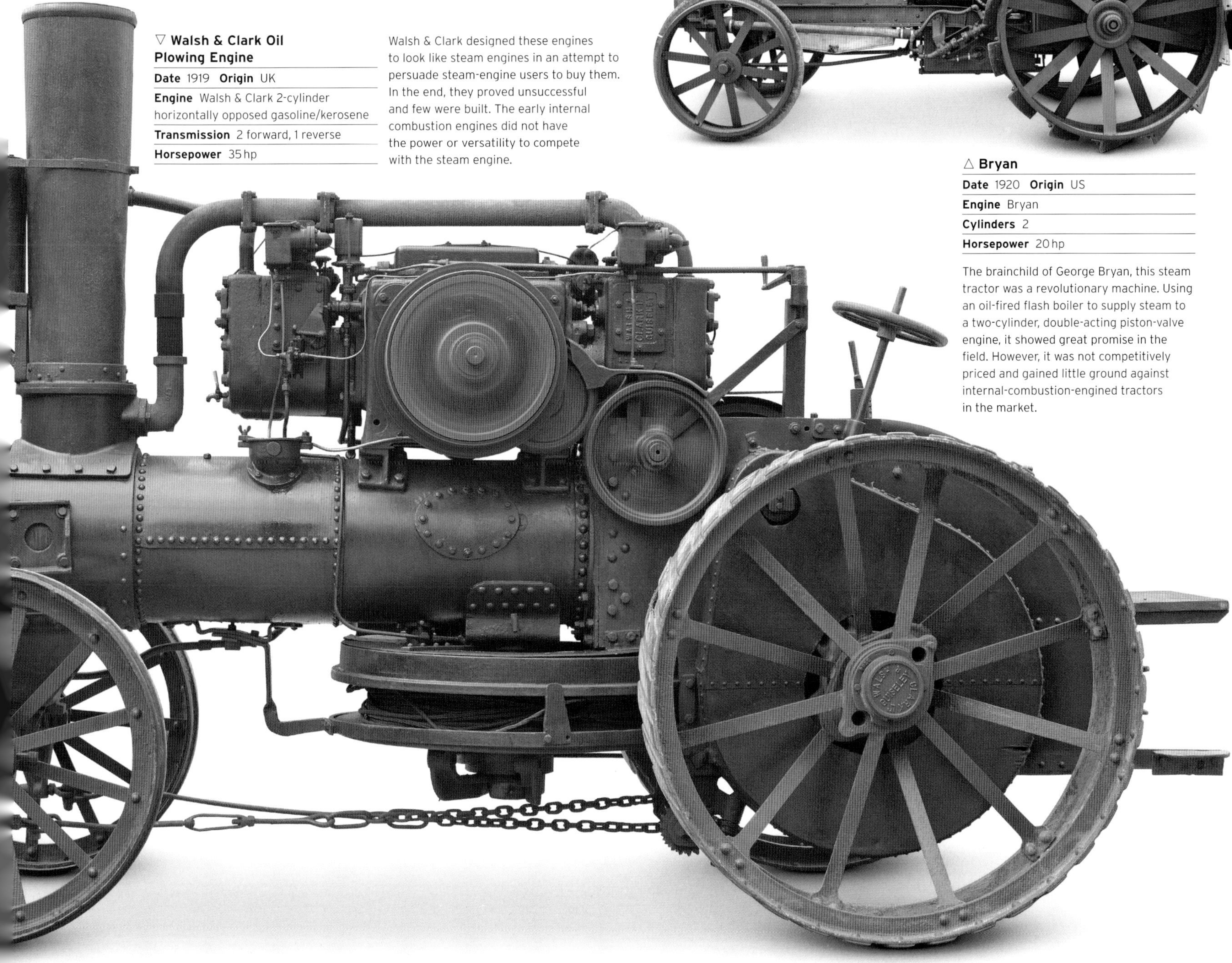

△ Bryan

Date 1920 **Origin** US	
Engine Bryan	
Cylinders 2	
Horsepower 20 hp	

The brainchild of George Bryan, this steam tractor was a revolutionary machine. Using an oil-fired flash boiler to supply steam to a two-cylinder, double-acting piston-valve engine, it showed great promise in the field. However, it was not competitively priced and gained little ground against internal-combustion-engined tractors in the market.

ORD. DEPT
FIVE TON
ARTILLERY TRACTOR
MOD. 1917

The US Army's Artillery Tractor

As stalemate settled over the trenches during World War 1, and the Western Front became a sea of mud, the need for crawler tractors to handle the heavy guns became acute. The British, French, and Russian governments placed orders for various sizes of artillery tractors, including 75-hp and 120-hp models, with the Holt Manufacturing Company. After the US entered the war in 1917, Holt was asked to develop an armored 45-hp tractor for the US Army Ordnance Department. To ease supply problems, production of this machine, officially designated the "Artillery Tractor 5 Ton Model 1917," was licensed to two motor car builders, Reo of Lansing and Maxwell of Detroit.

DEMONSTRATING THE TRACTOR

Holt's artillery tractor was demonstrated at Rock Creek Park on the outskirts of Washington, DC, on June 3, 1918. During the trial, the tractor was used to haul a 4.7-in (11.9-cm) field howitzer and lumber. Attending the demonstration were Newton Baker, the US secretary of war (seen in the photograph arriving in his Model T Ford), and General Peyton March, the US Army's chief of staff. The armored "5 Ton" was released for production soon afterward, and 1,100 were shipped to Europe before the Armistice was signed in November 1918. A civilian version was built from 1919 to 1922.

US Army engineers at Rock Creek, Washington, DC, await the arrival of Secretary Baker for the demonstration of the Holt "5-Ton" artillery tractor.

Beating the U-Boat

The demand for supplies to the European Western Front, including ordnance, horses, and foodstuffs, was seriously underestimated by the British government in World War I. This reached a crisis point by mid-1916, with farms struggling to combat the disruption of food imports caused by the attacks on shipping by Germany's U-boats. The British government turned to the US to supply tractors to replace the farm horses lost to the army. Large orders were also placed with Holt for the supply of tractors to the army for transportation duties.

△ Holt 75

Date 1918 **Origin** US

Engine Holt 4-cylinder gasoline

Horsepower 75hp

Transmission 2 forward, 1 reverse

The Holt 75 was the Allies' standard heavy artillery tractor, with 1,651 eventually being delivered from 1915 to November 1918. In the atrocious mud on the Western Front, it was the only tractor capable of hauling the heavy guns into position. As conditions worsened, the 75 was also used to haul supply trains bringing ammunition and other essentials up to the front lines.

△ Mogul 8-16

Date 1915 **Origin** US

Engine International single-cylinder horizontal gasoline/kerosene

Horsepower 25hp

Transmission Epicyclic single-speed forward and reverse

The 8-16 was simple, dependable, and rugged. Being equipped with low-tension ignition and a total-loss lubrication system, there was little that could go wrong. Around 500 units were delivered between 1915 and 1918, most to fill the UK Ministry of Munitions' orders.

◁ International Junior 8-16

Date 1919 **Origin** US

Engine International 4-cylinder gasoline/kerosene

Horsepower 16hp

Transmission 3 forward, 1 reverse

This was a popular tractor on farms in the early 1920s. Featuring a water-washer air cleaner and a mid-mounted radiator, the still chain-driven Junior filled the gap between the "old" type of tractors, the Mogul and Titan, and the gear-drive 15-30 and 10-20 models.

△ Overtime Model R

Date 1916 **Origin** US/UK

Engine Waterloo twin-cylinder horizontal gasoline/kerosene

Horsepower 24hp

Transmission 1 forward, 1 reverse

Known in the US as the Waterloo Boy and in the UK as the Overtime, it was the first of the US-built tractors to appear on British farms as a result of government purchase orders. These tractors were brought in to help produce more food as WWI began affecting imports of foodstuffs.

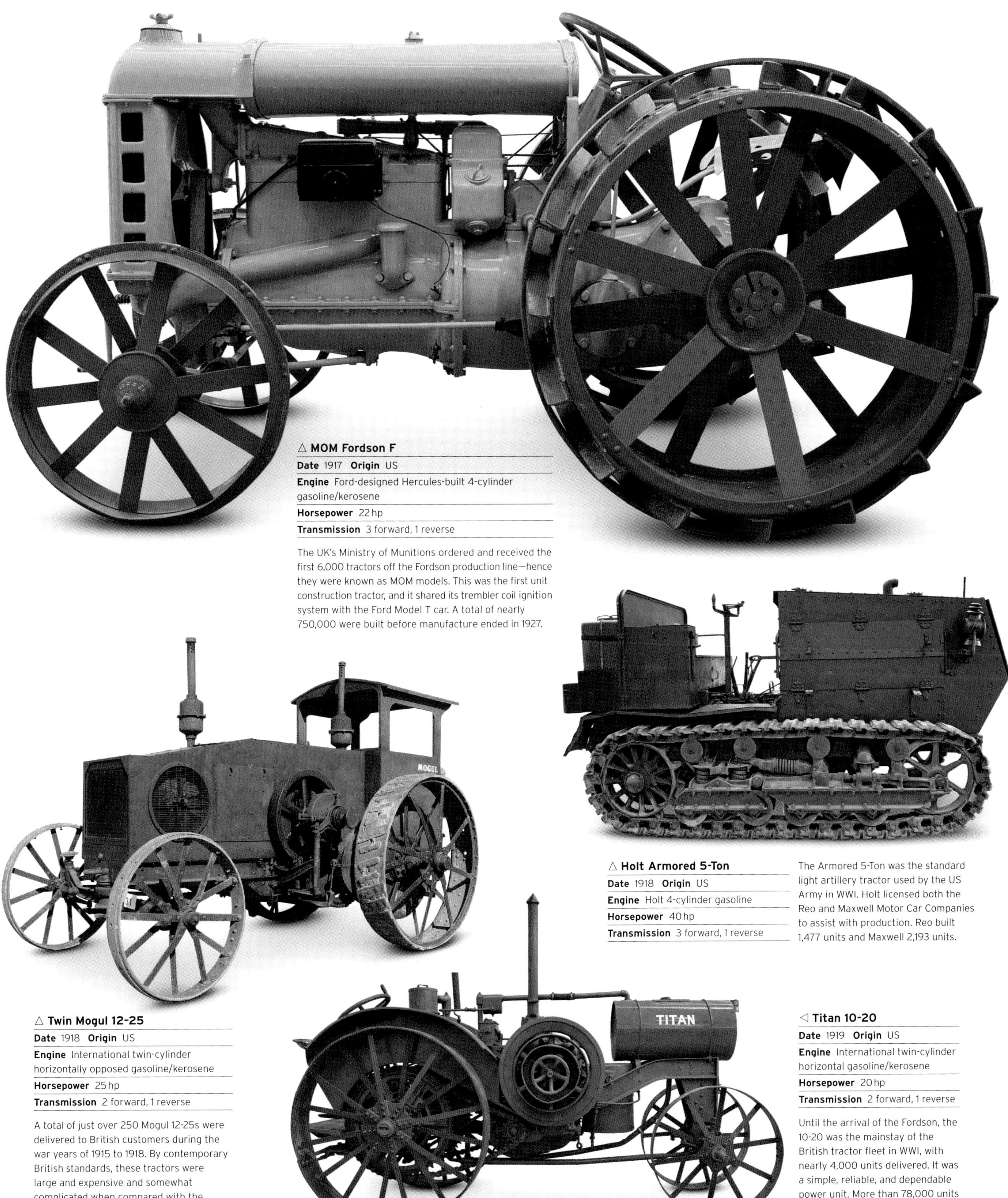

△ **MOM Fordson F**

Date 1917 **Origin** US

Engine Ford-designed Hercules-built 4-cylinder gasoline/kerosene

Horsepower 22 hp

Transmission 3 forward, 1 reverse

The UK's Ministry of Munitions ordered and received the first 6,000 tractors off the Fordson production line—hence they were known as MOM models. This was the first unit construction tractor, and it shared its trembler coil ignition system with the Ford Model T car. A total of nearly 750,000 were built before manufacture ended in 1927.

△ **Holt Armored 5-Ton**

Date 1918 **Origin** US

Engine Holt 4-cylinder gasoline

Horsepower 40 hp

Transmission 3 forward, 1 reverse

The Armored 5-Ton was the standard light artillery tractor used by the US Army in WWI. Holt licensed both the Reo and Maxwell Motor Car Companies to assist with production. Reo built 1,477 units and Maxwell 2,193 units.

△ **Twin Mogul 12-25**

Date 1918 **Origin** US

Engine International twin-cylinder horizontally opposed gasoline/kerosene

Horsepower 25 hp

Transmission 2 forward, 1 reverse

A total of just over 250 Mogul 12-25s were delivered to British customers during the war years of 1915 to 1918. By contemporary British standards, these tractors were large and expensive and somewhat complicated when compared with the other makes available at the time.

◁ **Titan 10-20**

Date 1919 **Origin** US

Engine International twin-cylinder horizontal gasoline/kerosene

Horsepower 20 hp

Transmission 2 forward, 1 reverse

Until the arrival of the Fordson, the 10-20 was the mainstay of the British tractor fleet in WWI, with nearly 4,000 units delivered. It was a simple, reliable, and dependable power unit. More than 78,000 units were produced from 1915 to 1922.

Holt 75 Gun Tractor

The Holt 75 became the standard artillery tractor for the Allied forces during World War I. A total of 1,810 military versions were built at the company's factory in Peoria, IL. Of these, 1,362 were ordered by the British War Department, arriving in the UK at Avonmouth Docks, where they were adapted to War Office requirements before being shipped to France. The military specification included a canopy and a capstan winch.

THE 75 TRACTOR was built by the Holt Manufacturing Company, the precursor of the Caterpillar organization, in Peoria and at the company's plant in Stockton, CA. The model was in production from 1914 to 1924, and 4,161 were made. Aside from those pressed into military service, most were supplied as agricultural tractors or for road-building.

The four-cylinder gasoline engine, developing 75 hp at a leisurely 550 rpm, was largely reliable, if a little outdated. It also had an incredible thirst for fuel, drinking gasoline like it was going out of style, and suffered from an inadequate cooling system. Most of the military tractors were equipped with an auxiliary water tank to feed the radiator. They were physically demanding machines to drive; changes of direction required numerous turns on the steering wheel while disengaging the appropriate clutch—and even then the tractor might continue in a straight line.

THE DETAILS

1. Holt name embossed on radiator header tank **2.** 4-cylinder gasoline engine has exposed push-rods **3.** Capstan winch for recovery work **4.** Bevel gears for steering column **5.** Driver's platform with levers for steering clutches and drive clutch **6.** Massive steering clutches are 3 ft (0.9 m) in diameter

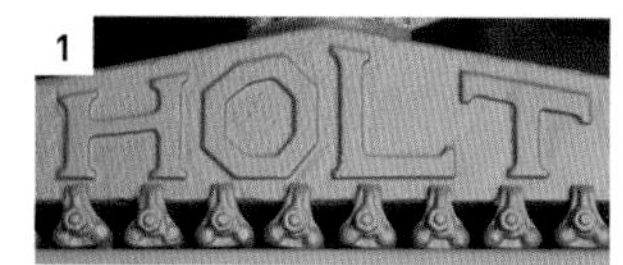

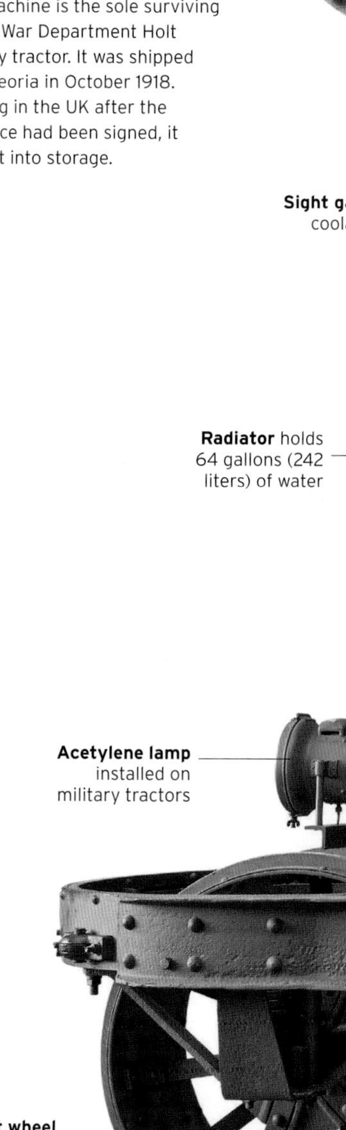

British Army Holt
This machine is the sole surviving British War Department Holt artillery tractor. It was shipped from Peoria in October 1918. Arriving in the UK after the armistice had been signed, it was put into storage.

SPECIFICATIONS	
Model	Holt 75 Gun Tractor
Built	1918
Origin	US
Production	4,161
Engine	75 hp Holt 4-cylinder gasoline
Capacity	1,400 cu in (22,900 cc)
Transmission	2 forward, 1 reverse
Top speed	3 mph (5 km/h)
Length	20 ft (6 m)
Weight	11.8 tons (10.7 metric tons)

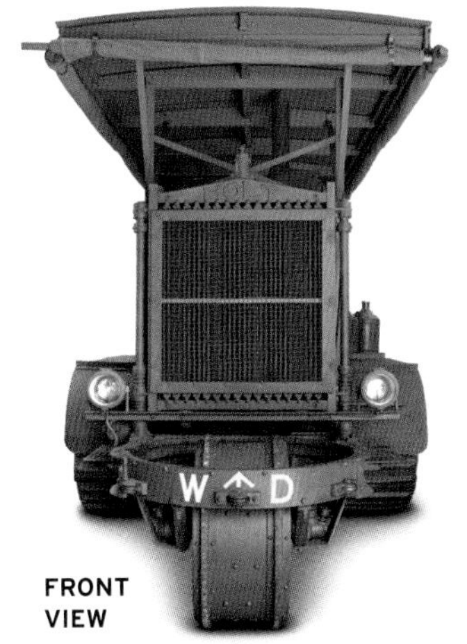

W↑D
FRONT VIEW

W↑D
REAR VIEW

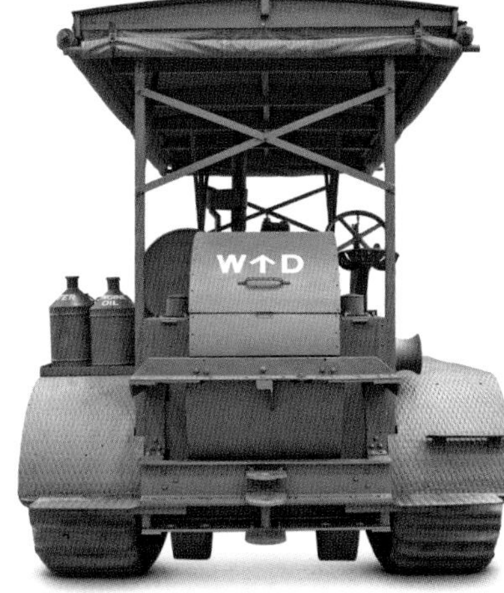

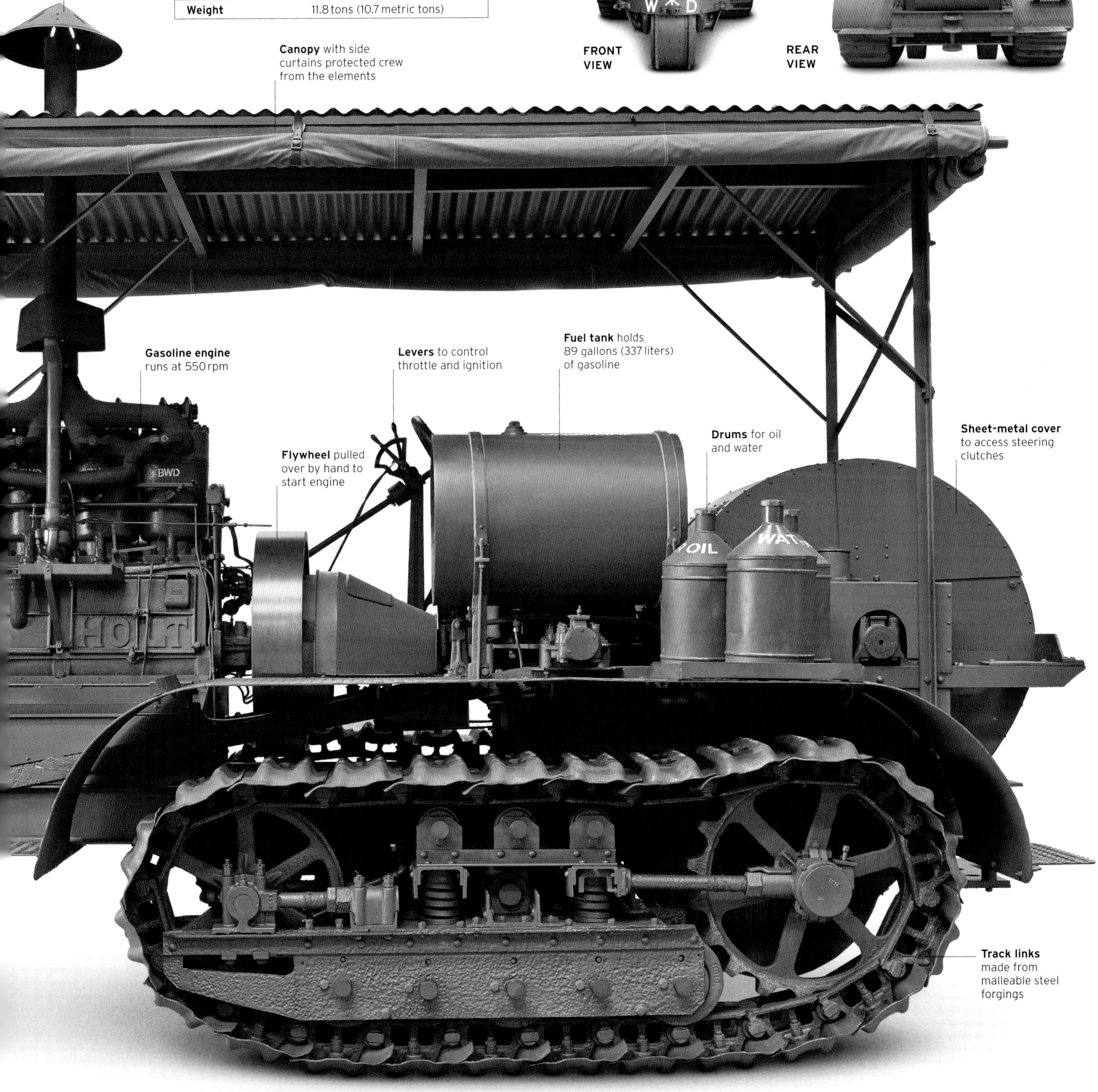
Conical top keeps rain out of exhaust pipe
Canopy with side curtains protected crew from the elements
Gasoline engine runs at 550 rpm
Levers to control throttle and ignition
Fuel tank holds 89 gallons (337 liters) of gasoline
Flywheel pulled over by hand to start engine
Drums for oil and water
Sheet-metal cover to access steering clutches
Track links made from malleable steel forgings
BWD
HOLT
OIL

Women at the Wheel

The early motor tractors captured the imagination of the public, and no more so than in the UK during World War 1, when increased mechanization was the answer to the German U-boat attacks that threatened to starve the country into submission. Tractors replaced horses that had been pressed into military service, but drivers were also needed at a time when most of the men were serving on the Western Front.

Women made up a third of all the auxiliary labor drafted onto the land during World War I, with the remainder being drawn from the Army Service Corps, prisoners of war, and boys too young for military service. The Women's Land Army (WLA) was not formed until February 1917, and most of its 23,000 members were given little opportunity to drive tractors. However, those who did proved to be the equal of the men they replaced, and were sometimes more proficient.

WOMEN'S LEGION AND ITS TRACTORS

The agricultural section of the Women's Legion, led by Miss Sylvia Brocklebank, raised money to buy tractors and implements, then leased them to farmers. Many of the tractors were US models purchased with donations from the US.

A woman at the wheel of a US Little Giant tractor illustrated in a British children's book from about 1917.

Tractor Development

Tractor evolution and production started in the US, with the UK close behind. Other European countries plus Canada and Australia followed, with Eastern European and South American manufacturers next on the list. Countries such as Japan, Turkey, India, and China came much later. The big influence before 1920 was the impact of World War I, which brought large-scale tractor production in the US and sped up the tractor-powered farm mechanization process that eventually replaced both horses and steam.

◁ Champeyrache

Date c. 1910 **Origin** France
Engine Chapuis Dornier 4-cylinder gasoline
Horsepower 12 hp
Transmission 1 forward, 1 reverse

This is possibly the only surviving Champeyrache tractor, a make for which few records exist. The rough finish on some of the parts suggests it may have been a prototype version. An unusual feature is a power-operated winch.

△ Pavesi America

Date 1913-17 **Origin** Italy
Engine OTO 4-cylinder gasoline
Horsepower 50 hp
Transmission 2 forward, 2 reverse

As Europe prepared for WWI, the Pavesi company realized there could be a demand for military tractors to replace horses for pulling heavy loads. The result was the US model, and after the war some of these ex-army tractors were sold for farm work.

▷ McDonald Imperial EB

Date 1912 **Origin** Australia
Engine McDonald 2-cylinder gasoline/kerosene
Horsepower 20 hp
Transmission 3 forward, 1 reverse

Australia's tractor industry started in 1908 when the first McDonald Imperial EA series tractor was built. About 13 were sold before a new model arrived in 1912—the Imperial EB. Design improvements included more efficient engine cooling.

△ Minerva

Date 1914 **Origin** Belgium
Engine Minerva 2-cylinder gasoline
Horsepower 25 hp
Transmission 3 forward, 1 reverse

The Minerva company made luxury cars and trucks before WWI. It also built this load-carrying utility vehicle for the Belgian army. When the war ended, some of the tractors were bought by farmers who used them for haulage work.

◁ Chapron CR

Date c. 1919 **Origin** France
Engine Chapuis Dornier 4-cylinder gasoline
Horsepower 12 hp
Transmission 2 forward, 1 reverse

This was a vineyard tractor built in a major wine-producing area by a company specializing in equipment for the wine industry. The tractor was designed to work with implements attached to the front or the rear, or both together.

▷ Pavesi P4

Date 1917 **Origin** Italy
Engine Pavesi 4-cylinder gasoline
Horsepower 40 hp
Transmission 3 forward, 1 reverse

This was another of the Pavesi tractors designed for military use with articulated steering, plus four-wheel drive for pulling heavy guns and equipment on the battlefield. Production ended in 1942 after Pavesi was taken over by Fiat.

◁ Scemia Universal

Date 1919 **Origin** France
Engine Saunderson 2-cylinder gasoline/kerosene
Horsepower 25 hp
Transmission 3 forward, 1 reverse

While Britain's Saunderson company faced increasing financial problems in its home market after WWI, a version of the Saunderson Universal tractor was built under license in France by Scemia. It was a popular choice in the medium-power sector.

▷ Fiat 702

Date 1919 **Origin** Italy
Engine Fiat 4-cylinder gas/kerosene
Horsepower 20 hp
Transmission 3 forward, 1 reverse

Fiat used engines from its truck range for its first tractor, the 702, which was available from 1919. Fiat started with a 5.6-liter engine, but within 18 months the output was increased to 30 hp when equipped with a 6.2-liter truck engine.

▷ Renault HO

Date 1921 **Origin** France
Engine Renault 4-cylinder gasoline
Horsepower 20 hp
Transmission 3 forward, 1 reverse

Renault's first tractor was a tracklayer based on a small army tank the company built in WWI. It was joined in 1921 by this HO wheeled tractor, which shared the GP's styling with the radiator behind the engine, but the power was reduced.

Early Tracked Machines

The years from 1915 to 1922 were a period of experimentation in the development of crawler tractors. The designs were wide and varied; however, most led to dead ends, because of either mechanical failures or the excessive cost of production. British manufacturers produced some very promising designs that were well built, of advanced design, but expensive to build. All eventually succumbed to the relentless commercial pressure of the early 1920s. The US manufacturers were better placed—they had a vast home market to satisfy and were not as reliant as the British on the old construction methods inherited from the days of steam.

△ **Cleveland Cletrac Model H**

Date 1917 **Origin** US
Engine Weidley 4-cylinder gasoline/kerosene
Horsepower 20 hp
Transmission 1 forward, 1 reverse

The Model H was one of the company's early designs that was marketed in the UK by H.G. Burford. The tractor was steered by the somewhat unconventional method of a steering wheel connected to the controlled differential by a series of links.

△ **Holt Experimental Model T16**

Date 1917 **Origin** US
Engine Holt 4-cylinder gasoline
Horsepower c. 30 hp
Transmission 3 forward, 1 reverse

This experimental Holt tractor, one of a batch of seven, was the result of a request by the US military to Holt for the design of a medium-sized artillery tractor. Although a compact and neat design emerged, the project was not continued.

△ **Bullock Creeping Grip 12-20**

Date 1917 **Origin** US
Engine Waukesha 4-cylinder gasoline
Horsepower 35 hp
Transmission 2 forward, 1 reverse

The 12-20 Creeping Grip was another early type of crawler tractor imported into the UK. It was exhibited at the Highland Tractor Trials in 1917, but it was expensive and only a few were sold. This very unusual machine has the engine placed behind the transmission.

△ **Bates Steel Mule Model F**

Date 1921 **Origin** US
Engine Midwest 4-cylinder gasoline/kerosene
Horsepower 38 hp
Transmission Fully-enclosed running in oil

Produced from 1921 to 1937, the Model F was built with three different engines. The Midwest was followed in 1926 by a Beaver, and in 1928 by a Le Roi, both four-cylinder units. As is usual with a half-track design, the tractor was steered by the front wheels. Throughout the production of their tracked machines, Bates always referred to these tracks as "crawlers," in an attempt to avoid any connection with the registered trademark of Caterpillar.

◁ **Clayton Chain Rail**

Date 1918 **Origin** UK
Engine Dorman 4JO 4-cylinder gasoline/kerosene
Horsepower 35 hp
Transmission 2 forward, 1 reverse

Produced by established steam engine manufacturers Clayton & Shuttleworth, Clayton Chain Rail was one of the earliest successful designs of British crawlers. Steering was by clutch and brake, the brakes being operated by foot pedals, but the clutches were actuated by a linkage system connected to a steering wheel.

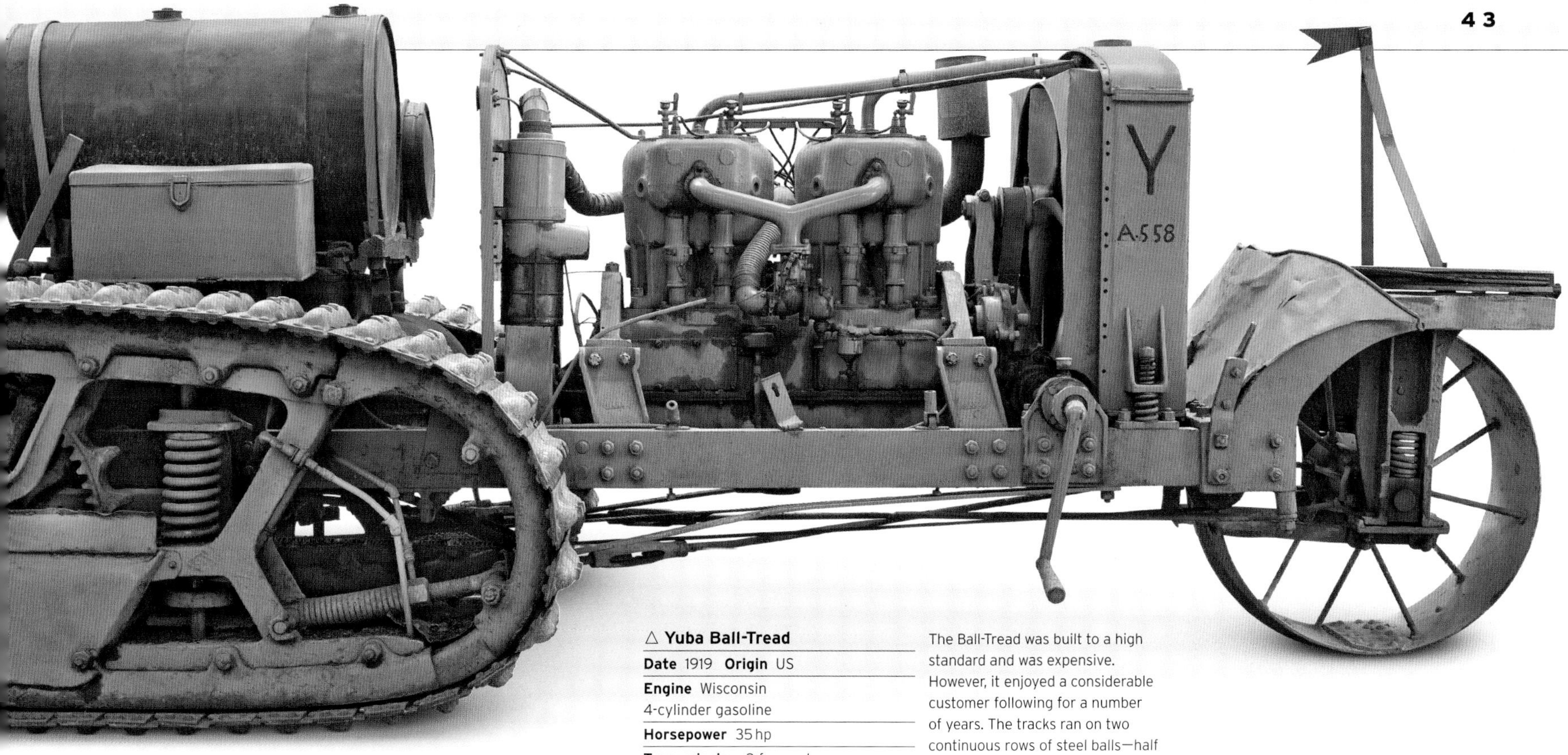

△ **Yuba Ball-Tread**

Date 1919 **Origin** US

Engine Wisconsin 4-cylinder gasoline

Horsepower 35 hp

Transmission 2 forward, 2 reverse

The Ball-Tread was built to a high standard and was expensive. However, it enjoyed a considerable customer following for a number of years. The tracks ran on two continuous rows of steel balls—half the race formed in the track rails and the other half in the track frames.

◁ **Blackstone Track Tractor**

Date 1919 **Origin** UK

Engine Blackstone 3-cylinder kerosene (lamp oil)

Horsepower 25 hp

Transmission 3 forward, 1 reverse

The Blackstone was an advanced and complex machine for its day. The engine was designed to start from cold on lamp oil. A simple braked differential was used to steer the tractor. The Blackstone was very expensive and few were sold.

△ **Best 60**

Date 1919 **Origin** US

Engine Best 4-cylinder gasoline

Horsepower 60 hp

Transmission 2 forward, 1 reverse

A milestone in crawler development, the basic design of Best 60 became the standard for mainstream track-type tractor evolution. When tested in Nebraska, it could produce only 56 hp, but the engine timing details were reworked and the output increased to 60 hp.

◁ **Renault Model HI**

Date 1920 **Origin** France

Engine Renault 4-cylinder gasoline

Horsepower 34 hp

Transmission 3 forward, 1 reverse

The distinctive features of the HI, as with other Renault crawlers, were the angled center-mounted radiator, the track suspension spring across the front of the chassis, and the tiller steering. It was exported to places such as Australia, New Zealand, and Russia. Total sales amounted to only 610 units.

Model A row-crop tractor in 1936

Great Manufacturers
John Deere

Today, John Deere is the world's biggest farm equipment manufacturer, with best-selling tractor and combine harvester ranges. Yet its success story had a much more modest start in 1825, when a 21-year-old blacksmith opened a small tool shop in Illinois, repairing and manufacturing agricultural tools.

JOHN DEERE LEARNED his trade as a blacksmith in Vermont, but it was in the state of Illinois in 1837 that he created the first of many pieces of farm equipment that would bear his name: the cast-steel plow. The tool featured steel moldboards that improved wear resistance, and a more efficient self-cleaning action. It was ideal for the tough soil of the Midwestern plains, and it was soon joined by other products in a fast-expanding machinery range. Although tractors were still proving difficult to market, many firms had attempted to sell them. However by World War I, John Deere was one of the few big US farm equipment companies not selling tractors.

John Deere (1804-1886)

Elsewhere, in 1892, John Froelich, seeking a replacement for the costly and time-consuming steam tractor, designed and built a gas-powered engine. The Froelich Gasoline Traction Engine was probably the first tractor with a reverse gear, and, unlike steam engines, it did not demand copious coal or time spent lighting fires or heating water. Other contractors liked the machine, and the Waterloo Gasoline Traction Engine Company was formed to produce it.

The Froelich machines were not successful, but their engines were in high demand, and from 1895 the newly renamed Waterloo Gasoline Engine Company focused on engine building. In 1912 the company returned to the tractor market, selling the Waterloo Boy tractors in time to benefit from the wartime boom, although engines remained its most popular product.

The John Deere Company had been developing tractors since 1912, but it was not until 1918 that it firmly resolved to enter the tractor market by acquiring Waterloo. The Iowa company's tractors had enjoyed a good reputation, but their design was dated and relied on a two-cylinder engine at a time when four cylinders were increasingly popular. However, when John Deere took control, it kept the two-cylinder horizontal engine design for its new tractors, a setup that continued to power almost all John Deere machines until 1960. It became one of the most successful tractor engines ever built.

Plow that broke the plains
In this painting entitled "Reviewing Performance" by Walter Haskell Hinton, farmers are witnessing a demonstration of the improved moldboards on John Deere's new plow.

> "I will **never** put my **name** on a **product** that doesn't have in it the **best** that is in me."
>
> JOHN DEERE

Following the takeover, John Deere improved the Waterloo Company's existing Model N tractor and retained the Waterloo Boy name while it began designing a new Model D two-cylinder tractor. The first Model D to carry the John Deere name arrived in 1923, the first in a long series of Model Ds available in various versions until 1953. Later models included the General Purpose row-crop tractor, the first tricycle-style John Deere, in 1929.

Major developments during the 1930s included the introduction of a hydraulically operated implement lift option on the row-crop Model A. Then in 1938 John Deere joined the styling trend, employing the leading industrial designer Henry Dreyfuss to give its tractor range a new look, starting with the Model A and Model B tractors.

Diesel arrived with John Deere's Model R tractor in 1948. The engine was the familiar two-cylinder design, using an electric motor to engage the small gasoline engine, which then started the larger diesel. With up to 51 hp available, the Model R was the most powerful John Deere yet. Model identification changed from letters to numbers in 1952 when a new tractor range was announced, and in 1956 John Deere's all-green paint color was replaced by a more eye-catching green and yellow livery on the new 20 series.

Four-cylinder engines began to take over after John Deere announced its "New Generation of Power" in 1960.

Waterloo Boy at work
This photograph shows a Waterloo Boy tractor towing a disc plow in July 1918, the year that its manufacturer, the Waterloo Gasoline Engine Co., was bought by John Deere.

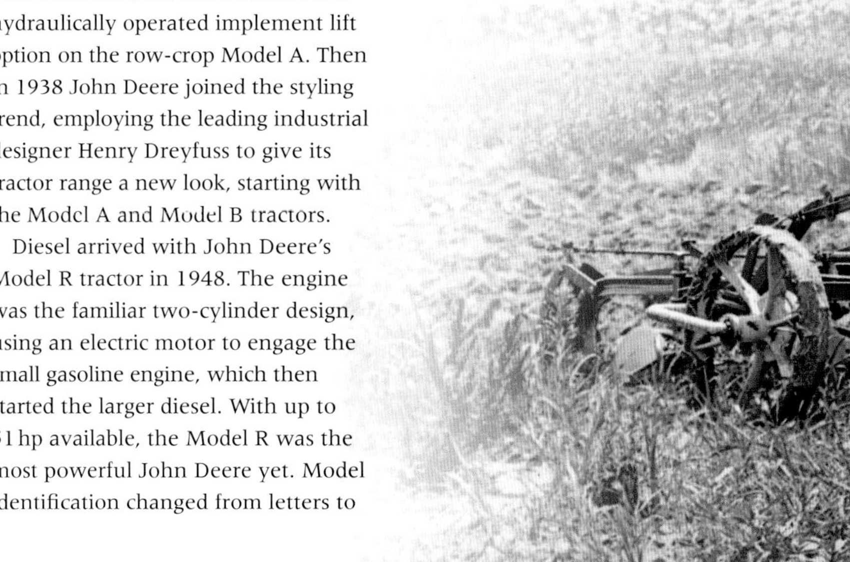

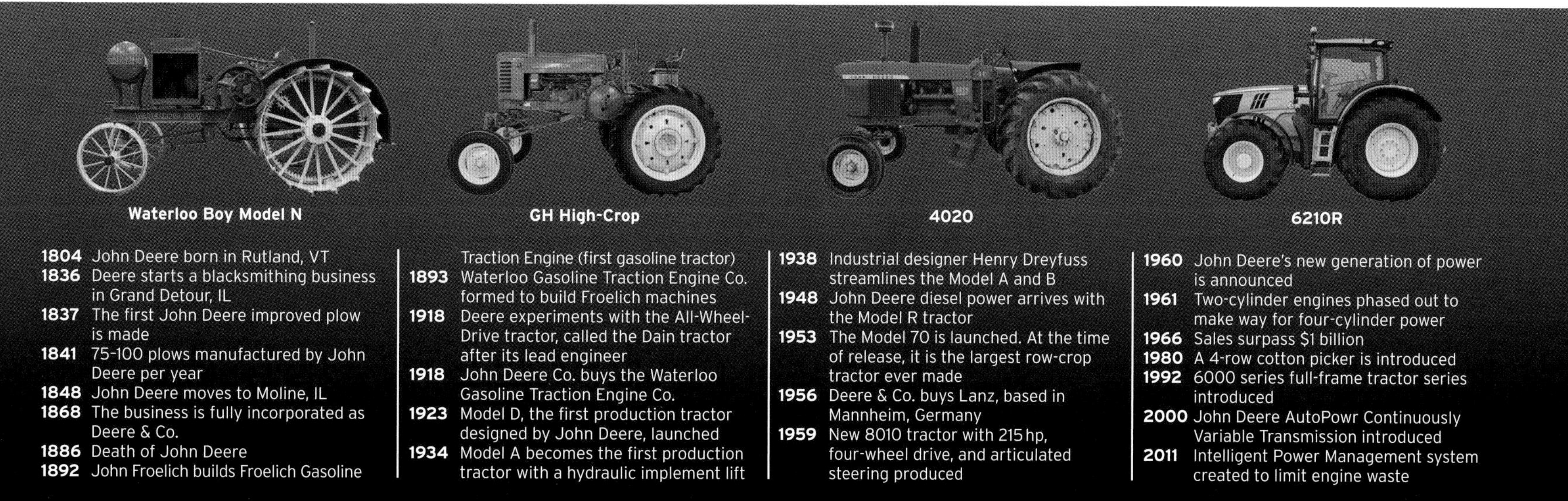

1804 John Deere born in Rutland, VT
1836 Deere starts a blacksmithing business in Grand Detour, IL
1837 The first John Deere improved plow is made
1841 75–100 plows manufactured by John Deere per year
1848 John Deere moves to Moline, IL
1868 The business is fully incorporated as Deere & Co.
1886 Death of John Deere
1892 John Froelich builds Froelich Gasoline Traction Engine (first gasoline tractor)
1893 Waterloo Gasoline Traction Engine Co. formed to build Froelich machines
1918 Deere experiments with the All-Wheel-Drive tractor, called the Dain tractor after its lead engineer
1918 John Deere Co. buys the Waterloo Gasoline Traction Engine Co.
1923 Model D, the first production tractor designed by John Deere, launched
1934 Model A becomes the first production tractor with a hydraulic implement lift
1938 Industrial designer Henry Dreyfuss streamlines the Model A and B
1948 John Deere diesel power arrives with the Model R tractor
1953 The Model 70 is launched. At the time of release, it is the largest row-crop tractor ever made
1956 Deere & Co. buys Lanz, based in Mannheim, Germany
1959 New 8010 tractor with 215 hp, four-wheel drive, and articulated steering produced
1960 John Deere's new generation of power is announced
1961 Two-cylinder engines phased out to make way for four-cylinder power
1966 Sales surpass $1 billion
1980 A 4-row cotton picker is introduced
1992 6000 series full-frame tractor series introduced
2000 John Deere AutoPowr Continuously Variable Transmission introduced
2011 Intelligent Power Management system created to limit engine waste

New and improved
Sales brochure for a later version of the Model D. By this time, John Deere had adopted its striking green and yellow livery.

Five models, beginning with the 36 hp 1010, were announced between 1961 and 1963. About 95 percent of the tractors' components were totally new, and the machines featured further styling by Dreyfuss and an increased emphasis on diesel power. More new models followed in 1963; the 130 hp 5020 was the top model, and the main technology advance was the introduction of powershift transmission.

Another new tractor generation arrived in 1972 when the 20 series made way for the 30 range. This time the emphasis was on cabs that ensured safety and driver comfort. The 6000 series tractor launch in 1992 brought another major development: the flexible full-frame structure. John Deere underwent another large-scale redesign in 2011 with the 7R series. Only the front axle was carried over from the older 7030; the engine, transmission, cab, and hydraulics were all upgraded.

Today John Deere is spearheading research into driverless tractors. Despite being an area of interest for manufacturers for many years, commercial progress has been almost nonexistent thus far. However, in 2011, John Deere demonstrated its new Machine Sync remote-control system, which allows combine harvester drivers to control the speed and steering of a tractor and trailer accurately and remotely while unloading the combine's grain tank. The device won a gold medal at 2011's Agritechnica show and is now increasingly available on a commercial basis.

Shipping out
Barge loads of new John Deere tractors are shipped along the Rhine from the company's Mannheim factory.

1921–1938

COMING OF AGE

CASE

RUMELY

COMING OF AGE

The arrival of the Fordson, with its engine, gearbox, and rear transmission forming the backbone of the tractor, ushered in a new era of unit construction that other manufacturers were quick to follow. A few firms persevered with outdated "framed" designs, but most agreed that Ford's concept was the way forward.

The tractor had come of age, and the manufacturers were eager to consolidate their position in the marketplace. However, the 1920s and '30s were marked by a severe worldwide economic depression, with agriculture falling into decline. The US manufacturers, having made greater use of mass-production techniques, were able to weather the storm much better than their European counterparts, and were soon dominating the global markets. Many of the British makers fell by the wayside, to the point where there was only one model of wheeled tractor, the Marshall 18/30, in production in the UK in 1932.

To encourage sales, the tractor industry explored new ideas and redefined the concept of power farming with specialty machines for particular markets. It was the era of the row-crop tractor and the heavy crawler. New features included power takeoffs and pneumatic tires. During this period, the diesel engine also gained a toehold in agriculture, especially in those European countries that did not have access to cheap oil supplies.

The most notable invention of the era was the introduction of Harry Ferguson's hydraulic lift and three-point linkage on his Type A tractor in 1936. Although few realized it at the time, this innovative concept would revolutionize tractor design for many years to come.

△ **Dagenham production**
The manufacture of Fordson's Model N tractor was moved to the UK in 1933 following the opening of Ford's Dagenham factory.

"I merely sat on a **sack-covered iron seat**, and was **carried** across the field by the power of **20 synthetic horses**."

HENRY WILLIAMSON (1895–1977)
BRITISH WRITER, NATURALIST, FARMER, AND PROLIFIC RURALIST

◁ **Progress in mechanization** replaces generations of toil on the land, according to this 1928 Advance-Rumely advertisement.

Key events

- ▷ **1921** International Harvester's 15-30 model is equipped with a drawbar, belt pulley, and power takeoff.
- ▷ **1921** German manufacturer Lanz introduces the 12 hp HL Bulldog.
- ▷ **1923** John Deere introduces its first two-cylinder tractor, the Model D. The German Benz-Sendling Type BK is the world's first tractor with a high-speed diesel engine.
- ▷ **1924** International Harvester launches the Farmall model, introducing the idea of a general-purpose row-crop tractor.
- ▷ **1927** The Soviet Politburo approves a factory to build Russian versions of the International 15-30 model in Stalingrad.
- ▷ **1929** The Wall Street Crash foreshadows the Great Depression, which led to a severe downturn in agriculture and a declining tractor market.
- ▷ **1930** Royal Agricultural Society of England and University of Oxford organize World Tractor Trials.
- ▷ **1931** Caterpillar launches its first diesel crawler tractor, the Diesel Sixty model.
- ▷ **1932** Allis-Chalmers offers pneumatic tires as standard on its Model U.
- ▷ **1936** Harry Ferguson's Type A is built at the David Brown factory, Yorkshire, UK.

△ **Hydraulic lift**
Harry Ferguson's Type A, the first production tractor to feature a hydraulic lift, was launched to great acclaim in 1936.

US Consolidation

This was a time for consolidation in the US. Some smaller manufacturers disappeared because of competition from the mass-produced Fordson or, in some cases, because the new Nebraska tractor test program exposed poor standards of performance and after-sales support. Frameless designs were taking over—another Fordson influence—and reliability improved. Transmission design advanced only in the 1930s, when inflatable rubber tires allowed faster travel speeds and made a greater choice of gears necessary.

▷ Avery 45-65

Date 1923 **Origin** US

Engine Avery 4-cylinder horizontally opposed gasoline/kerosene

Horsepower 69 hp

Transmission 2 forward, 1 reverse

Demand for heavyweight prairie giant tractors faded in the 1920s, but they were still available, and for farmers aiming to impress their neighbors, the Avery 45-65 might have been a good choice. It weighed almost 10 tons (9,072 kg), with a big engine developing its rated power at a leisurely 634 rpm.

△ Rumely OilPull Model M 20-35

Date 1924-27 **Origin** US

Engine Rumely 2-cylinder gasoline/kerosene

Horsepower 43 hp

Transmission 3 forward, 1 reverse

Advance-Rumely was one of the few steam traction engine companies that switched successfully to tractors. The 20-35 was among the smaller models, but it included familiar OilPull features such as a rectangular cooling tower and a big, slow-speed, two-cylinder engine.

△ Baker 22-40

Date 1926 **Origin** US

Engine Beaver 4-cylinder gasoline

Horsepower 40 hp

Transmission 2 forward, 1 reverse

Baker Co. was a steam traction engine builder and the 22-40 was its first tractor. It was a little more than a collection of proprietary parts with an engine from Beaver Manufacturing Co., Milwaukee. A 25-50 tractor with a Wisconsin engine followed in 1927. Rugged and powerful, they remained in production until the late 1930s.

△ Case 12-20

Date 1927 **Origin** US

Engine J.I. Case 4-cylinder gasoline/kerosene

Horsepower 25.5 hp

Transmission 2 forward, 1 reverse

The distinctive-looking Case "crossmount" series with its transverse engines is among the tractor industry's design classics. It was available from 1916 until the late 1920s and included the 12-20 model with pressed steel front and rear wheels.

▷ Allis-Chalmers 20-35

Date 1927 **Origin** US

Engine Allis-Chalmers 4-cylinder gasoline

Horsepower 35 hp

Transmission 2 forward, 1 reverse

The 20-35 was typical of many US tractors in the 1920s. It was sturdy and solidly built, with a big 7.2-liter engine producing a modest power output and earning it a reputation for reliability. The transmission was just a basic two-speed gearbox.

△ **John Deere General Purpose**

Date 1929 **Origin** US

Engine John Deere 2-cylinder horizontal gasoline/kerosene

Horsepower 25 hp

Transmission 3 forward, 1 reverse

John Deere's first row-crop tractor arrived in 1927 as the Model C and reappeared in 1928 as the General Purpose. It was one of the first tractors with a power-operated implement lift, but the rubber tires on this 1929 example would have been a later addition.

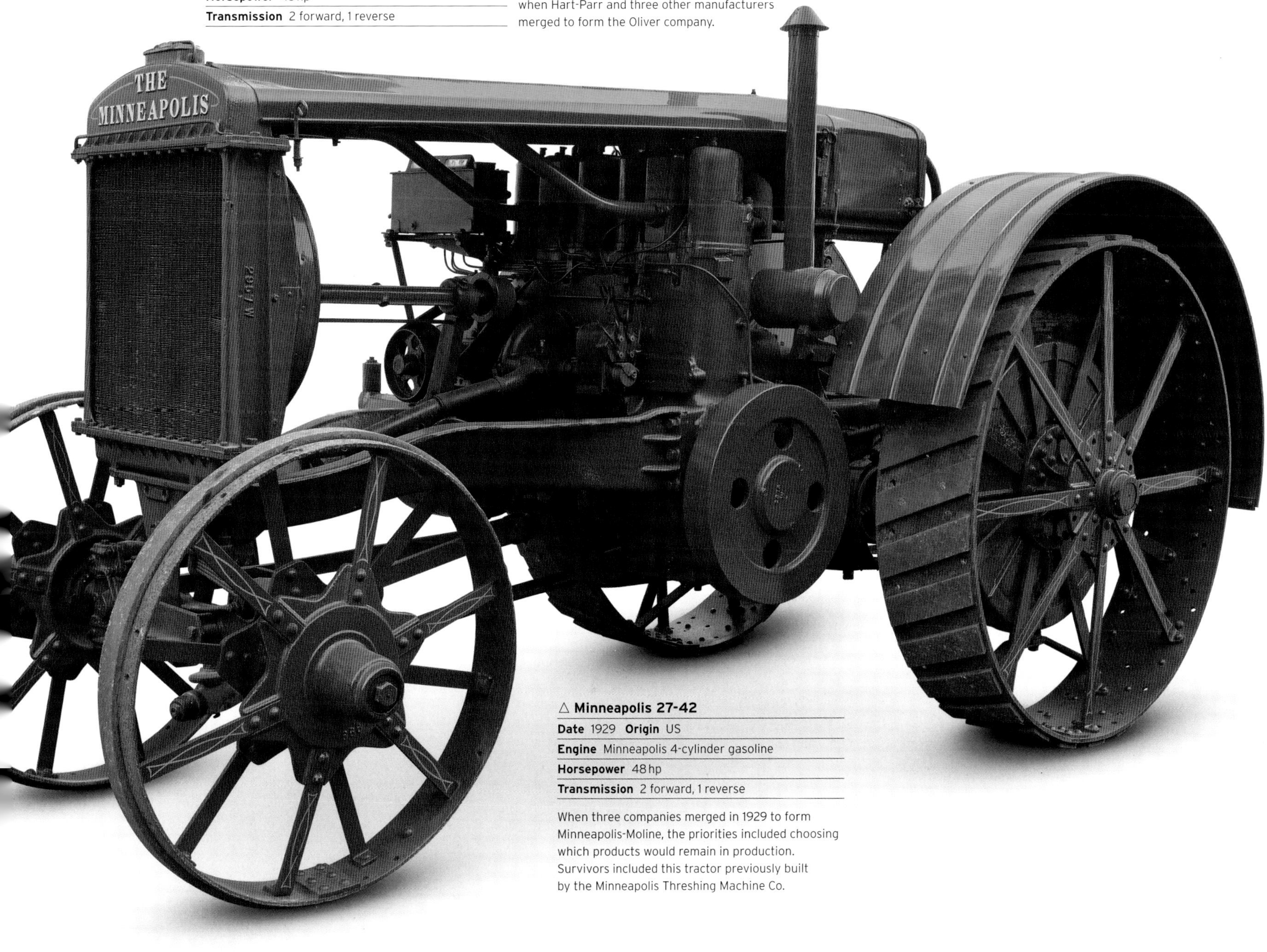

△ **Hart-Parr 18-36**

Date 1927 **Origin** US

Engine Hart-Parr 2-cylinder horizontal gas/kerosene

Horsepower 43 hp

Transmission 2 forward, 1 reverse

There were two versions of the 18-36, starting with the Model G with a two-speed gearbox and followed in 1928 by the three-speed Model H. The 18-36 was still in production in 1929 when Hart-Parr and three other manufacturers merged to form the Oliver company.

△ **Minneapolis 27-42**

Date 1929 **Origin** US

Engine Minneapolis 4-cylinder gasoline

Horsepower 48 hp

Transmission 2 forward, 1 reverse

When three companies merged in 1929 to form Minneapolis-Moline, the priorities included choosing which products would remain in production. Survivors included this tractor previously built by the Minneapolis Threshing Machine Co.

International 10-20 industrial tractor loading lumber in the 1930s

Great Manufacturers

International Harvester

For much of the 20th century, International Harvester dominated the North American market, and was the foremost "full-line" manufacturer of agricultural equipment. Living up to its name, the company was a truly international organization with factories across the US, Canada, the UK, France, Germany, Sweden, and even Russia.

INTERNATIONAL HARVESTER'S origins date back to 1809, when Virginia farmer Robert McCormick began tinkering with a mechanical reaper to harvest his grain. Robert's son, Cyrus Hall McCormick, later perfected the reaper and formed a partnership to manufacture the machine in Chicago.

McCormick's rival, William Deering, made grain reapers in Plano, IL, from 1870. In 1902 the Deering Harvester Company merged with the McCormick Harvesting Machine Company and three other firms to form the International Harvester Company (known as IH). The other constituents of the new organization were Warder, Bushnell & Glassner of Ohio, the Milwaukee Harvester Company, and the Plano Manufacturing Company.

Based in Chicago, IH grew to become North America's leading farm equipment manufacturer with subsidiaries in many countries. The old rivalries took a while to fade, and for a time the company operated separate product lines, with McCormick and Deering being the prominent brands.

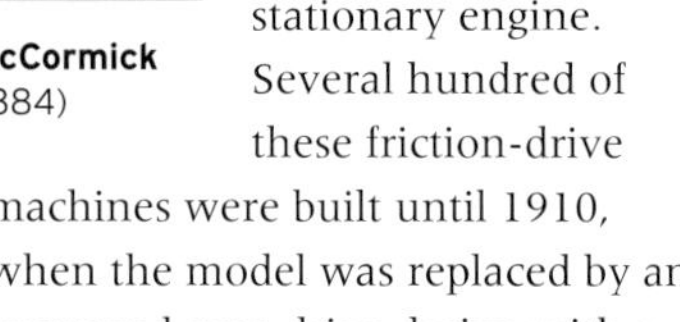

Cyrus Hall McCormick
(1809–1884)

International entered the tractor market in 1906 by adopting a proprietary chassis designed to take its single-cylinder "Famous" gasoline stationary engine. Several hundred of these friction-drive machines were built until 1910, when the model was replaced by an improved gear-drive design with a friction reverse.

By 1910, IH was the US's largest tractor manufacturer, followed by Rumely and Hart-Parr. The following year, it opened its new Tractor Works in Chicago, which became home to the Mogul line. These were very much a McCormick product, with the developments supervised by Edward A. Johnston, who had joined McCormick in 1894.

The Deering faction was responsible for the Titan line, produced at the Milwaukee Works, the former home of the Milwaukee Harvester Company. The move toward a single product line was finally prompted by competition in the marketplace from Henry Ford and his cut-price Fordson Model F. IH countered Ford's attack with a new, lightweight model, the Junior 8-16, based on automotive components. The two industry giants then clashed in the so-called "tractor war," with each drastically slashing prices. Ford eventually withdrew from the US market after IH introduced its superior new "gear-drive" tractors—the 15-30 launched in 1921 and the 10-20 two years later.

The company strengthened its hold on the industry during the 1920s and '30s, with several innovations. In 1924 the Farmall concept of a general-purpose row-crop machine was born, sending ripples through the industry. To cope with demand for this new model, IH opened the Farmall Works in Rock Island, Illinois, which later became the company's main tractor production center.

Crawlers were added to the line in 1931, and the company began developing a gasoline-start diesel engine. In November 1936, International brightened its image by changing the color of its tractors from gray to red—ostensibly to

Triple power
One of the first tractors to come with a power takeoff, the 10-20 model was sold under the "Triple Power" slogan. More than 200,000 were built.

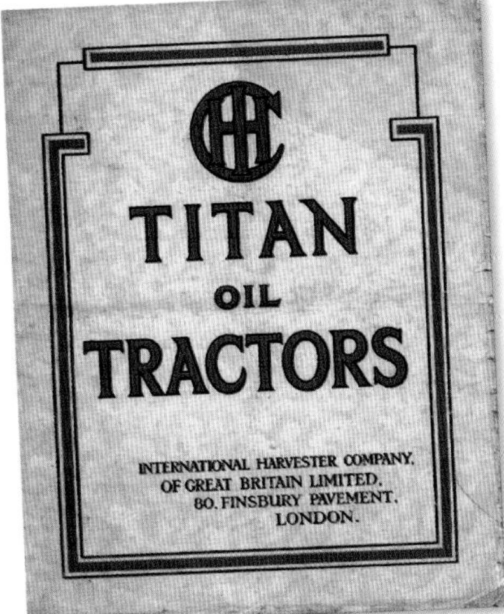

Titan in the UK
To aid in the Ministry of Munitions' plowing campaign, nearly 2,000 Titan 10-20 tractors were supplied to the International Harvester Co. of Great Britain during WWI.

International at war
This Industrial I-4 tractor is moving a B-25 Mitchell bomber at North American Aviation's Kansas City, KS, facility in 1942. Much of IH's production capacity was diverted to munitions work during World War II.

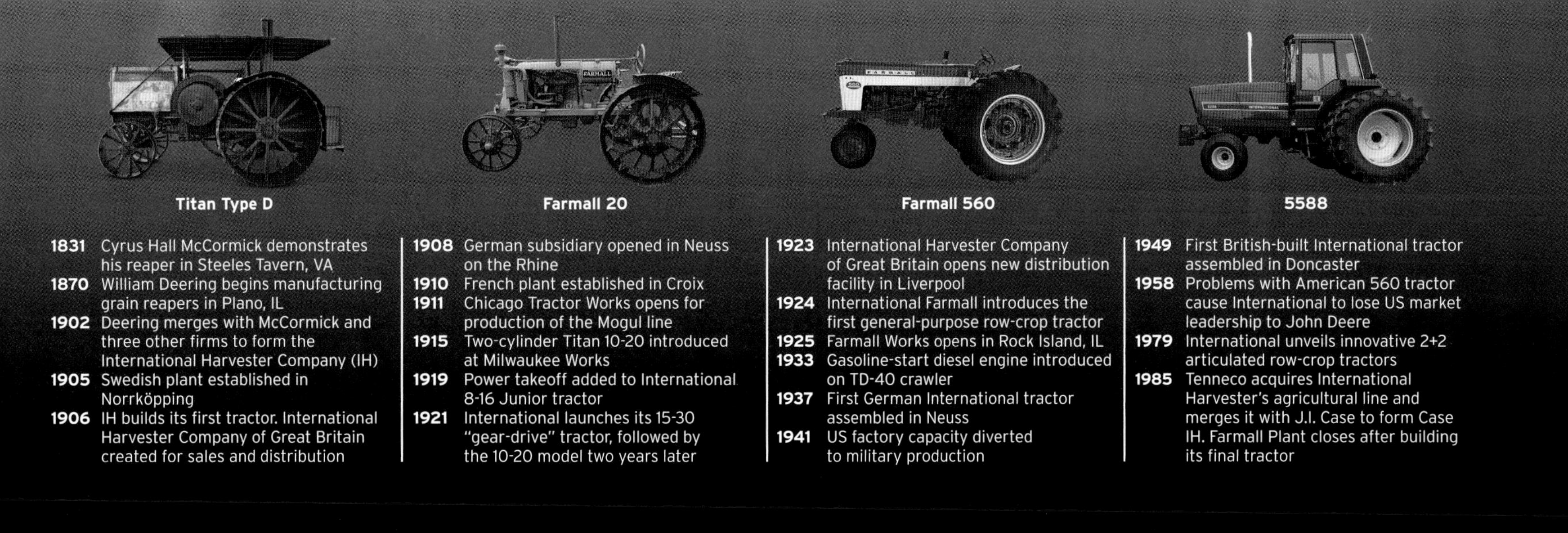

1831 Cyrus Hall McCormick demonstrates his reaper in Steeles Tavern, VA
1870 William Deering begins manufacturing grain reapers in Plano, IL
1902 Deering merges with McCormick and three other firms to form the International Harvester Company (IH)
1905 Swedish plant established in Norrköpping
1906 IH builds its first tractor. International Harvester Company of Great Britain created for sales and distribution
1908 German subsidiary opened in Neuss on the Rhine
1910 French plant established in Croix
1911 Chicago Tractor Works opens for production of the Mogul line
1915 Two-cylinder Titan 10-20 introduced at Milwaukee Works
1919 Power takeoff added to International 8-16 Junior tractor
1921 International launches its 15-30 "gear-drive" tractor, followed by the 10-20 model two years later
1923 International Harvester Company of Great Britain opens new distribution facility in Liverpool
1924 International Farmall introduces the first general-purpose row-crop tractor
1925 Farmall Works opens in Rock Island, IL
1933 Gasoline-start diesel engine introduced on TD-40 crawler
1937 First German International tractor assembled in Neuss
1941 US factory capacity diverted to military production
1949 First British-built International tractor assembled in Doncaster
1958 Problems with American 560 tractor cause International to lose US market leadership to John Deere
1979 International unveils innovative 2+2 articulated row-crop tractors
1985 Tenneco acquires International Harvester's agricultural line and merges it with J.I. Case to form Case IH. Farmall Plant closes after building its final tractor

make them safer on the road. A modernized range of "styled" tractors arrived soon afterward, in 1939.

In 1941, many of IH's US factories went onto a military footing to support the war effort. Its wartime products included military vehicles, tanks, torpedoes, guns, and artillery shells. During the postwar years, International expanded its empire still further by establishing tractor production in the UK and France and reinstating operations in Germany. International's dominance in the US was unchallenged until the late 1950s, when badly conceived and rushed product releases allowed John Deere to eat into its market share. By the 1970s, IH was overstretched. Its unwieldy product line included trucks, construction machinery, and even domestic refrigerators.

The company fought back with an ambitious worldwide program, with tractors and components being manufactured in the US, UK, France, Germany, Japan, India, Australia, and Mexico. But problems with labor relations in the US plunged the business into difficulty. The end came in 1985 after Tenneco acquired IH's agricultural line and merged it with J.I. Case to form Case IH. On May 14, the Farmall Plant closed after the final tractor, a 5488, rolled off the line.

Doncaster production
Brochure for the British International 85 Series, introduced in Doncaster, Yorkshire, in 1981. The range covered six tractors from 53 hp to 85 hp, and included a hydrostatic model.

Britain Between the Wars

The 1920s and 1930s were a difficult time for the UK's tractor industry, with many leading companies failing to survive. One of the problems was strong competition from North American imports, while slim profit margins made many UK farmers reluctant to invest in new equipment. The good news included export success, and the decision to move Fordson production to England was a major boost to the UK's tractor industry. There were also important technical developments, including progress in diesel engine design, plus the launch of the Ferguson System for implement attachment and control.

△ Vickers Aussie

Date 1925-33 **Origin** UK
Engine Vickers 4-cylinder gas/kerosene
Horsepower 30 hp
Transmission 2 forward, 1 reverse

Perhaps with an eye to export sales, Vickers called its new tractor launched in 1925 the Aussie, although the name was dropped in 1926. A special feature was the patented self-cleaning design of the rear wheels developed in Australia.

△ British Wallis

Date 1921 **Origin** UK
Engine Ruston 4-cylinder gasoline/kerosene
Horsepower 25 hp
Transmission 2 forward, 1 reverse

This machine, built by Ruston & Hornsby when they switched from making steam engines, was based on an agreement with Wallis Tractor Co., a subsidiary of J.I. Case Plow Works, to build tractors under license. It was available for about 10 years from 1919.

▽ Saunderson

Date 1922 **Origin** UK
Engine Saunderson V-twin overhead-valve gasoline/kerosene
Horsepower 20 hp
Transmission 2 forward, 1 reverse

Competition from Fordson in the early 1920s, plus government tractor sales from the WWI food production campaign, caused problems for Saunderson. This new lightweight model in 1922 failed to attract enough customers, in spite of a three-year warranty.

▷ Peterbro

Date 1925 **Origin** UK
Engine Peterbro 4-cylinder overhead-valve gasoline/kerosene
Horsepower 30 hp
Transmission 2 forward, 1 reverse

The Peterbro was built in Peterborough by Peter Brotherhood. It was available from 1920 with a power unit developed by Harry Ricardo, a leading engine designer, but in spite of this, and the addition of a half-track version in 1928, sales were disappointing.

◁ Garrett

Date 1933 **Origin** UK
Engine Gardner 4-cylinder diesel
Horsepower 38 hp
Transmission 3 forward, 1 reverse

The Agricultural & General Engineers group introduced its Garrett tractor with a choice of Aveling & Porter or Blackstone diesel engines at the 1930 World Tractor Trials. Just 12 were built before the AGE group collapsed in 1932. This is a replica of proposals for a version with a Gardner engine.

◁ Marshall 15/30

Date 1930 **Origin** UK
Engine Marshall single-cylinder two-stroke horizontal diesel
Horsepower 30 hp
Transmission 3 forward, 1 reverse

When Marshall developed its new tractor in the late 1920s, it used a single-cylinder diesel engine—a surprise choice when diesel power was almost unknown outside Germany and Italy and most British and US tractor companies used multi-cylinder engines.

△ Rushton

Date c. 1929 **Origin** UK
Engine AEC 4-cylinder gasoline/kerosene
Horsepower 20 hp
Transmission 3 forward, 1 reverse

Launched in Britain in 1929 as a competitor to the Fordson in an effort to break the US dominance of the tractor market, the Rushton copied much of the Fordson design. It was as an offshoot product of the Associated Equipment Company (AEC), which built London's buses.

▷ Fordson Model N

Date 1933 **Origin** UK
Engine Fordson 4-cylinder gasoline or gasoline/kerosene
Horsepower 23 hp (kerosene)
Transmission 3 forward, 1 reverse

The move of Fordson production from Ireland to the Ford factory in the east London suburb of Dagenham boosted the UK tractor industry and added to its export success. The tractor here was the first Dagenham-built Fordson, with rubber tires from 1935.

◁ Ferguson Type A

Date 1937 **Origin** UK
Engine Coventry Climax 4-cylinder gasoline/kerosene
Horsepower 20 hp
Transmission 3 forward, 1 reverse

The principal features of Harry Ferguson's three-point linkage implement attachment and control system were developed by 1930. The first production tractor with the equipment was the Ferguson Type A or "Ferguson-Brown" available from 1936.

Ferguson Type A

Harry Ferguson's first tractor, introduced in 1936, resulted from a manufacturing agreement with David Brown. Officially called the Type A, it was more commonly known as the "Ferguson-Brown." The tractor featured Ferguson's revolutionary hydraulic system, with three-point linkage and automatic depth control. But farmers were put off by the idea of buying special implements, and only 1,350 Type A models were made before production ended in 1939.

WHAT MADE Ferguson's design unique was its combination of draft control—using the tractor's hydraulics automatically to control the depth of the implement—with a converging three-point linkage. The linkage lifted the plow out of the ground, and both pulled and carried it while working, transferring its weight onto the tractor's rear wheels to aid traction.

The line of pull of the active implement extended to a theoretical hitch point just behind the tractor's front axle. This created strong downward thrust on the front wheels to keep the tractor stable, particularly on hillsides. The converging linkage also ensured that the plow followed a straight course behind the tractor.

The pump powering the hydraulic system was driven from the gearbox. The top link, in compression during use, sent signals to the hydraulic system, opening and closing a valve to regulate working depth.

SPECIFICATIONS			
Model	Ferguson Type A	**Capacity**	133 cu in (2,175 cc)
Built	1937	**Transmission**	3 forward, 1 reverse
Origin	UK	**Top speed**	4.9 mph (8 km/h)
Production	1,350	**Length**	9 ft 5 in (2.9 m)
Engine	20 hp Coventry Climax 4-cylinder gasoline/kerosene	**Weight**	0.9 tons (0.8 metric tons)

Revolutionary tractor
The Type A's integrated linkage made it easier and simpler to attach implements, but the special mounted equipment proved too expensive for many farmers.

FRONT VIEW

REAR VIEW

Two-furrow plow

The Ferguson-Brown with its plow
The Ferguson Type A tractor and the Type B two-furrow mounted plow were designed as an integral unit. The plow was attached to the tractor by a triangulation of hitch points, consisting of two converging lower links and a single top link. Ferguson's concept was a total farming system, and the product line of matched implements included a tiller, ridger, and cultivator.

THE DETAILS

1. Radiator top casting made from aluminum alloy **2.** Toolbox mounted on nearside of engine **3.** Ignition via magneto with impulse coupling **4.** Quadrant for lever controlling hydraulic system **5.** Coil-spring absorbs draft forces acting on top link

Gasoline compartment for starting fuel

Main fuel tank holds 10.8 gallons (41 liters) of kerosene

Water temperature gauge

Vaporizer for running on kerosene, made by Gladwell & Kell, offered as an accessory from 1938

Starting handle

Transmission housing cast from aluminum alloy for lightness

Exhaust pipe underslung for neatness

Pneumatic tires were optional from 1937

European Expansion

Some tractor development in European countries during the 1920s came from established names such as Fiat in Italy, Renault in France, and the German Benz company, but there were also numerous small start-ups. The newcomers made a significant contribution to Europe's tractor production, particularly in sectors such as vineyard tractors, but the failure rate was high. Technical progress in mainland Europe included adopting diesel and hot-bulb engines for tractors. There was a British influence too, as the Austin and Saunderson Universal tractors were built in France.

▷ Fiat 703

Date 1923 **Origin** Italy
Engine Fiat 4-cylinder gasoline/kerosene
Horsepower 35 hp
Transmission 3 forward, 1 reverse

The 703 was an improved version of the original 702, Fiat's first production tractor. Both were powered by a 6.2-liter Fiat truck engine available in gasoline and gasoline/kerosene versions, but the 703 engine was uprated to further increase the power output.

▽ Renault PE

Date 1927 **Origin** France
Engine Renault 4-cylinder gasoline
Horsepower 20 hp
Transmission 3 forward, 1 reverse

A distinctive Renault feature was placing the radiator and cooling fan at the rear of the engine compartment. In front, instead of a radiator, the Renault had an enormous air cleaner. In 1933 the PE became the first French-built tractor available with rubber tires.

◁ Lanz HL12

Date 1925 **Origin** Germany
Engine Lanz single-cylinder hot-bulb
Horsepower 12 hp
Transmission 1 forward, 1 reverse

Heinrich Lanz was Germany's leading manufacturer of hot-bulb or semi-diesel-powered tractors, starting with the Bulldog HL12 designed by Dr. Fritz Huber. A total of 6,000 HL12 tractors were produced between 1921 and 1929.

▽ Latil KTL

Date 1929 **Origin** France
Engine Latil 4-cylinder gasoline
Horsepower 20 hp
Transmission 6 forward, 2 reverse

The Latil KTL was an unconventional transport tractor with advanced features. The specification included four-wheel drive through equal-sized wheels, a six-speed gearbox with a top speed of 17 mph (27 km/h), and the rare advantage of four-wheel braking.

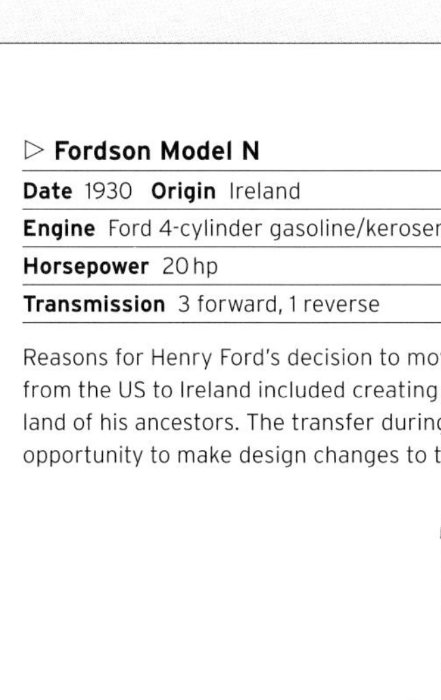

▷ **Fordson Model N**

Date 1930 **Origin** Ireland

Engine Ford 4-cylinder gasoline/kerosene

Horsepower 20 hp

Transmission 3 forward, 1 reverse

Reasons for Henry Ford's decision to move tractor production from the US to Ireland included creating employment in the land of his ancestors. The transfer during 1928–29 was also an opportunity to make design changes to the 12-year-old Fordson.

△ **Austin**

Date 1928 **Origin** France

Engine Austin 4-cylinder gasoline/kerosene

Horsepower 15 hp

Transmission 3 forward, 1 reverse

Britain's Austin car company began making tractors in 1918, but UK sales were disappointing and production ended in 1927. Austins assembled in France were popular with French farmers, and production continued in France, which exported Austins to the UK.

△ **Grillet**

Date 1930 **Origin** France

Engine Ford Model A 4-cylinder gasoline

Horsepower 24 hp

Transmission 2 forward, 1 reverse

Built in Bordeaux, a major wine-producing center, the Grillet was designed for vineyard work. Its compact size and single front-wheel steering made it suitable for small areas, and it could operate equipment attached at the front or the rear.

▽ **Deutz MTZ320**

Date 1934 **Origin** Germany

Engine Deutz 2-cylinder diesel

Horsepower 40 hp

Transmission 3 forward, 1 reverse

Germany led the way with diesel tractor development during the 1920s and 1930s, and the Deutz MTZ320 was an example. As well as the diesel engine with its decompression device for starting, the tractor featured an 11-mph (17.5-km/h) transport speed.

△ **Tirtou**

Date 1934 **Origin** France

Engine Briban single-cylinder gasoline

Horsepower Not known

Transmission 3 forward, 1 reverse

This three-wheeled tractor, with handlebar steering just in front of the fuel tank, appears to have been designed by a motorcycle enthusiast. Surprisingly perhaps, it was meant for general farm work, and photographs from the 1930s show it pulling a plow.

Steam engines and threshing machines were built at Lanz's Mannheim factory

Great Manufacturers

Lanz

One of the major European agricultural engineering firms of the early 20th century, Lanz became famous for its simple and reliable Bulldog tractors. John Deere began investing in the firm in 1956 and owned it outright by 1960. Today the US firm's European factory remains at the same Mannheim site where Lanz began.

THE SON OF A farm machinery importer, Heinrich Lanz was born in Mannheim, Germany, in 1838. He joined the family business in his early twenties, selling British-built equipment to the region's farmers. Popular items included Clayton & Shuttleworth threshing machines and Fowler steam engines.

It was not long before Lanz had opened his own business importing, retailing, and repairing farm implements, and by 1867 he had grown it from a small company with just two employees into a full-scale manufacturing operation that produced various pieces of field and barn equipment.

Heinrich Lanz
(1838–1905)

Twelve years later, in 1879, Lanz's company built its first steam power unit, and the construction of new machinery replaced sales and repairs as the company's principal activity.

By the turn of the century, Lanz could comfortably lay claim to being Europe's largest farm equipment manufacturer, with around 3,000 people producing a range of both non-powered and steam-driven implements, including threshing machines, at the company's base in Mannheim.

More than **200,000** Bulldogs had been built in **Mannheim** when production **ended** in 1956.

Heinrich Lanz died in 1905, but his company continued under the management of family and employees. The product that was to make the firm globally famous was designed and launched 16 years later, in 1921. Dr. Fritz Huber, Lanz's chief engineer, was responsible for the development of what was to become the company's best-known product: the Bulldog tractor. First branded the HL Landbaumotor, or Heinrich Lanz agricultural engine, the Bulldog moniker had its origins in the likeness of the hot bulb located on the head of the cylinder to the British breed of dog.

To start the tractor, the bulb was heated with a blowtorch, after which the detachable steering wheel was used to crank the flywheel. This fired up the single-cylinder two-stroke engine, which relied on heat, rather than compression, for combustion. The versatile engine could run on

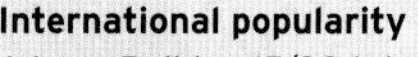

International popularity
A Lanz Bulldog 15/30 takes part in the 1930 World Tractor Trials in Oxfordshire. It was entered by the British agents, the Locomobile Engineering Co. of London.

Steam

HC L2

HR2 Gross

John Deere-Lanz 2416

1838 Heinrich Lanz is born, son of a farm machinery importer based in Mannheim, Germany
1858 Lanz joins the family business, selling products including Clayton & Shuttleworth threshing machines and Fowler steam engines
1860 Lanz founds a farm machinery repair business, serving farmers in the vicinity of Mannheim
1867 The Lanz company begins producing equipment to its own designs

1879 Lanz designs and manufactures its first steam power unit
1880 Steam engines and threshing machines comprise the main part of Lanz's output
1900 Lanz employs 3,000 staff in its Mannheim factory and is said to be Europe's largest farm equipment maker
1905 Death of Heinrich Lanz, aged 67
1921 Launch of the Lanz HL12, the first Bulldog tractor
1923 Four-wheel-drive Bulldog HP model with articulated steering is introduced

1929 A more powerful HR2 model is introduced, producing 22–30 hp
1942 Death of Dr. Fritz Huber, designer of the Bulldog
1950 Introduction of the D5506, the first Lanz tractor to be built after WWII
1951 Launch of the Alldog A1205 tool carrier, with 12-hp air-cooled single-cylinder gasoline engine
1952 Hot bulb replaced with flat-top cylinder head with ignition chamber on new D5506

1955 Launch of the D1616 and D2016 with full diesel engines
1956 Bulldog production ends, with more than 200,000 models having been built in Mannheim
1956 Deere & Co. buys a majority share in Lanz; subsequent tractors badged John Deere-Lanz
1960 Deere acquires remaining Lanz shares
1965 Lanz name disappears from tractor manufacturing with the launch of the John Deere 20 series

Lanz All Purpose Tractor
This pamphlet illustrates the Bulldog's inner workings. Simplicity was the tractor's strength, and even crude oil could be used as fuel.

less-refined fuels, such as kerosene and creosote, in addition to diesel. While the units had just one cylinder, the displacement of the models offered over the tractor's life span covered a range from 6.2 to 14.1 liters. However, in the machine's most common early format it produced just 12 hp. Transmission was simply a chain drive to the rear wheels, with a lever-type clutch, and reversing the tractor was achieved by stopping the engine and restarting it with the flywheel running in the opposite direction.

Later HL models saw engine speed increase from 420 to 500 rpm and a corresponding jump in power to 15 hp. Introduced in 1923, an innovative four-wheel-drive variant, the Bulldog HP, featured articulated steering, with front wheels larger than those at the rear. Over the following decades, developments included a crawler version of the Bulldog.

The company delivered its 100,000th tractor in 1942, having coped with the various engineering demands placed on it by World War II. The Nazi government had dictated that diesel and other liquid fuels were to be conserved for military use. To provide an alternative, Lanz issued kits to equip its tractors with a gas generator. These heavy and unwieldy additions converted other fuels, typically wood, into gas.

Further challenges came with the death of Huber in 1942 and significant war damage to Lanz's factory. The company struggled to get full production back on stream until the early 1950s, when it introduced a complete six-model range. The series spanned 17–36 hp, with a six forward and two reverse gearbox and pneumatic tires. By 1950 a new 55-hp D5506 model was added to the top of the line, with some significant revisions. This marked the beginning of the end for some of Lanz's best-known design features, with the hot bulb replaced in 1952 by a flat-top cylinder head incorporating an ignition chamber. In 1955 Lanz switched from its single-cylinder engine design to multi-cylinder diesels on the D1616 and D2016.

By 1956 Bulldogs were being built under license around the globe, but Mannheim was outdated and overstaffed. Lanz's product lines were in dire need of rationalization. The answer was John Deere, which, seeking a European tractor production base, became majority shareholder in Lanz in 1956, completing its purchase four years later, in 1960.

Pioneer Bulldog
When model P and S Bulldogs were introduced, maximum power had increased to 55 hp. They featured optional electrical starting.

Deere's first products from Mannheim, the 300 and 500, co-branded as John Deere-Lanz, were among a new breed of multi-cylinder machines. The Lanz name was swept aside in 1965 by the John Deere 20 series, but its legacy remains. Bulldog is a common word for tractor across Germany today.

John Deere-Lanz
The number of co-branded John Deere-Lanz Bulldogs was reduced from 19 to 14 in 1956 and then to 12 in 1957.

World Tractor Trials

Held in the UK in 1930, this was one of the most important machinery events of the time. The trials, open to any machines manufactured in any country, with no restrictions as to weight or horsepower, attracted entries from the UK, Ireland, the US, France, Germany, Hungary, and Sweden. The results achieved global recognition, and the information provided proved invaluable to tractor makers, farmers, and users for many years.

TESTING THE TRACTORS

The trials were organized by the Royal Agricultural Society of England in conjunction with the Institute of Agricultural Engineering at the University of Oxford. The testing, done between June and September in Wallingford, Oxfordshire, included various trials to determine the power and performance of the tractors under different loads and conditions.

Drawbar tests were carried out using a dynamometer car. The car's rear wheels were connected to a generator, which provided the load. The load could be varied by altering the resistance of the generator. Sometimes a spare tractor was hitched to the rear of the car to provide an additional load. Belt and plowing tests were also carried out as part of the trials.

One of the British entries, a Vickers tractor, is being connected to the dynamometer car with a Lanz 18/30 adding to the load.

North American Row Crops

The vast size and diversity of the farming industry in the US and Canada created a demand for specialty tractors, including the popular row-crop models designed for working between rows of corn and other vegetables. Essential features included space for carrying implements under the tractor, a good view forward to allow accurate steering between the rows and, for some crops, twin or single front wheels for increased maneuverability.

▷ **Twin City KT**

Date 1930 **Origin** US

Engine Minneapolis-Moline 4-cylinder gasoline/kerosene

Horsepower 26 hp

Transmission 3 forward, 1 reverse

The company making Twin City tractors joined others in a 1929 merger to form Minneapolis-Moline, and the Twin City name continued for about three years. The KT or Kombination Tractor was designed for versatility with a standard front axle and high clearance for row crops.

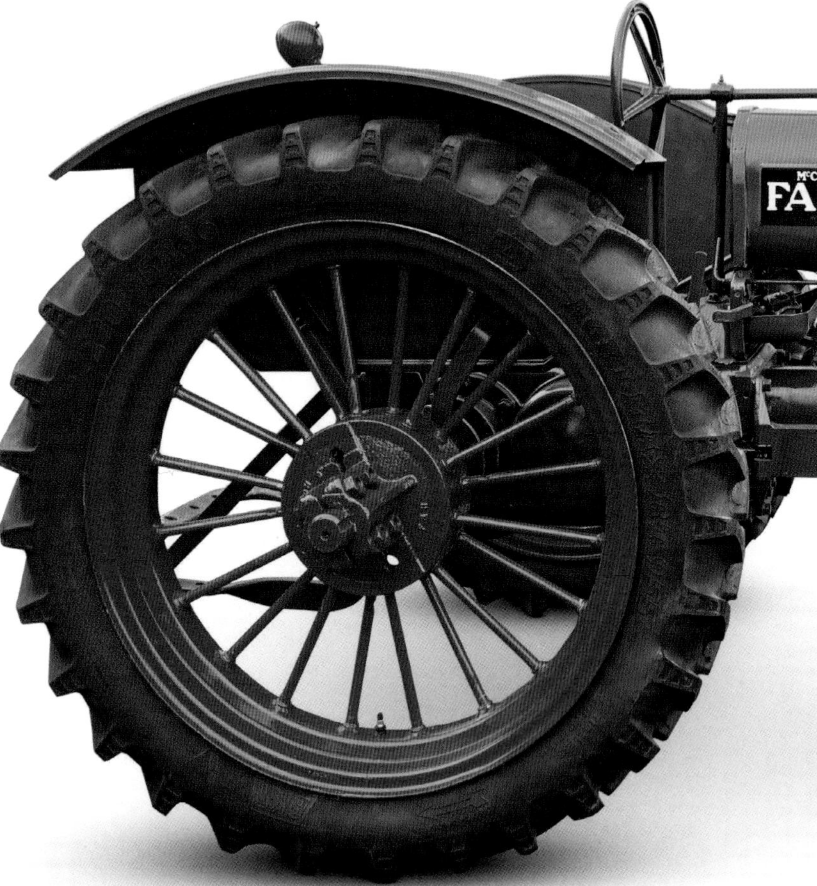

△ **International Farmall F-12**

Date 1935 **Origin** US

Engine International 4-cylinder gasoline/kerosene

Horsepower 16 hp on gasoline

Transmission 3 forward, 1 reverse

The row-crop tractor idea came from Bert R. Benjamin of International, who designed the first Farmall. Production started in 1924 with output totaling 100,000 by 1930, and new models introduced in 1932 included the entry-level F-12 for small acreages.

△ **Allis-Chalmers WC**

Date 1936 **Origin** US

Engine Allis-Chalmers 4-cylinder kerosene engine; gasoline version was available

Horsepower 21 hp

Transmission 4 forward, 1 reverse

The WC was one of the most popular of the row-crop models during the 1930s and 1940s, with more than 170,000 sales. Allis-Chalmers was the company that introduced rubber tires on tractors, in 1931, and in 1934 the WC became the first rubber-tired tractor tested in Nebraska.

▷ **International Farmall F-20**

Date 1933 **Origin** US

Engine International 4-cylinder gasoline/kerosene

Horsepower 23 hp

Transmission 4 forward, 1 reverse

The F-20 was the model that replaced Bert R. Benjamin's original Farmall. It was available from 1932 with a number of design improvements introduced later, including a power increase to 28 hp and the option of rubber tires.

◁ John Deere Model B

Date 1936 **Origin** US
Engine John Deere 2-cylinder horizontal gasoline/kerosene
Horsepower 16 hp (first production version)
Transmission 4 forward, 1 reverse

The B was one of the models chosen to introduce John Deere's new design look in 1938, and this tractor is a "styled" example. The Model B is also an example of faster engine speeds as power outputs increased, and with 1,150 rpm rated speed, it was the first John Deere engine to exceed 1,000 rpm.

△ Oliver 70 Row Crop

Date 1937 **Origin** US
Engine Oliver 6-cylinder gasoline/kerosene
Horsepower 30 hp
Transmission 4 forward, 1 reverse

Oliver was formed in 1929 in a merger that included Hart-Parr, one of the pioneer US tractor companies. Tractors built after the merger, including the early 70 row-crop models, carried the Oliver Hart-Parr brand name at first, but by 1937 this was shortened to Oliver.

△ Silver King

Date 1937 **Origin** US
Engine Hercules 4-cylinder gasoline
Horsepower 25 hp
Transmission 4 forward, 1 reverse

Silver King tractors from the Fate-Root-Heath company included this 3-wheel model, a classic row-crop design with generous ground clearance and a tricycle-style single front wheel. A special design feature was the unusually high 25 mph (40 km/h) top speed for tractors equipped with rubber tires.

△ Case CC

Date 1937 **Origin** US
Engine Case 4-cylinder gasoline/kerosene
Horsepower 30 hp
Transmission 3 forward, 1 reverse

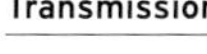

Other manufacturers reacted to the success of the Farmall by introducing row-crop models of their own. The Case version was based on their small Model C tractor, which was modified with twin front wheels, adjustable wheel track settings, and extra ground clearance.

▷ Eagle 6B

Date 1938 **Origin** US
Engine Hercules 6-cylinder gasoline
Horsepower 40 hp maximum
Transmission 3 forward, 1 reverse

The Eagle company offered their six-cylinder tractor in two versions during the 1930s. The 6C was the general-purpose version, and the 6B was basically the same tractor in a row-crop model with tricycle-style dual front wheels and increased ground clearance.

Massey-Harris GP

The Massey-Harris General Purpose, or GP, was not the earliest four-wheel-drive tractor, but it was the first to go into mass production. The design offered near-perfect weight distribution, a good turning circle, exceptional clearance for row-crop work, and unbeatable traction. The model lived up to its name in every respect but one—it was short on power, and was given a disappointing 22-hp drawbar rating when tested in Nebraska.

MASSEY-HARRIS was a Canadian concern, but the GP, introduced in 1930, was built in the United States at the former Wallis Tractor Company factory in Racine, WI. The GP's four-wheel-drive concept was ahead of its time, and with a little more power, the model might have fared better in the marketplace.

The drive from the four-cylinder Hercules engine was taken from beneath the front axle to a three-speed gearbox. A gearset on top of the gearbox transmitted the drive to the front and rear differentials. The front drive-steer axle had a brake band on each differential shaft to aid steering. The drive to the rear axle was via an enclosed torque tube, which swiveled to allow the axle to oscillate and follow the contours of the ground. The GP was listed by Massey-Harris from 1930 to 1938.

FRONT VIEW

REAR VIEW

The GP export model
GP models exported to the UK were painted green with red wheels instead of the gray and red livery used for the US market. The tractor was available in various tread widths; this is the standard 66-in (167.6-cm) version.

Final drives running in an enclosed oil bath

Pedestal for clutch and gear levers

SPECIFICATIONS	
Model	Massey Harris GP
Built	1932
Origin	US
Production	approx. 3,000
Engine	24 hp Hercules 4-cylinder gasoline/kerosene
Capacity	226 cu in (3,703 cc)
Transmission	4 forward, 1 reverse
Top speed	4 mph (6 km/h)
Length	9 ft 11 in (3 m)
Weight	2 tons (1.8 metric tons)

THE DETAILS

1. Massey-Harris name adorns top of Modine radiator **2.** Front axle steered via ball joints **3.** Pinion on axle shaft meshes with a ring gear **4.** Belt pulley was optional **5.** Levers on steering wheel column controlling clutch and gear selection

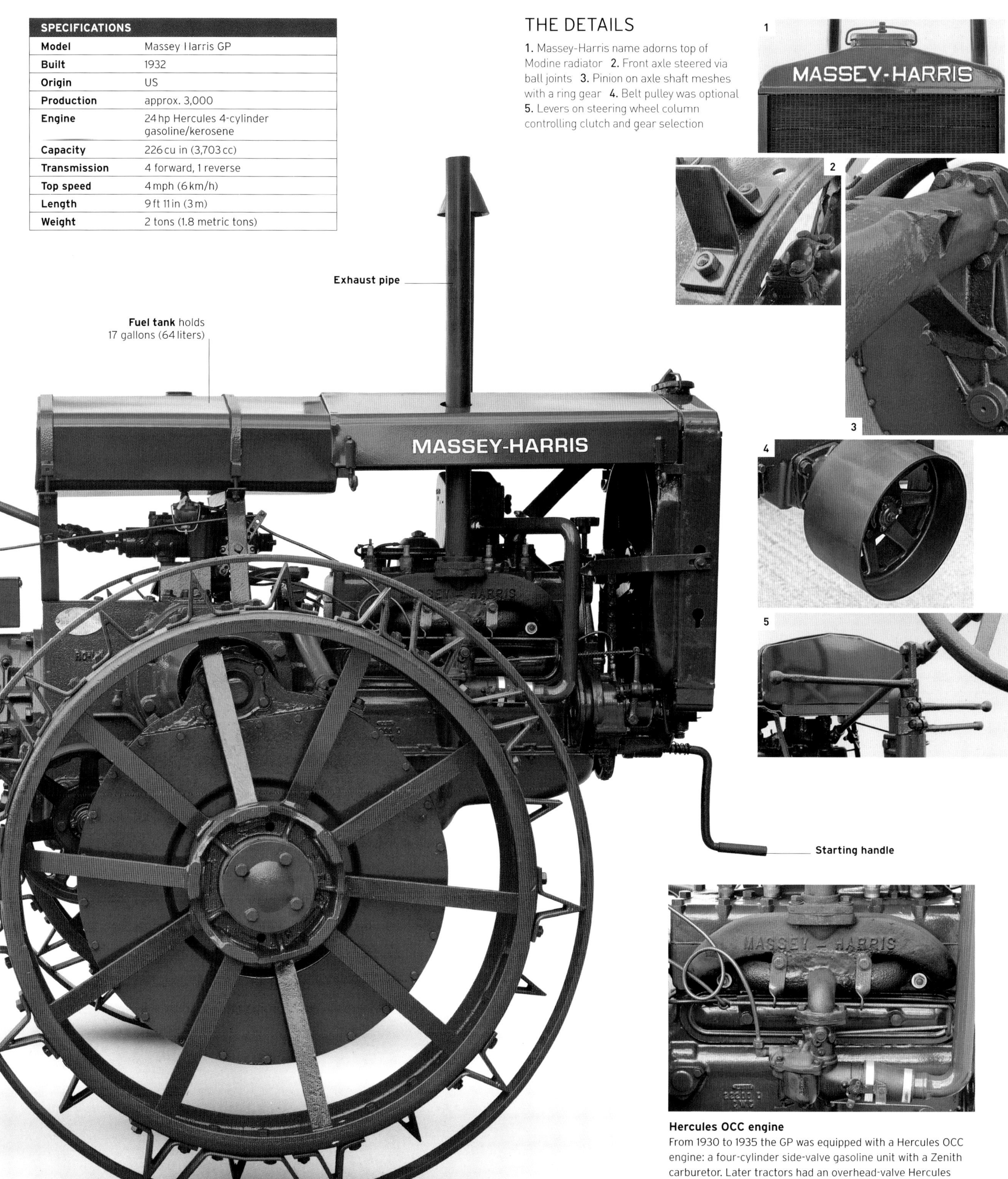

Hercules OCC engine
From 1930 to 1935 the GP was equipped with a Hercules OCC engine: a four-cylinder side-valve gasoline unit with a Zenith carburetor. Later tractors had an overhead-valve Hercules engine, but the power was unchanged at 24 hp.

General Purpose

Power farming advanced rapidly during the 1920s and 1930s as large numbers of working horses and mules were replaced by tractors. US and Canadian farms were at the forefront of the trend, creating a big demand for versatile, general-purpose tractors designed to operate a wide range of machinery: many were equipped with a belt pulley for stationary operation. Popular models were mechanically simple, usually with four-cylinder gasoline/kerosene engines and outputs of around 30 hp to 45 hp.

▽ International 10-20

Date 1930 **Origin** US

Engine International 4-cylinder gasoline/kerosene

Horsepower 20 hp

Transmission 3 forward, 1 reverse

International's McCormick-Deering 10-20 became a tractor industry classic. Production totaled more than 200,000 spread over 16 years, with success based on a reputation for long-term reliability. It was among the first tractors with a power takeoff (PTO).

▷ Rumely 6-A

Date 1931 **Origin** US

Engine Waukesha 6-cylinder gasoline/kerosene

Horsepower 48 hp

Transmission 3 forward, 1 reverse

This was the last new model from the Advance-Rumely Thresher Co. It was designed as a high-specification, six-cylinder tractor with a premium price, but demand was disappointing, and in 1931 the company was taken over by Allis-Chalmers.

◁ Massey-Harris GP

Date 1932 **Origin** US

Engine Hercules 4-cylinder gasoline/kerosene

Horsepower 24 hp

Transmission 4 forward, 1 reverse

Four-wheel drive made a big contribution to tractor efficiency, but it was still a novelty when Massey-Harris included it on their GP model in 1930. Farmers were concerned about the extra cost and sales were disappointing.

◁ **Allis-Chalmers Model U**

Date 1933 **Origin** US
Engine Allis-Chalmers 4-cylinder gas/kerosene
Horsepower 34 hp
Transmission 3 forward, 1 reverse

The Model U was an ordinary tractor, but in 1932 it became the first production model with inflatable rubber tires. It was a huge breakthrough, as replacing the old steel wheels allowed faster speeds and avoided road damage.

△ **Case Model C**

Date 1934 **Origin** US
Engine Case 4-cylinder gas/kerosene
Horsepower 27 hp
Transmission 3 forward, 1 reverse

The Model C and the larger Model L were the first Case tractors with a more conventional longitudinal engine position instead of the transverse layout used previously. The Model C also featured the Case final drive, using a rugged spur gear and chain.

△ **International W-12**

Date 1935 **Origin** US
Engine International 4-cylinder gasoline/kerosene
Horsepower 15 hp
Transmission 3 forward, 1 reverse

The W-12 was the general-purpose version of the F-12, International's smallest Farmall row-crop model. To gain extra sales there was also an O-12 orchard model and the I-12 industrial version, all sharing the same mini-sized mechanical specification.

▷ **Graham-Bradley 104**

Date 1938 **Origin** US
Engine Graham-Paige 6-cylinder gasoline/kerosene
Horsepower 32 hp
Transmission 4 forward, 1 reverse

Graham-Paige Motors built upscale cars in Detroit, but falling sales persuaded them to add high specification Graham-Bradley tractors to their range in 1937. In spite of the 104's streamlined styling and 20 mph (32 km/h) top speed, the tractors were not a success.

▽ **Huber Modern Farmer Model L**

Date 1937 **Origin** US
Engine Waukesha 4-cylinder gasoline/kerosene
Horsepower 43 hp
Transmission 3 forward, 1 reverse

Huber made farm machinery in the 1860s and progressed to tractors, but they never achieved big sales. New tractors announced in 1937, including the Model L, were the last to carry the Huber name; the factory closed in the early 1940s.

◁ **Oliver 80**

Date 1940 **Origin** US
Engine Oliver 4-cylinder gasoline/kerosene
Horsepower 36 hp
Transmission 3 forward, 1 reverse

The 80 models were Oliver's midrange tractors, available as a row-crop or with the standard or general-purpose layout as shown. A diesel 80 available from 1940 was among the first general-purpose tractors with this engine type.

Fowler Gyrotiller

The legendary Gyrotiller, or "rotary plow," as it was sometimes known, was originally developed for the cultivation of sugarcane in the West Indies. Manufactured by John Fowler of Leeds, the first machine was completed in 1927 according to the patents of Norman Storey, an estate manager in Puerto Rico. The Gyrotiller was built in several sizes, but just 67 of the largest versions were made before production ended in 1935.

STOREY DEVELOPED the concept of using two contra-rotating rings of tines for the deep cultivation of sugarcane, and then assigned manufacturing rights to Fowler. Early machines had thirsty 225-hp Ricardo gasoline engines, which were soon replaced by 150-hp industrial diesel engines sourced from MAN in Germany. Subsequent models all had MAN diesel engines, eventually up-rated to 170 hp.

Around half of the larger Gyrotillers built were supplied to British contractors, who used them for arable cultivations and land reclamation. The machines were often unreliable and costly to operate, using almost 10 gallons (36 liters) of fuel per hour. Farmers felt they were not getting their money's worth unless the ground was cultivated as deeply as possible, which destroyed the soil structure and gave the Gyrotillers an unfair reputation for damaging the land.

Deep cultivator
The only remaining 170-hp machine still in running order is this Gyrotiller, which was delivered new to A.J. Ward & Son of Egham, Surrey, on April 6, 1935. The purchase price was reputed to be £6,000 (about $29,000).

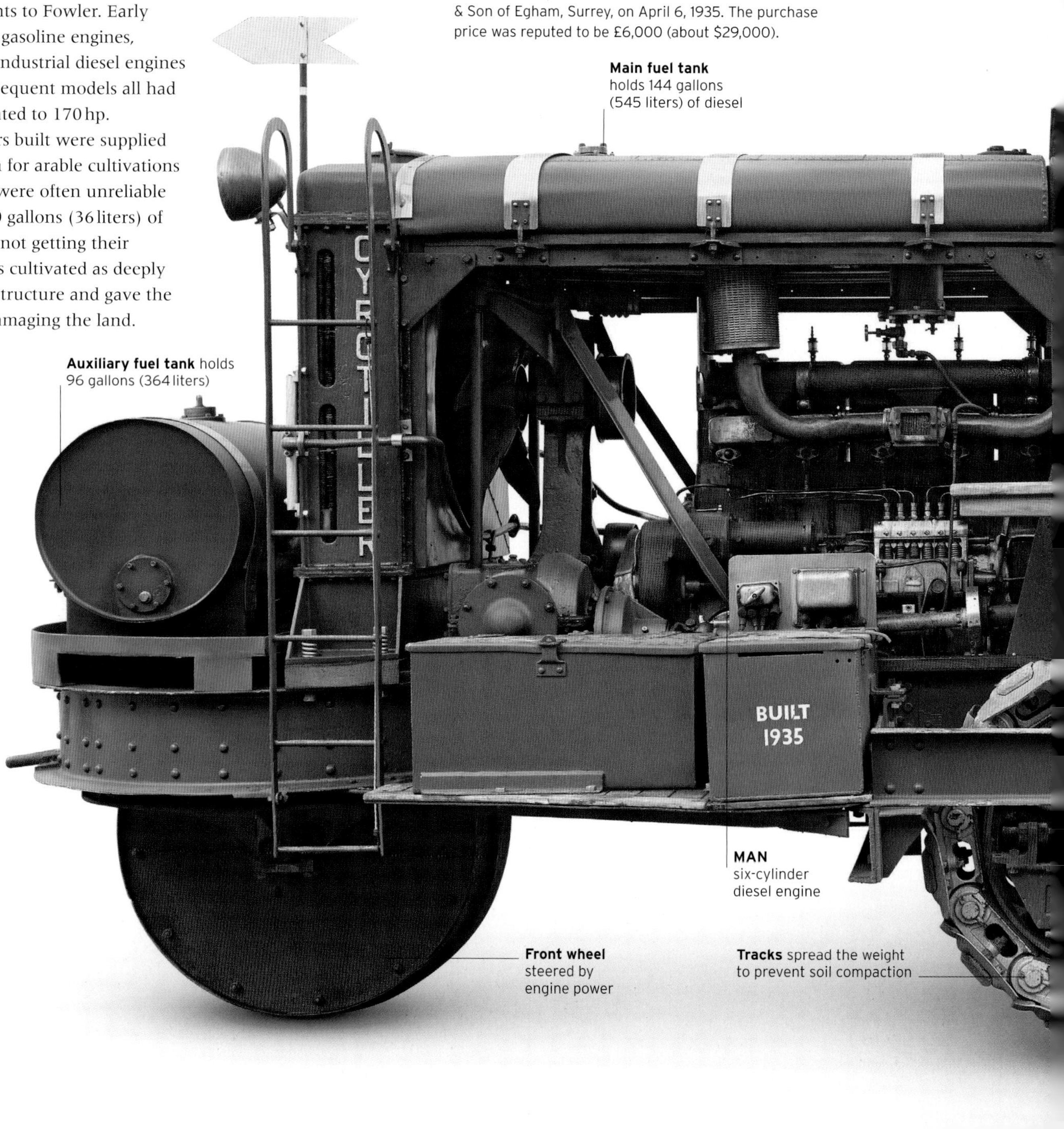

Main fuel tank holds 144 gallons (545 liters) of diesel

Auxiliary fuel tank holds 96 gallons (364 liters)

GYROTILLER

BUILT 1935

MAN six-cylinder diesel engine

Front wheel steered by engine power

Tracks spread the weight to prevent soil compaction

FRONT VIEW

REAR VIEW

SPECIFICATIONS	
Model	Fowler Gyrotiller 170 hp
Built	1935
Origin	UK
Production	67
Engine	170 hp MAN 6-cylinder diesel
Capacity	1,937 cu in (31,750 cc)
Transmission	2 forward, 1 reverse with high/ low gearbox
Top speed	2 mph (3 km/h)
Length	26 ft 3 in (8 m)
Weight	28 tons (25.4 metric tons)

THE DETAILS

1. Directional arrow shows position of front wheel **2.** Tachometer indicates engine revolutions **3.** CAV Bosch Model PE fuel injection pump for MAN diesel engine **4.** Control levers operating steering clutches, brakes, transmission, and rotors **5.** Contrarotating rotor rings each equipped with four tines

Lever to raise rotary tillers

Control levers for steering clutches and brakes

Operator's seat

Lift for tillers is chain-operated

Compensating springs allow rotary tillers to "float" while working

Tines can cultivate to a depth of 20 in (50.8 cm)

Crawler Market Grows

The crawler tractor market in the UK between the wars was dominated by imported machines, mainly from the US. Several old British steam engine companies produced some promising track-type designs, but most failed to make any impact on the market. The Fowler Gyrotiller was the exception; it was found to be the ideal machine for reclaiming land that had fallen derelict during the Depression. The German machines from Lanz and Hanomag were successful in their home markets, but struggled abroad against the US tractors such as Caterpillar, International, Cletrac, and Allis-Chalmers.

◁ **Georges Vidal Vineyard**

Date 1925 **Origin** France
Engine Baudoin 4-cylinder gasoline
Horsepower 30 hp
Transmission 2 forward, 1 reverse

Vidal started building tractors in 1920, its intended market being the vineyards in the Languedoc-Roussillon region of France. A version of the tractor was built to run on producer gas, in this case generated from wood, from the generator built onto the back of the tractor. Only 20 hp was available when run on this fuel.

▷ **International T20 TracTracTor**

Date 1933 **Origin** US
Engine International 4-cylinder gas/kerosene
Horsepower 28 hp
Transmission 3 forward, 1 reverse

The T20 was the smallest of International's crawlers up to 1939. It shared its engine with International's wheeled tractors of similar horsepower. The design of the transmission layout was unique to International Trac-Tractors: it allowed access to the steering clutches from the rear of the machine without having to dismantle the transmission case.

▽ **Case Model L Roadless**

Date 1938 **Origin** UK
Engine Case 4-cylinder gasoline/kerosene
Horsepower 42 hp
Transmission 3 forward, 1 reverse

The addition of Roadless tracks to the Case Model L was an attempt to improve its tractive capabilities. The result did not produce a popular machine. The ratios of the three-speed gearbox were not ideally suited to crawler work, and the design of the Roadless tracks meant that the tractor pitched fore and aft on anything but level ground. A waterproofed version was built for launching lifeboats.

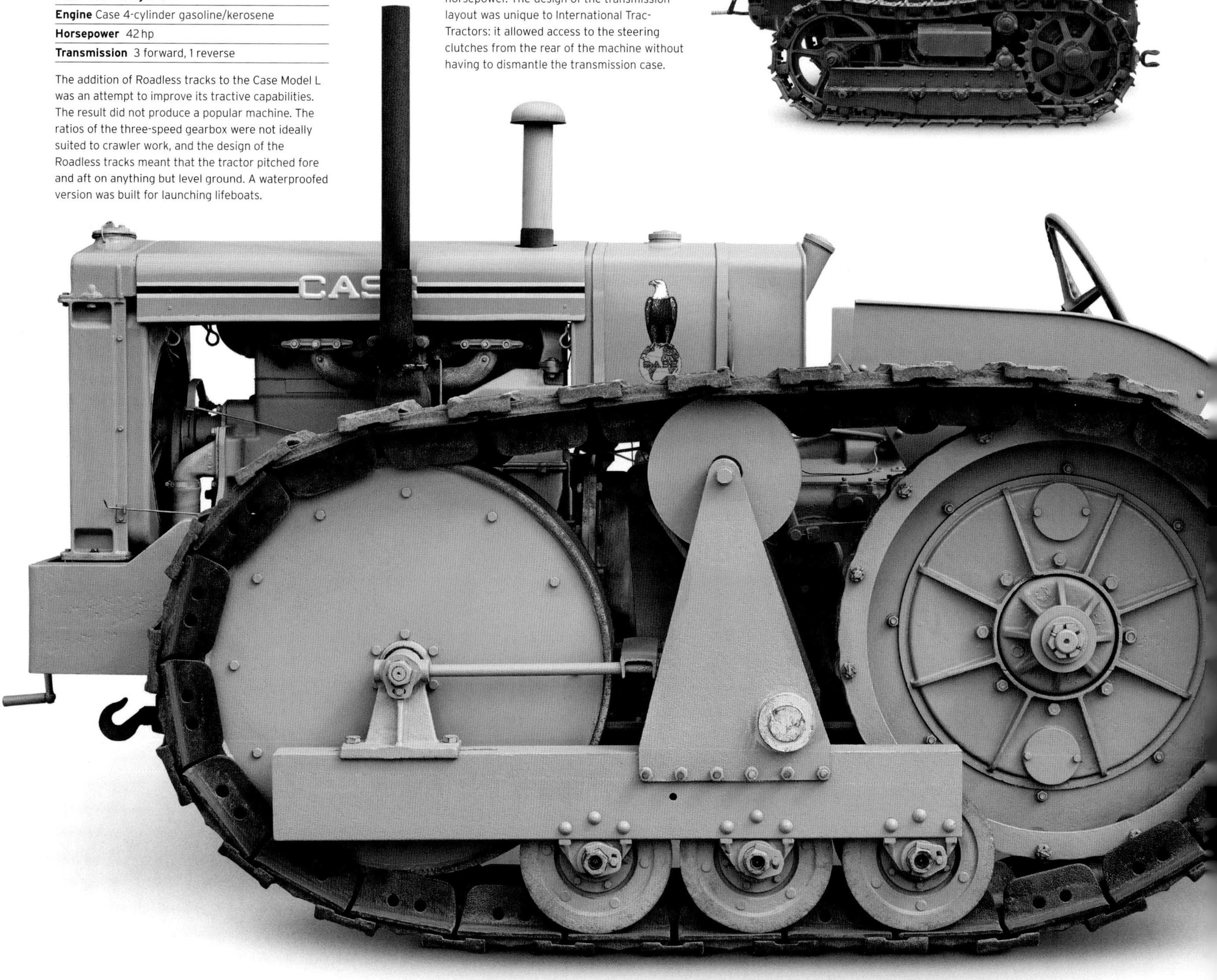

▷ **Fowler 75**

Date 1935 **Origin** UK
Engine Fowler 6A 6-cylinder diesel
Horsepower 75 hp
Transmission 5 forward, 1 reverse

These tractors were produced from 1931 to c. 1937, and initially nearly all had a Gyrotiller attachment. The tractor shown here is equipped with a cable winch and spent its entire working life at the Daventry Wireless Mast Establishment. It is the only one known to exist that was sold new as a tractor and not a Gyrotiller.

△ **Fowler Gyrotiller**

Date 1935 **Origin** UK
Engine MAN 6-cylinder diesel
Horsepower 170 hp
Transmission 2 forward, 1 reverse with high/low gearbox

The first Gyrotiller, powered by a 225 hp Ricardo gasoline engine, was shipped to Cuba in 1927. Three more were built with this engine, but because they consumed 17 gallons (64 liters) of gasoline per hour, they were soon replaced with MAN diesels. These large Gyrotillers cost £6,000 ($29,000) each; 67 were built.

▷ **Caterpillar 15**

Date 1931 **Origin** US
Engine Caterpillar 4-cylinder gasoline
Horsepower 25 hp
Transmission 3 forward, 1 reverse

Introduced in 1928, the Caterpillar 15 was at that time a modern and up-to-date tractor. In 1931 the color of all Caterpillar tractors was changed from battleship gray to "Hi-Way" yellow. A small number came to the UK when new.

◁ **Lanz Bulldog HR8 Model D9550**

Date 1935 **Origin** Germany
Engine Lanz single-cylinder 2-stroke hot bulb semi-diesel
Horsepower 38 hp
Transmission 6 forward, 2 reverse

This crawler rightly earned a reputation as a simple, reliable, and long-lived machine. Apart from the inconvenience of having to use a blowtorch to start the hot-bulb engine, the single-cylinder engine was foolproof. Clutch and brake steering meant there was no loss of power during turns.

▷ **International TD35 Trac-Tractor**

Date 1937 **Origin** US
Engine International 4-cylinder gasoline-start diesel
Horsepower 42 hp
Transmission 5 forward, 1 reverse

International introduced its range of diesel crawlers in the early 1930s. The design of the diesel engines was unusual in that they started on gasoline and, when warmed up, turned over to diesel. On the TD35 the changeover occurred automatically; on later models this was done manually.

△ **Caterpillar R2 4J**

Date 1940 **Origin** US
Engine Caterpillar 4-cylinder gasoline/kerosene
Horsepower 27 hp
Transmission 5 forward, 1 reverse

The five-speed R2 was the spark-ignition version of the diesel D2. Apart from their engines, the two tractors were identical; both were offered as a 44-in (112-cm) standard gauge, or a 50-in (127-cm) wide gauge tractor. Production of the R2 ceased in 1942 to make room for increased production of tractors required by the Allies for WWII. Much of the production run was shipped to the UK.

HENGIST

Aircraft-towing Tractors

One of the tractor's more interesting applications was the ground movement of both military and civilian aircraft. The early commercial "airports" were little more than grass airfields. When the long-range routes were opened up and the airliners became larger and heavier, tractors were pressed into service to handle the aircraft on the ground.

As aviation came of age, and grass aprons gave way to tarmac runways, the requirements for aircraft movement became more specialized, and companies like County, David Brown, Douglas, and Reliance-Mercury built dedicated aircraft tugs. Civil airports also used tractors for snow clearance and deicing runways.

IMPERIAL AIRWAYS

The UK's Imperial Airways, formed in 1924, was one of the first airlines to use tractors for ground movement. In the 1930s, Imperial bought a Rushton Roadless for its flying boat base at Hythe on Southampton Water to move and launch Short seaplanes. A second Roadless, this time based on a Fordson, was bought for Imperial's landplane operations in Croydon, Surrey.

Both tractors were full-track crawlers equipped with Roadless rubber-jointed tracks for silent running. Roadless, based in Hounslow, Middlesex, focused on building specialized tractors for specific purposes, and several of its machines also had military applications.

A Fordson Roadless at Croydon airport handling a Handley Page HP42W aircraft, which was in service on Imperial's long-haul routes from 1931 to 1937.

The First Diesels

The advantages of the diesel engine for tractors had been recognized by the mid-1920s, but manufacturing units for the field raised problems. The diesel engine was more fuel-efficient and it burned a cheaper fuel oil that took less refining than gasoline. However, engineers had to make engines that were light and cheap enough to install. The fuel injection system components needed to be reliable and adjustment-free in the field. The single-cylinder, hot-bulb engines, favored by European manufacturers, went some way toward addressing these issues, since they needed only one set of injection equipment per tractor.

◁ Avance Super Crude Oil Tractor

Date 1927 **Origin** Sweden

Engine Avance 2-cylinder 2-stroke hot bulb semi-diesel

Horsepower 35 hp

Transmission 3 forward, 1 reverse

The Avance tractor had unique features. To start the engine, the ignition plugs for each cylinder had first to be heated. An electrical system was provided to do this, which required the tractor to have a battery and a generator. Early examples suffered from broken crankshafts and had to be rebuilt using modified components.

△ Bubba UT6

Date 1930 **Origin** Italy

Engine Bubba single-cylinder horizontal 2-stroke hot-bulb semi-diesel

Horsepower 40 hp

Transmission 3 forward, 1 reverse

The company claimed that Bubba tractors could burn oils derived from shale, heavy naphtha, and tar distillation. This was possible as hot-bulb engines have relatively low engine speed and low injection pressures, allowing heavy fuel to be burned slowly.

△ HSCS K40

Date 1935 **Origin** Hungary

Engine HSCS single-cylinder horizontal hot-bulb semi-diesel

Horsepower 40 hp

Transmission 3 forward, 1 reverse

HSCS was founded in the early 20th century as Hofherr-Schrantz-Clayton-Shuttleworth. It was a partnership between a group of Hungarian businessmen and Clayton & Shuttleworth of the UK. The arrangement continued until 1921, when Clayton relinquished its interest. HSCS built its first tractor in 1923 following the Lanz single-cylinder design. Production continued into the 1950s.

▷ Caterpillar Sixty Atlas

Date 1928 **Origin** US

Engine Atlas-Imperial 4-cylinder diesel

Horsepower 65 hp

Transmission 3 forward, 1 reverse

By the late 1920s Caterpillar was facing demands from its customers for a diesel track-type tractor. One or two frustrated customers even started experiments of their own, and one of the results is seen here. In 1928, Henry Kaiser of the Kaiser Paving Co. and Fletcher Walker of the Red River Lumber Co. converted a handful of these Caterpillars and Monarch 75s using Atlas marine diesel engines. These engines were complicated and very heavy, even for a track-type tractor, and the conversions met with limited success.

△ **Bolinder-Munktell 22HK**

Date 1921 **Origin** Sweden
Engine Munktell 2-cylinder 2-stroke hot bulb semi-diesel
Horsepower 26 hp
Transmission 3 forward, 1 reverse

This company started tractor production in 1913 and continued into the 1970s. The 22HK was designed with unit construction, with the engine, gearbox, and rear axle forming the frame of the tractor. As an unusual feature, it had an oil bath air cleaner.

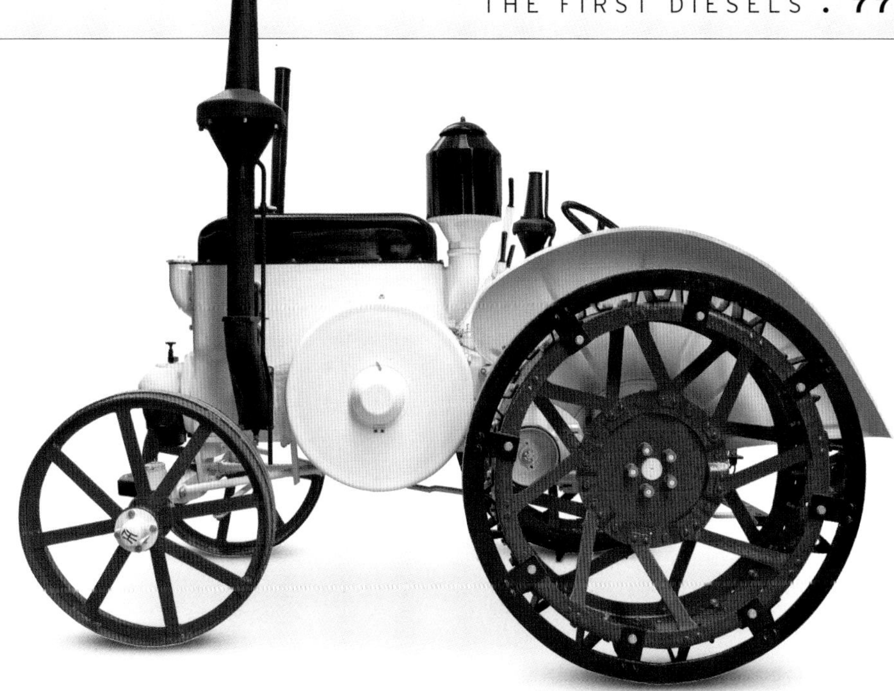

◁ **Vierzon H2**

Date 1936 **Origin** France
Engine Vierzon single-cylinder horizontal 2-stroke hot bulb semi-diesel
Horsepower 50 hp
Transmission 3 forward, 1 reverse

Before starting to produce tractors in 1931, Vierzon manufactured agricultural machinery. Production of tractors continued until the late 1950s, when Case bought an interest in the company. Case at one time built a few tractors for the French market in the Vierzon factory. The tractors followed the general design of the German Lanz.

△ **Lanz HR2 Gross "Bulldog"**

Date 1926 **Origin** Germany
Engine Lanz single-cylinder horizontal 2-stroke hot bulb semi-diesel
Horsepower 22 hp
Transmission 3 forward, 1 reverse

The Lanz was copied by many manufacturers of this type of tractor. The HR2 was the earliest of the larger machines, with the name "Bulldog" becoming famous. A feature of the Lanz tractors was that the steering wheel was removed from the column and attached to the flywheel to start the engine.

▷ **Marshall 12-20**

Date 1938 **Origin** UK
Engine Marshall single-cylinder horizontal 2-stroke diesel
Horsepower 20 hp
Transmission 3 forward, 1 reverse

The 12-20 was the first successful model produced by Marshall, being lighter and cheaper than its earlier models. This was the forerunner of the famous Field-Marshall line, all of which were very economical to operate. Marshall adopted the design of the German Lanz, but opted for a full diesel. The 12-20 was excellent for driving threshing machines.

◁ **McDonald TWB Imperial**

Date 1938 **Origin** Australia
Engine McDonald single-cylinder horizontal 2-stroke hot bulb semi-diesel
Horsepower 35 hp
Transmission 3 forward, 1 reverse

The TWB tractor was produced from 1931 to 1944. The transmission and chassis was a close copy of a Rumely design for which McDonald adapted its "T" type engine. As with all McDonald products, it was a rugged and simple design with the engine following the Lanz principle. McDonald also built road-rollers in large numbers, and later imported rollers, road-building machines, and general farm and road machinery.

Benjamin Holt and British Major General E.D. Swinton meeting at Holt's Stockton plant on April 22, 1918

Great Manufacturers

Caterpillar

A world leader in its field, Caterpillar Inc. is a major manufacturer supplying customers in more than 180 countries from 93 principal facilities. Today Caterpillar Inc. builds in excess of 300 products, from construction and mining equipment to diesel and natural gas engines, industrial gas turbines, and diesel-electric locomotives.

THE STORY OF CATERPILLAR began in the early years of machinery development on the West Coast of the US. It can be traced back to the Holt brothers, whose Stockton Wheel Company was established to supply wooden wagon wheels, poles, and axles, and Daniel Best who began manufacturing farm machinery in 1869 before later moving on to building a number of steam and gas traction engines.

Charles Holt and his inventor brother Benjamin started the Stockton Wheel Company in 1863, after moving from New Hampshire to Stockton, CA. In 1886 they sold their first combine harvester, a machine that incorporated their "Link Belt Drive" invention. This mechanism protected the machine from damage caused by overspeeding if the 30 or more mules that were required to haul the machine bolted. The Holts built their first steam traction engine in 1890, and two years later they formed the Holt Manufacturing Company.

The soil in the Stockton district was deep, black peat, which could not support the weight of a steam-traction engine. To overcome this, Holt tried putting wide wheels on its machines, which in some cases made the tractor more than 30 ft (9 m) wide. These wheels were only partially successful, so Holt looked for a better solution. Although not a new idea, it decided to add a set of tracks to one of its steam traction engines. The machine was tested in November 1904 and proved to be an immediate success; it was around this

Caterpillar 2-ton tractor, c. 1927
Probably staged to demonstrate ease of use, this Caterpillar is shown pulling a mole-draining plow by the British dealers Tractor Traders Ltd. of Westminster Square, London.

Early field trials
Holt's new gasoline-powered, track-type tractor is trialed in 1908 over the deep, peaty soils found near the Holt plant.

Best 60

60 Atlas

D7

Challenger 65

1863 Holt Brothers form the Stockton Wheel Company, incorporated as the Holt Manufacturing Company in 1892
1871 Daniel Best forms his farm equipment company, which he later sells to Holt
1910 C.L. Best Gas Traction Company formed
1925 Caterpillar Tractor Company formed on April 15 from the merger of Holt Manufacturing Company and C.L. Best Gas Traction Company
1927 First Caterpillar-designed model, the 20, is launched
1928 Caterpillar Tractor Company purchases Russell Graders
1931 First diesel—Caterpillar 60 Atlas—enters the market
1942 Caterpillar enters military production
1946 Largest expansion program in the company's history announced
1951 Caterpillar Tractor Company Ltd., first overseas facility, formed in England
1955 Caterpillar D9 introduced
1956 Plans announced to establish production in Glasgow, Scotland
1964 Sales exceed a billion dollars
1972 Caterpillar 225, a 360-degree excavator, is introduced
1976 Sales exceed $5 billion
1977 High-Drive D10 tractor introduced.
1982 $180 million loss reported; first loss since 1932
1986 Company reorganized under its present name, Caterpillar Inc.
1987 Caterpillar Challenger 65 introduced. Plans announced to close Glasgow
1988 New Caterpillar trademark introduced
1995 *Chief Executive Magazine* names Caterpillar board in the top five in US
1998 Caterpillar acquires Perkins Diesel Engines
1999 Caterpillar becomes the world's largest producer of diesel engines
2002 Agricultural equipment assets sold to AGCO
2011 Caterpillar Inc. acquires Bucyrus International for $8.8 billion, the biggest acquisition in the company's history

The 1944 annual report
Caterpillar produces an annual report giving a perspective of the company's performance over the previous 12 months. This one focuses on the war effort for World War II.

time that the name "Caterpillar" was first used to describe the Holt track-type tractors. By 1906 the commercial production of these machines was underway, and two years later, Holt was ready to place its first internal-combustion-engine, track-type tractor on the market. Holt moved east and set up a manufacturing facility in Peoria, IL, which is now the site of the world headquarters of Caterpillar Inc.

The Holt Manufacturing Company continued to develop its track-type tractors and became the market leader for this type of machine. With the coming of World War I in 1914, the Allies began to order considerable quantities of Holt tractors. Holt concentrated on these orders during the war, but when peace came, it was left with huge numbers of machines and no immediate market for them. Holt therefore supplied equipment to the US Army from its earliest days, and Caterpillar continues to supply this market today.

Daniel Best, like the Holt brothers, built combine harvesters and progressed to steam traction engine production in 1889, but he sold out to the Holt Manufacturing Company in 1908. Daniel's son, Clarence Leo Best, Holt's arch rival in the production of track-type tractors, formed the C.L. Best Gas Traction Company in 1910 and moved into his father's old premises in San Leandro, CA, in 1916. From here Best built high-quality tractors using the latest technology and materials. Although Holt remained twice the size of the C.L. Best Gas Traction Company, court cases over patent rights and subsequent loss of market share had severely weakened Holt.

In 1919 Best released its Model 60, one of the most significant landmarks in crawler tractor design. This brought renewed pressure on Holt to settle with Best, and on April 15, 1925, the two companies merged to form the Caterpillar Tractor Company. Their product line was rationalized with Best machines at the forefront. Caterpillar then bought the Russell Grader Manufacturing Company in 1928. The new company now had the resources and expertise to develop the much-needed diesel tractor to add to its product line. The first Caterpillar diesel finally appeared on the market in late 1931, followed in 1933 by a range of new diesel tractors. This placed the company ahead of the competition, and by 1938 it claimed to be the largest producer of diesel engines in the world. During World War II, Caterpillar focused on supplying the requirements of the Allies. In 1946 the company announced the largest expansion program in its history. Since then, continuous, well-funded research and development programs have resulted in advanced machine design, such as the rubber Mobil-trac system developed for the Caterpillar 65, launched in 1987, and the Hi-Drive steel-tracked tractors. These technological advances have led to continued customer demand and confidence in the product. Caterpillar's machines are backed up by an extensive spare parts availability and supply system. Its product line today consists of more than 300 different machines.

Technical support
An example of one of the many different styles of maintenance guides issued for customer information.

Advanced technology
This cutaway of the Caterpillar's revolutionary Challenger 65 shows the engine, gearbox, and hydraulic-over-mechanical steering system.

BROWN

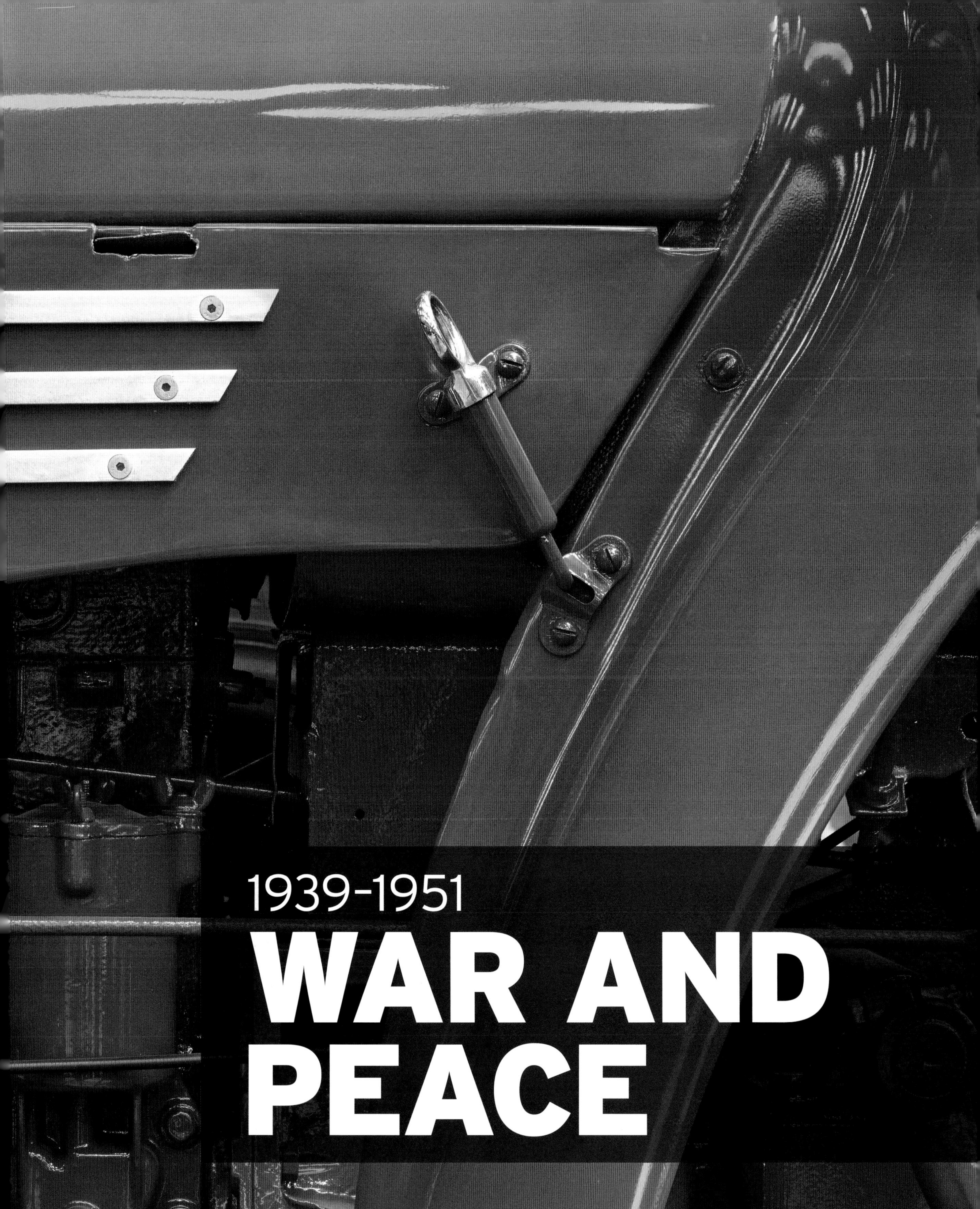

1939–1951

WAR AND PEACE

557
CWT

WAR AND PEACE

The tractor had a significant role to play in World War II, not just in mobilizing the farms and bringing more land under the plow, but also in a military capacity: moving aircraft and artillery, establishing beachheads, and constructing airfields.

During the war, the UK tractor population rose from 55,000 in 1939 to 140,000 in 1944. Most were supplied by Ford's Dagenham plant, which had been awarded the Ministry of Supply contract for tractor production. At its peak, it was building a Fordson tractor every 17 minutes and 36 seconds. By the end of World War II, more than 19 million acres (7.7 million hectares) of land were in arable production in the UK.

△ **New styling**
International's TracTracTor crawler range was revamped in 1939 with a line of new models featuring styling by Raymond Loewy.

The tractor industries of almost all of the world's nations were affected, either directly or indirectly, by the conflict. The dark days of war stifled development, and most manufacturers soldiered on with outdated designs that were little changed from those of the previous decade. In Germany, fuel shortages were a problem, and tractors were converted to run on producer gas, generated from burning wood or coke. The manufacturing shortfall in the UK was made up by the mighty US tractor industry. Between 1941 and 1945, some 30,000 wheeled tractors and 5,000 crawlers were shipped to the UK under the wartime Lend-Lease program.

US manufacturers had moved into a new era of styled tractors finished in vivid hues, bringing a welcome splash of color to the austerity years. Postwar food shortages drove the agricultural recovery, and the tractor industry flourished as never before. By 1949 as mechanization became universal, the number of tractors on US farms was double the 1940 tally of 1.5 million.

"Indeed, we **have** now, by a good margin, the most highly **mechanized agriculture** in the world."

ROBERT SPEAR HUDSON, UK MINISTER OF AGRICULTURE & FISHERIES, 1940-45

◁ **Fordson E27N Major tractor production** in Dagenham in 1948, from a painting by Terence Cuneo.

Key events

- ▷ **1939** Ford's UK general manager persuades the Ministry of Agriculture to establish a pool of tractors for the War Agricultural Committees.
- ▷ **1940** UK's Air Ministry asks David Brown to supply aircraft-towing tractors to the RAF for grass airfields.
- ▷ **1941** Tractor production in Dagenham disrupted by Luftwaffe bombing raids.
- ▷ **1941** US Congress passes the Lend-Lease Act.
- ▷ **1942** The Cleveland Tractor Co. is awarded contract to supply military crawlers to the US armed forces.
- ▷ **1942** US Army employs Caterpillar, International, and Allis-Chalmers heavy crawlers to build the Ledo Road.
- ▷ **1944** Lanz's Mannheim plant severely damaged by Allied bombing.
- ▷ **1945** The Stalingrad tractor plant in Russia is awarded the Order of the Patriotic War First Class for the mass heroism of its staff during World War II.
- ▷ **1946** The Ferguson TE-20 goes into production in Coventry, UK. The British Nuffield tractor is announced, but not launched until 1948.
- ▷ **1947** The Polish Ursus factory, badly damaged during WWII, starts tractor production with the C-45 model.
- ▷ **1949** International Harvester Co. of Great Britain begins tractor production at the new Doncaster factory.

△ **Building the Ledo Road**
US crawlers worked on military construction projects including building the Ledo Road, a 1,079-mile (1,736-km) military supply route from India to China.

Tractors with Styling

The emphasis during the early years of tractor development was on performance and reliability, not appearance. The situation started to change during the 1920s as the tractor market became more competitive, and from the mid-1930s onward, appearance moved up the priority list. Designers, particularly in the US, were influenced by the car industry, where the old boxy shapes were being replaced by the new streamlined look. Subdued grays and greens were often replaced by brighter, more eye-catching paint finishes.

△ Minneapolis-Moline UDLX

Date 1938 **Origin** US

Engine Minneapolis 4-cylinder gasoline

Horsepower 46 hp

Transmission 5 forward, 1 reverse

Minneapolis-Moline decided incorrectly that there was a market for a comfortable tractor with a steel cab, padded seats, a 40 mph (64 km/h) top speed, a heater, and even an ashtray. The UDLX ended production with only 125 made.

◁ International W-4

Date 1940 **Origin** US

Engine International 4-cylinder gasoline only or gasoline/kerosene

Horsepower 22 hp on gasoline, 24 hp on kerosene

Transmission 4 forward, 1 reverse

Demand for diesel tractors in the US was almost nonexistent in the late 1930s, so the W-4 engine was available in gasoline-only and gasoline/kerosene versions. When tested in Nebraska, the gasoline-fueled engine produced 10 percent more power output.

▷ Minneapolis-Moline GT

Date 1941 **Origin** US

Engine Minneapolis 4-cylinder gasoline/kerosene

Horsepower 40 hp

Transmission 4 forward, 1 reverse

If there were a prize for the most eye-catching 1930s tractors, Minneapolis-Moline's bright yellow paint finish would probably win. The GT with its "five-plow" US power rating was the most powerful model in the 1938 batch of new tractors.

◁ Massey-Harris 102 Senior

Date 1941 **Origin** US

Engine Continental 6-cylinder gasoline/kerosene

Horsepower 47 hp

Transmission 4 forward, 1 reverse

The 102 is the gasoline/kerosene version of the 101 gasoline model, and both are available in four-cylinder Junior and six-cylinder Senior versions. Massey-Harris used the US "three-plow" power rating to describe the output.

◁ **John Deere Model D**

Date 1945 **Origin** US
Engine John Deere 2-cylinder horizontal gasoline/kerosene
Horsepower 40 hp
Transmission 2 forward, 1 reverse up to 1934; 3 forward from 1935

The first production tractor designed by John Deere was the Model D. Various versions were available for more than 30 years, all with the simple, two-cylinder engine design that became one of the tractor industry's most successful power units.

△ **Cockshutt 30**

Date 1948 **Origin** Canada
Engine Buda 4-cylinder gasoline
Horsepower 30 hp
Transmission 3 forward, 1 reverse

As well as the Cockshutt 30 identity, this tractor was also available as the Co-op E-3 and the Farmcrest 30. None of the three brands achieved big sales and the Cockshutt name disappeared after being taken over in 1962.

△ **Case Model DC**

Date 1950 **Origin** US
Engine Case 4-cylinder gasoline
Horsepower 33 hp
Transmission 3 forward, 1 reverse

Case tractors changed color in 1939 when the previous gray was replaced by a bright orange the company called Flambeau Red. This is the DC-3 row-crop version with extra features including adjustable wheel spacing and special steering settings.

◁ **John Deere AN**

Date 1942 **Origin** US
Engine John Deere 2-cylinder horizontal gasoline/kerosene
Horsepower 38 hp
Transmission 4 forward, 1 reverse

The Model A was built from 1934 to 1952, and during that time there were numerous design changes and special versions. The N model shown is a "styled" version with the new look introduced in 1938—N denotes a single front wheel.

△ **Case Model DO**

Date 1951 **Origin** US
Engine Case 4-cylinder gasoline
Horsepower 35 hp
Transmission 3 forward, 1 reverse

Tractors for orchard work are special, and this orchard "O" version of a Case Model D is an example. Removing the upright exhaust pipe avoids damaging branches, and even the steering wheel and rear wheels are shielded to avoid tree damage.

PIONEERS

The Makeover

Industrial styling became fashionable in the 1930s as manufacturers turned to specialists to give products more visual appeal. The stylist for the John Deere tractor makeover was the US industrial designer Henry Dreyfuss (1904–72). His "Dreyfuss look" was revealed when new John Deere A and B models were announced at the end of 1937. Traces of his influence remained until a completely new tractor range was introduced in 1961.

Practical designer Henry Dreyfuss designed everything from telephones and vacuum cleaners to the casing for the NYC Hudson locomotive.

MM UDLX

Epitomizing the new era of "styled" tractors appearing on the US market at the end of the 1930s, the UDLX Comfortractor was something of a sensation, offering a level of luxury previously unheard of in agricultural circles. Launched in 1938, it was sold as a practical farm tractor in which the farmer and his wife could ride in superb comfort, but few could afford the purchase price.

THE UDLX was special in every sense, with flowing lines, enclosed bodywork, and a well-appointed cab. The powerful, high-compression gasoline engine and five-speed transmission achieved speeds of up to 40 mph (64 km/h).

The All-Weather cab, entered by the large rear door, contained a passenger seat, rubber mats, a heater, and a glove compartment. There was every modern convenience including oil, temperature, and fuel gauges, ammeter, speedometer, clock, lighter, ashtray, sun visor, rear-view mirror, electric horn, and radio. The tractor also had electric start, headlights, brake lights, and windshield wipers.

Introducing a new concept to the marketplace was certainly a bold move for the Minneapolis-Moline Power Implement Company, and the UDLX definitely turned heads, but the $2,155 price tag put it beyond the reach of most farmers.

Minneapolis Moline's Comfortractor
The UDLX's bumper, hubcaps, and hood ornament were chrome-plated, and the model was finished in MM's bright Prairie Gold livery, which the company had introduced during 1938 for its new Visionlined range of styled tractors.

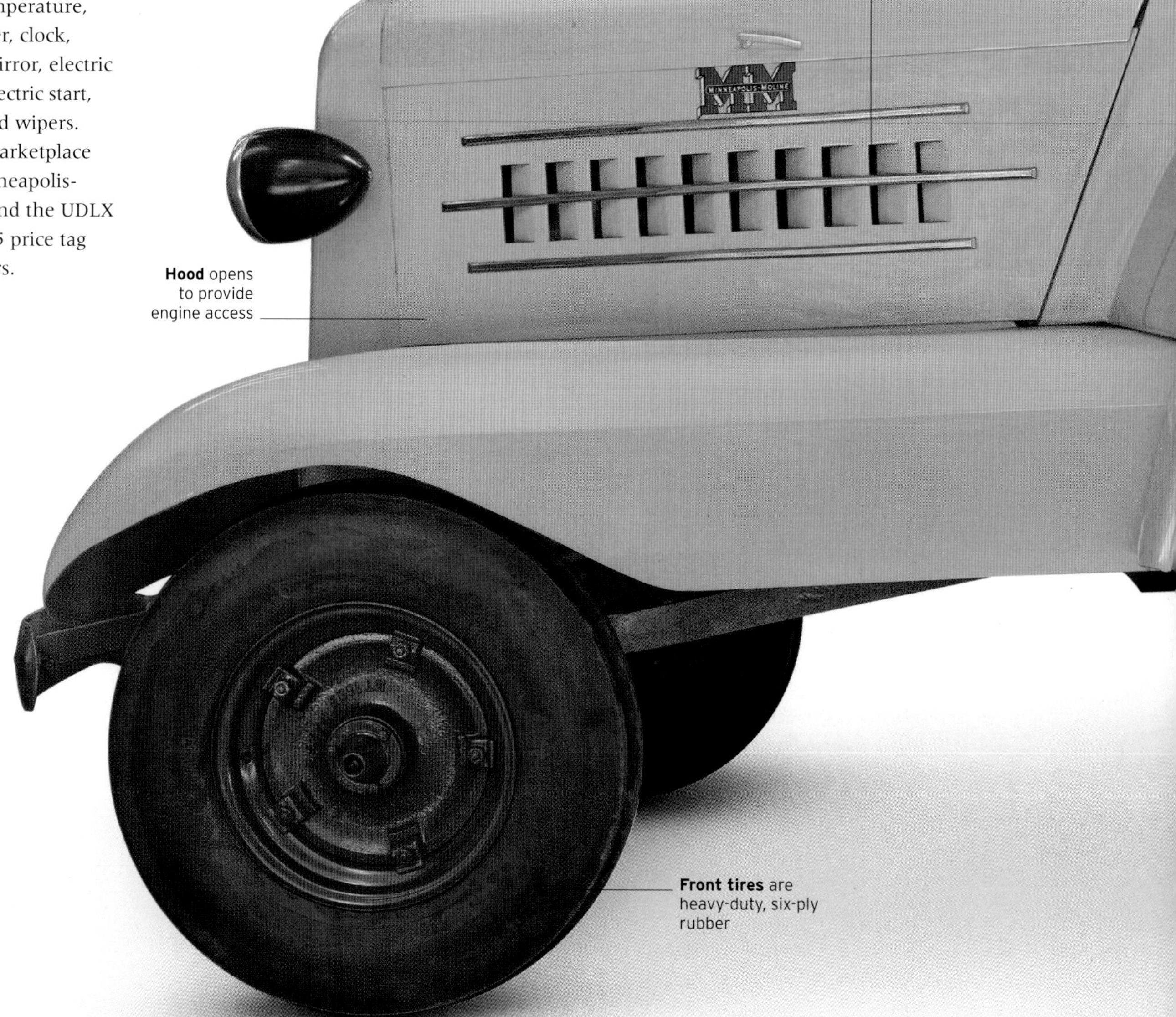

Louvers dissipate heat from enclosed engine

MM hood ornament plated in chrome

Hood opens to provide engine access

Front tires are heavy-duty, six-ply rubber

FRONT VIEW

REAR VIEW

THE DETAILS

1. MM logo adorns the radiator grille **2.** Full lighting was provided as part of the Delco-Remy electrical system **3.** Automobile-style controls include foot pedals for the clutch, brake, and throttle **4.** Four-cylinder high-compression KED gasoline engine with electric start

Windshield made from safety glass

4-PAGE COLOR BROCHURE, 1938

Rear wheels have chrome hubcaps

SPECIFICATIONS	
Model	Minneapolis-Moline UDLX
Year	1938
Origin	US
Production	125
Engine	46 hp Minneapolis-Moline 4-cylinder gasoline
Capacity	283 cu in (4,637 cc)
Transmission	5 forward, 1 reverse
Top speed	40 mph (64 km/h)
Length	11 ft 9 in (3.58 m)
Weight	3.2 tons (2.9 metric tons)

US Crawlers Move On

Machines designed and built in the US have dominated the crawler market since their introduction in the early 1900s. The heyday of the type began in the early 1930s when the first diesel-engined machines appeared in the market, and continued into the early 1960s. The products of Caterpillar Inc., Allis-Chalmers, International, and Cletrac saw service all over the world on farms and construction sites, and in theaters of war.

◁ Cleveland Cletrac BD

Date c. 1940 **Origin** US
Engine Hercules 6-cylinder diesel
Horsepower 38 hp
Transmission 4 forward, 2 reverse

The Hercules diesel engine installed in the Cletrac BD crawler was one of the most advanced engine designs available. Like all Cletrac crawlers, the BD had controlled elliptical differential steering, which was known in the US as "Cletrac steering."

◁ Caterpillar D2

Date 1942 **Origin** US
Engine Caterpillar 4-cylinder diesel
Horsepower 32 hp
Transmission 5 forward, 1 reverse

This D2 has a Killifer power lift toolbar and could be equipped with a range of tools for both row-crop and general field work. The toolbar could be removed, allowing the tractor to be used for plowing and cultivating.

▽ Caterpillar D7

Date 1948 **Origin** US
Engine Caterpillar 4-cylinder diesel
Horsepower 92 hp
Transmission 5 forward, 4 reverse

Being robust and reliable to a fault, the D7 saw worldwide service. A total of 49,110 were produced under 7M, 1T, 3T, 4T, and 6T serial numbers. Like all Caterpillar tractors of this era, starting was by a two-cylinder auxiliary gasoline engine.

△ International TD14

Date 1944 **Origin** US
Engine International 4-cylinder gasoline-start diesel
Horsepower 64 hp
Transmission 6 forward, 2 reverse

The starting method used for the TD series of International Trac-Tractors was somewhat complicated, using an extra valve in each cylinder. The TD14 was never seen in large numbers outside the US.

▽ Oliver HG

Date 1944 **Origin** US
Engine Hercules 4-cylinder gasoline
Horsepower 18 hp
Transmission 3 forward, 1 reverse

The Oliver Corporation acquired the Cleveland Tractor Company in 1944 and continued to produce the then-current range of crawler tractors. The only noticeable difference was the change of name and color from orange to Oliver green. The HG was intended for row crops.

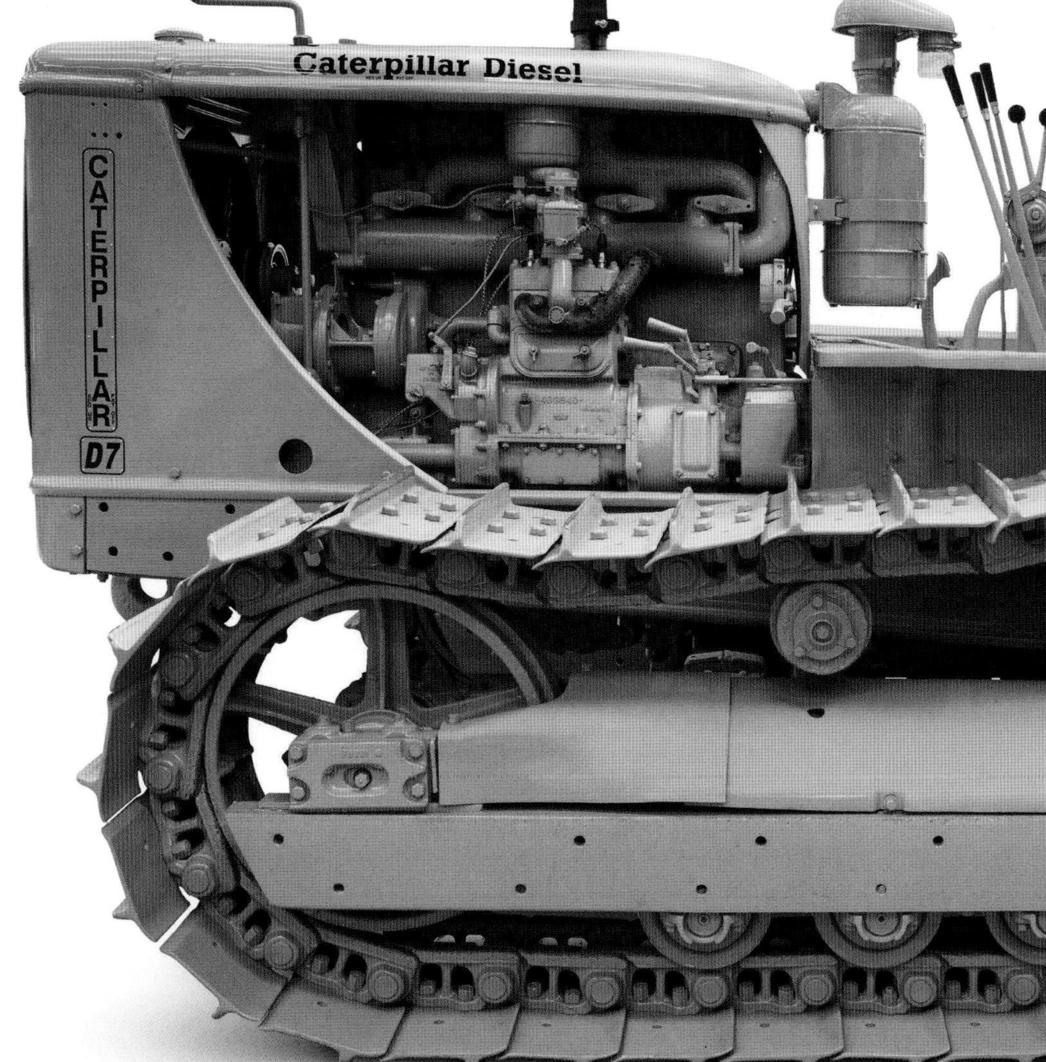

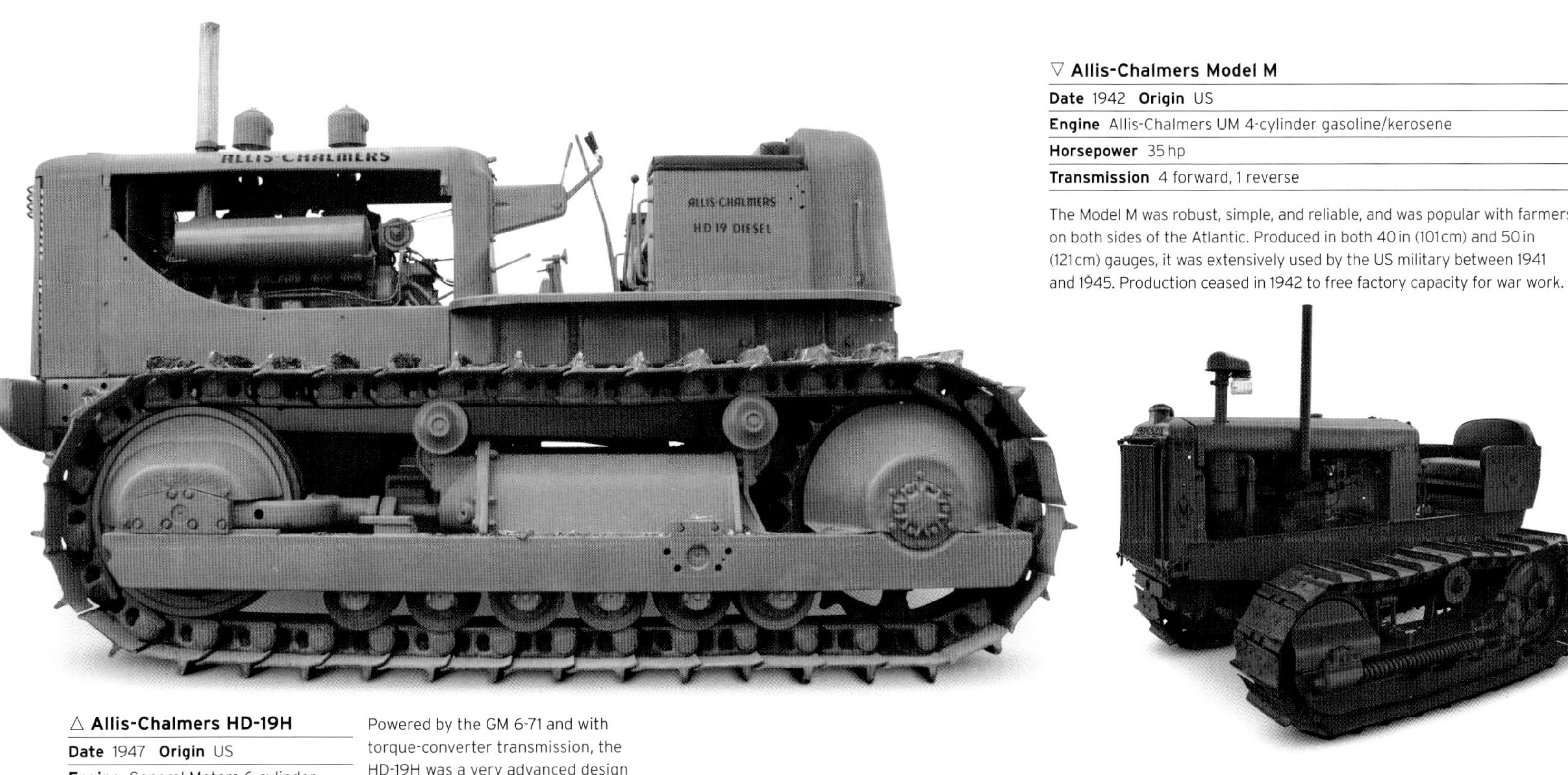

▽ Allis-Chalmers Model M

Date 1942 **Origin** US
Engine Allis-Chalmers UM 4-cylinder gasoline/kerosene
Horsepower 35 hp
Transmission 4 forward, 1 reverse

The Model M was robust, simple, and reliable, and was popular with farmers on both sides of the Atlantic. Produced in both 40 in (101 cm) and 50 in (121 cm) gauges, it was extensively used by the US military between 1941 and 1945. Production ceased in 1942 to free factory capacity for war work.

△ Allis-Chalmers HD-19H

Date 1947 **Origin** US
Engine General Motors 6-cylinder 71 series 2-stroke diesel
Horsepower 129 hp
Transmission 2 forward, 1 reverse torque converter

Powered by the GM 6-71 and with torque-converter transmission, the HD-19H was a very advanced design at the time of its introduction. Diesel fuel was used as the fluid medium for the torque converter. This arrangement was soon replaced with a converter that used oil.

▽ John Deere MC

Date 1950 **Origin** US
Engine John Deere 2-cylinder gasoline
Horsepower 18 hp
Transmission 4 forward, 1 reverse

This small, compact machine was the first John Deere-designed crawler tractor. Skid units were sent from Dubuque, Iowa, to the Lindeman factory in Washington, which John Deere had bought in 1946. The MC was not imported into the UK.

▷ John Deere BO-Lindeman

Date 1942 **Origin** US
Engine John Deere 2-cylinder horizontal gasoline/kerosene
Horsepower 14 hp
Transmission 4 forward, 2 reverse

The BO-Lindeman was built to satisfy customer demand from the west coast of the US, with 2,000 BO tractors being converted by Lindeman. Late models could be equipped with a hydraulic attachment to control a range of different implements.

Lightweight Tractor Power

In the early 1900s, small tractors provided a first stepping stone from animal power to powered farming. Replacing work animals offered plenty of sales potential. Official census figures for the number of horses and mules on US farms peaked at more than 26 million in 1920, but there were still about 8 million in 1950. Some of the small tractors, particularly those from big, established companies, were well-built, but first-time buyers with little knowledge of tractor design were often victims of salesmen offering badly designed tractors with little or no after-sales support.

△ Cleveland Cletrac General GG

Date 1948 **Origin** US

Engine Hercules IXA3 4-cylinder gasoline

Horsepower 19 hp maximum

Transmission 3 forward, 1 reverse

The General was built by the Leader Tractor Co., Cleveland, OH, specialty manufacturer of the Cletrac range of crawler tractors. The General GG was the first Cletrac wheeled tractor, and there was a tracklaying version called the HG. In 1941 production was transferred to the implement manufacturer B.F. Avery.

△ Thieman

Date 1941 **Origin** US

Engine Ford Model A car engine preferred, other car engines supplied could be used

Horsepower 40 hp

Transmission 3 forward, 1 reverse

Thieman tractors were for cost-conscious customers. There were three versions: one sold complete and ready to work; a cheaper, self-assembly kit version with a Ford car engine, and—the cheapest option—a kit with no engine for customers who could source their own.

▷ International Farmall Cub

Date 1948 **Origin** US

Engine International 4-cylinder gasoline

Horsepower 9 hp

Transmission 3 forward, 1 reverse

The success of International's 18 hp Model A encouraged it to offer an even smaller model, the 9 hp Farmall Cub with a 980 cc engine. The design included plenty of ground clearance to avoid crop damage, and the offset seat and steering wheel allowed good forward visibility.

◁ Leader Model D

Date 1947–49 **Origin** US

Engine Hercules IXB5 4-cylinder

Horsepower 22 hp

Transmission 3 forward, 1 reverse

Leader tractor history is confusing because the same brand name was used by at least three different US manufacturers. The Model D was built in small numbers during the tractor boom following WWII, with a specification that included a belt pulley and power takeoff.

◁ Brockway 49G

Date 1949 **Origin** US

Engine Continental F-162 4-cylinder gasoline

Horsepower 35 hp

Transmission 4 forward, 1 reverse

The family-owned Brockway company built tractors in Chagrin Falls, OH. Its 49 model was available in three versions, all powered by Continental engines. The diesel model was the 49D, the 49K engine used kerosene, and this 49G was the gasoline version.

◁ Gibson Model H

Date 1949 **Origin** US

Engine Hercules IXB3 4-cylinder gasoline

Horsepower 24.5 hp

Transmission 4 forward, 1 reverse

Gibson tractors were available briefly from the late 1940s in a four-model range powered by one-, two-, four-, and six-cylinder engines. The Model H, available in tricycle and standard-wheel versions, was given a "two-plow" rating (it could turn two furrows).

△ Allis-Chalmers Model G

Date 1949 **Origin** US

Engine Continental AN-62 4-cylinder gasoline

Horsepower 10.3 hp

Transmission 4 forward, 1 reverse

The Model G is a tool carrier, a special type of lightweight tractor developed for working in row crops. The rear engine and the skeleton frame at the front allowed excellent visibility for positioning hoes accurately to avoid damaging the crops.

▽ Newman AN3

Date 1950 **Origin** UK

Engine Coventry Victor air-cooled twin-cylinder horizontally opposed gasoline

Horsepower 10.75 hp

Transmission 3 forward, 1 reverse

Price was a problem for the Newman AN3. In 1949 the 10.75 hp three-wheeler cost more than a Fordson Major E27N with four wheels and three times as much power.

△ Indian Motorcycle Tractor

Date 1952 **Origin** US

Engine Indian V-twin gasoline

Horsepower 24 hp

Transmission 3 forward, 1 reverse

This unique tractor was built by Leonard Kanner, an Ohio engineer, using an Indian motorcycle engine plus WWII government surplus items. The tractor was used to prepare the site for a house Kanner built and after that for garden work and snow clearance.

Mobilizing the Women's Land Army

When conflict seemed inevitable, even before the UK had declared war on Germany, moves were made to mobilize the Women's Land Army (WLA). The organization was re-formed on July 1, 1939, and more than 1,000 volunteers joined almost immediately; by December there were 4,544 "Land Girls."

TRAINING THE LAND ARMY

The women of the WLA carried out a myriad of tasks in every sphere of agriculture—arable, livestock, and dairy farming, land reclamation, market gardening, fruit picking, and even rat-catching. Many trained as tractor drivers, and some were so adept at their job that they were sent out to instruct farmers on the care and use of their machinery. Many found the tractors uncomfortable—"Fairly shakes the inside out of you" is a typical quote—and the work exhausting.

The women were trained by Ministry of Agriculture machinery instructors. Most of the tractors were Fordsons, and training courses were organized by the Henry Ford Institute of Agricultural Engineering in Essex. Land Girls learned about lubrication and maintenance and were taught how to service tractors. In short, they knew as much as the men, if not more.

Members of the Women's Land Army learn to plow with a Fordson tractor at the Henry Ford institute on the Boreham House estate in Essex.

CKX
904

Military Might

The tractors supplied to the military during World War II fell into two categories: they were either built to standard civilian specifications or adapted for specific tasks. For example, tractors used to tow aircraft required different transmissions and extra weight to cope with heavy starting loads; tracked machines were favored for towing. The ability to start instantly was important, so many ran on pure gasoline. Civilian models that were normally fueled by kerosene after warm-up had modified manifolds for continuous gasoline running.

▷ Caterpillar D8 8R

Date 1943 **Origin** US

Engine Caterpillar 6-cylinder diesel

Horsepower 120 hp

Transmission 6 forward, 2 reverse

The Caterpillar D8 was used in the US and UK to prepare the ground for airfields and other facilities in WWII. It was usually fitted with LeTourneau equipment such as a bulldozer or, as shown here, with a cable control unit (CCU) for operating scrapers and rippers.

△ Cletrac Medium M2 High-Speed Tractor

Date 1942 **Origin** US

Engine Hercules 6-cylinder WXLC3 gasoline

Horsepower 150 hp

Transmission 4 forward, 1 reverse

The Cletrac M2 was designed to tow US Air Force heavy bombers. It was equipped with planetary steering, a 10,000 lb (4,500 kg) pull winch, a high-pressure air compressor, 24- and 110-volt auxiliary generators, and nitrogen bottles for fire control. From 1942 to 1945, 8,510 were built.

◁ Brons Model EAT

Date 1940 **Origin** Holland

Engine Brons 2-cylinder diesel

Horsepower 40 hp

Transmission 3 forward, 1 reverse

The mid-mounted radiator added the unusual layout of this rare tractor. The total production of the model EAT, in both two- and three-cylinder sizes, was 49 between 1935 and 1959. The three-cylinder version produced 60 hp. The engine used in the tractor was also offered for stationary and marine application.

△ Roadless Half-Track

Date c. 1940 **Origin** UK

Engine Fordson 4-cylinder gasoline

Horsepower 25 hp

Transmission 3 forward, 1 reverse

The Fordson was the basis for a number of applications, including this Roadless Half-Track, which was specially built for aircraft haulage with the RAF. The front forecarriage made it stable and easier to steer as well as providing a mounting platform for the chain-driven Hesford winch.

Farming Front

When war broke out in September 1939, the UK was well prepared from an agricultural point of view. Under an arrangement with the Ford Motor Co. a pool of new model Ns had been built up, ready for allocation to the county-based War Agricultural Executive Committees (WAEC). Some tractors from other British manufacturers were available and machines were still being imported from the US, but German U-boats sank many of the ships. When the US entered the war in 1941, supplies were curtailed initially, but resumed under the Lend–Lease arrangements.

◁ David Brown VAK1

Date 1941 **Origin** UK

Engine David Brown 4-cylinder gasoline/kerosene

Horsepower 26 hp

Transmission 4 forward, 1 reverse

The VAK1 was announced in 1939. At the time it was considered an advanced design, including the then novel feature of dished wheel centers to adjust the track width. The basic model had steel wheels and magneto ignition. Rubber tires and coil ignition were available at extra cost.

◁ David Brown VIG1/100

Date 1941 **Origin** UK
Engine David Brown 4-cylinder gasoline
Horsepower 26 hp
Transmission 4 forward, 1 reverse

David Brown Tractors Ltd. were awarded the only contract for the design and production of heavy-wheeled aircraft towing and supply tractors for the RAF. The result was the VIG1/100, weighing nearly 4 tons (3,500 kg). It could be ballasted to over 7 tons (7,000 kg) to handle heavier loads.

▽ Fordson Model N Industrial

Date 1944 **Origin** UK
Engine Fordson 4-cylinder gasoline
Horsepower 25 hp
Transmission 3 forward, 1 reverse

The Industrial Model N was the tractor favored for hauling bombs out from the dump to the waiting aircraft. It had an oil-immersed clutch, which made it practically impossible to burn out under the heavy starting loads sometimes imposed on the transmission of these tug tractors while in service.

△ Ford 9N Moto-Tug B-NO-25

Date 1941 **Origin** US
Engine Ford 4-cylinder gasoline
Horsepower 23 hp
Transmission 3 forward, 1 reverse

The Moto-Tug was a light machine with a pull of about 2,500 lb (1,100 kg). As such, it could handle only light loads such as fighter planes and their ammunition supply carts on the airfield and concrete hangar standings. The Moto-Tug saw extensive service on US aircraft carriers and at overseas civilian airports.

△ Ford 9N

Date 1941 **Origin** US
Engine Ford 4-cylinder gasoline
Horsepower 23 hp
Transmission 3 forward, 1 reverse

The result of the famous handshake between Henry Ford and Harry Ferguson, the 9N (or "Ford-Ferguson") went on sale mid-1939. Some 10,000 were shipped to the UK under the Lend-Lease program; most went to the Women's Land Army, a British civilian organization of women working in agriculture.

◁ Fordson Model N

Date 1944 **Origin** UK
Engine Fordson 4-cylinder gasoline/kerosene
Horsepower 25 hp
Transmission 3 forward, 1 reverse

Known as the "Standard" Fordson, this tractor was the backbone of the British agricultural tractor fleet during WWII. Production peaked at 2,500 a month in 1943. The model N was available with the low-ratio "Red Spot" and the higher-ratio "Green Spot" gearbox options. Almost all of the wartime production was delivered on steel wheels.

Roadless Half-Track

Half-tracks based on Fordson tractors were supplied to the UK's Royal Air Force and other armed services during World War II for recovery work, mowing airfields, moving aircraft, and hauling bomb trolleys or fuel pumps. Several were used during the evacuation of Dunkirk in June 1940. The forecarriage arrangement, which supported the front axle and wheels, also provided a suitable platform for mounting a winch or crane.

ROADLESS TRACTION constructed tracked versions of everything from wheelbarrows to steam rollers. The company's "elastic girder" tracks had flexible rubber blocks that acted as frictionless joints between the metal plates to eliminate shocks. The tracks were silent in operation and maintenance was reduced to a minimum.

In 1936 the Air Ministry placed an order for Roadless full-track conversions of Fordson tractors. Problems with the tractors pitching at high speed led to the development of the forecarriage arrangement, which improved stability and turned the crawlers into half-tracks. The front wheels controlled the steering so the tractors were easier to drive.

Military applications
Wartime Roadless half-tracks, based on the Fordson Model N, saw service with the Royal Air Force. Equipped with a front-mounted winch, the tractor was used for aircraft recovery.

Control levers for winch

Driver's seat

Drive sprocket

Tracks are rubber-jointed

Bottom rollers

FRONT VIEW

REAR VIEW

SPECIFICATIONS	
Model	Roadless Half-Track
Built	c.1940
Origin	UK
Production	Not known
Engine	25 hp Fordson 4-cylinder gasoline
Capacity	267 cu.in (4,380 cc)
Transmission	3 forward, 1 reverse
Top speed	$4^{1}/_{2}$ mph (7 km/h)
Length	10 ft 8 in (3.35 m)
Weight	3.6 tons (3.3 metric tons)

THE DETAILS

1. Fordson radiator header tank **2.** Roundel for Royal Air Force **3.** Winch could carry 300 ft (92 m) of wire cable **4.** Chain-drive for winch taken from tractor's pulley shaft **5.** Lubrication plate for track running gear **6.** Driver's platform, steering assisted by independent foot brakes for each track unit

Postwar Variations

There was a considerable variation in tractor design around the world after the end of World War II. The older designs of single-cylinder, hot-bulb machines remained popular where the quality of the available liquid fuels had not yet returned to a standard specification. Gasoline as a fuel was still in use, but the price, as well as the greater fuel consumption, put it at a disadvantage. There was a huge demand for tractors to get the farms in war-torn areas back in production. Customers were easy to find and this supported a large and varied number of tractor manufacturers.

△ **HSCS R50**

Date 1947 **Origin** Hungary
Engine HSCS single-cylinder 2-stroke hot-bulb semi-diesel
Horsepower 50 hp
Transmission 3 forward, 1 reverse

The R50 was the largest size of these Lanz models produced by HSCS just after WWII. These tractors were widely exported, with some examples carrying the name "Le Robuste," which they certainly were. It was a perfectly functional machine, but lacked refinement of detail.

▽ **Hürlimann 40 DT70G**

Date 1945 **Origin** Switzerland
Engine Hürlimann 4-cylinder diesel
Horsepower 65 hp
Transmission 5 forward, 1 reverse

In the true Hürlimann tradition, this tractor was a high-quality machine, and every part was produced in-house. Designed specially for road haulage and timber work, it was supplied with a full air-braking system on all four wheels. A top speed of 25 mph (40 km/h) was available years before it became a standard feature on modern tractors.

△ **Super Landini**

Date 1946 **Origin** Italy
Engine Landini single-cylinder 2-stroke hot-bulb semi-diesel
Horsepower 48 hp
Transmission 3 forward, 1 reverse

Giovanni Landini founded Landini SpA in 1884 to manufacture agricultural equipment; he started tractor production in 1925. The Super Landini was based on the Lanz single-cylinder, hot-bulb design. It was a finely engineered and refined machine. Landini tractors are still produced, Landini now being part of Argo SpA.

▽ **KL Bulldog**

Date 1949 **Origin** Australia
Engine KL Tractors single-cylinder 2-stroke hot-bulb semi-diesel
Horsepower 42 hp
Transmission 3 forward, 1 reverse

Kelly & Lewis (formerly the agent for Lanz Bulldogs in Australia) formed KL Tractors Ltd. to produce copies of the 40-hp Lanz Bulldog. Announced in 1947, the KL Bulldog was produced until 1954, but because there were now better tractors on the market, it took several years to clear the inventory.

△ **Hürlimann D100**

Date 1948 **Origin** Switzerland
Engine Hürlimann 4-cylinder diesel
Horsepower 45 hp
Transmission 5 forward, 1 reverse

This was the first tractor in the Hürlimann range to use a new design of direct injection diesel engine that had been in development since 1944. The engine—with a 24-volt starter, a newly developed Bosch injection system, and a pneumatic governor—would start easily in very cold weather.

▷ **ECO Type N**

Date 1948 **Origin** France
Engine Poyaud 2-cylinder diesel
Horsepower 40 hp
Transmission 6 forward, 2 reverse

Founded by Eugene Mignot in 1920, ECO sold its tractors from offices in Paris, but assembly was contracted to outside firms. Its first tractors appeared in 1932. The Type N, introduced at the end of WWII, was manufactured at the French government's armaments factory in Le Havre, using either Panhard gasoline or Poyaud diesel engines.

▽ **Motomeccanica Balilla B50 Stradale**

Date 1948 **Origin** Italy
Engine Motomeccanica 4-cylinder gasoline
Horsepower 10 hp
Transmission 6 forward, 2 reverse

The Balilla, Italy's first compact tractor, was introduced in 1931 by Motomeccanica of Milan. Various models, including a B50 tracked version, were built until 1952. The tractor was also sold in France under the Alfa Romeo name.

◁ **Bolinder-Munktell BM20**

Date 1949 **Origin** Sweden
Engine Bolinder W5 2-cylinder hot-bulb semi-diesel
Horsepower 41 hp
Transmission 5 forward, 1 reverse

Bolinder-Munktell was the result of a merger between the two companies in 1932. The BM20 was one of the first tractors whose specifications included use of water ballast in the rear tires. Along with wheel weights, the tractor could be increased in total weight from 5,840 lb to 7,715 lb (2,650 kg to 3,500 kg).

▷ **Orsi Argo**

Date 1950 **Origin** Italy
Engine Orsi single-cylinder 2-stroke hot-bulb semi-diesel
Horsepower 55 hp
Transmission 6 forward, 1 reverse

The Orsi Argo was yet another tractor whose design was based on the Lanz Bulldog. Its simple and rugged construction appealed to customers, keeping it in production for some 12 years. The belt pulley had its separate clutch mounted in the flywheel, a feature that allowed it to be operated from ground level.

German Engineering

From low-output tractors used by small farmers to large and sophisticated models, tractor production was in full swing in Germany before the outbreak of World War II. Practically all tractors at this time were designed to run on either diesel or some heavy oil. With the outbreak of war, fuel ran short; instead, producer gas was used, requiring the addition of generators to the tractors, making them unwieldy and heavy. As soon as fuel became available again, these were discarded.

▷ **Hermann Lanz Aulendorf D40**

Date 1938 **Origin** Germany
Engine Deutz F2M414 2-cylinder diesel
Horsepower 22 hp
Transmission 4 forward, 1 reverse

A prewar design, the D40 was produced from 1937; production ceased in 1942 as diesel became practically impossible to obtain for civilian use. The company then went on to produce tractors that could run on producer gas, generated mainly from wood.

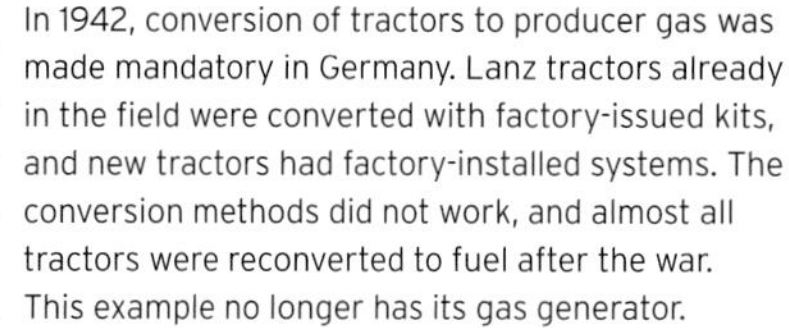

▷ **Hanomag RL 20**

Date 1939 **Origin** Germany
Engine Hanomag D19 4-cylinder diesel
Horsepower 20 hp
Transmission 4 forward (optional 3-speed), 1 reverse

Primarily a road tractor, the RL 20 was built from "off-the-shelf" parts; it used a car engine, hence the front styling. It had hydraulically actuated brakes on all four wheels: in its standard form it had equal-sized wheels; larger rear wheels were available as an option.

▽ **Lanz Gas Producer Bulldog**

Date 1943 **Origin** Germany
Engine Lanz single-cylinder
Horsepower 35 hp on gas
Transmission 6 forward, 2 reverse

In 1942, conversion of tractors to producer gas was made mandatory in Germany. Lanz tractors already in the field were converted with factory-issued kits, and new tractors had factory-installed systems. The conversion methods did not work, and almost all tractors were reconverted to fuel after the war. This example no longer has its gas generator.

Built for Roads

Popular in continental Europe, tractors equipped with road-legal braking and lighting systems provided an alternative to trucks. A road tractor with several four-wheel trailers offered flexible transportation, since empty trailers could be left at a location to be loaded while the tractor delivered another full trailer elsewhere. If the terrain was difficult—for example, on a farm in wet weather—road tractors were capable of movement when a truck was not, and on the road they operated at speeds similar to the trucks of the day.

▷ **O&K SA 751**

Date 1939 **Origin** Germany
Engine O&K 17B2s 2-cylinder diesel
Horsepower 30 hp
Transmission 3 forward, 1 reverse

Orenstein & Koppel (O&K) was founded by Benno Orenstein and Arthur Koppel as a general engineering company in 1876. Starting with heavy equipment such as railroad locomotives and mining machinery, it began to produce tractors in the late 1930s. The SA 751 was aimed at the large customer base of small German farms. A simple and basic machine, the tractor sold in modest numbers.

▽ Deutz F1M414

Date 1949 **Origin** Germany

Engine Deutz single-cylinder diesel

Horsepower 11 hp

Transmission 3 forward, 1 reverse

First produced in 1936, the F1M414 remained in production until 1951 with practically no modifications to the design throughout its run. This tractor was responsible for the mechanization of large numbers of small German farms. It came with a power takeoff, a belt pulley, and a side-mounted mower drive as standard.

▷ Deutz F2M315

Date 1948 **Origin** Germany

Engine Deutz 2-cylinder diesel

Horsepower 28 hp

Transmission Standard 3 forward (optional 5 forward), 1 reverse

The F2M315 was first produced in 1933, and production ceased in 1942 due to the lack of liquid fuels. After the war, production was resumed with only a few modifications. It was built in three different versions: the basic agricultural model, the Universal model, and a road tractor version.

△ Allgaier R22

Date 1949 **Origin** Germany

Engine Kaeble single-cylinder diesel

Horsepower 22 hp

Transmission 4 forward, 1 reverse

The R22 was a very basic tractor—it could be described as a stationary engine mounted on wheels. The single-cylinder engine was hopper-cooled, with no radiator, fan, or water pump. The tractor came with a list of possible attachments, which included a side-mounted cutter bar, a rear power lift, an adjustable drawbar, and a canopy.

▷ Lanz Traffic Bulldog

Date 1941 **Origin** Germany

Engine Lanz single-cylinder hot-bulb

Horsepower 55 hp

Transmission 6 forward, 2 reverse

With a maximum speed of 14 mph (23 km/h), a sprung front axle, a driver's cab, and a full set of mudguards, the Traffic Bulldog was purpose-built for road use. The driver's cabin was available as a full cab or with just a roof and windshield. The seat was suspended and dampened by two shock absorbers. Standard equipment also included a tachometer and a directional indicator.

△ MAN AS-250

Date 1944 **Origin** Germany

Engine MAN DO534GS 4-cylinder diesel

Horsepower 50 hp

Transmission 5 forward, 1 reverse

In the true tradition of MAN, the AS-250 tractor was a very well-engineered machine. Its quality was reflected in the high purchase price. The AS-250 achieved a very low fuel consumption rating both in the field and on the road. A total of 1,323 were built between 1938 and 1944.

Harry Ferguson, left, and Sir John Black

Great Manufacturers

Ferguson

Harry Ferguson made a huge contribution to tractor history. Almost every modern wheeled tractor has the three-point linkage system for implement attachment and control that he developed, and the company he owned became one of the world's most successful tractor manufacturers.

HARRY FERGUSON'S affinity for the mechanical drove him from his father's Ulster farm and its horses when he was 17 years old. In 1902 he left for Belfast and quickly found work in a garage, where he could make the most of his ability.

Ferguson's mechanical engineering skill earned him attention, and he quickly went into business for himself. His success as a racing driver also won publicity for his garage.

World War I saw a huge increase in demand for tractors as the UK sought ways to improve its agricultural output in the face of German blockades. Ferguson seized the chance to expand his business and began selling the US-built Waterloo Boy tractors (known as the Overtime in the UK) through his garage, doing many of the sales demonstrations himself. His skill with a tractor brought an invitation to make a government-sponsored tour of tractor-owning farms in 1917.

The tour was intended to show farmers how to use their equipment more efficiently, but its most significant result was to convince Ferguson that using a chain to pull implements behind a tractor was slow, inefficient, and costly.

A salesman's working model of a TE-20 and plow.

Ferguson's attempts to overcome this issue resulted in his invention of the three-point linkage system with hydraulic draft control. Throughout the 1920s and early '30s, Ferguson worked with his garage staff to perfect his device. The linkage firmly attached the implement to the tractor, aiding traction by increasing the weight of the rear wheels. The system was controlled by the tractor's hydraulics, allowing the driver to raise and lower the attached implement as necessary. In 1933 Ferguson finalized his prototype, calling it simply and elegantly the Black Tractor.

The Black Tractor resulted in a partnership between Ferguson and David Brown, managing director of an engineering business, with Ferguson designing and marketing the tractors built by David Brown's company. Their tractor, officially called the Ferguson Type A, but often known as the Ferguson-Brown, was launched in 1936. It was based on Ferguson's Black Tractor, with an 18–20-hp Coventry-Climax engine powering the first 500, followed by a 2,010-cc David Brown power unit.

The new tractor performed well and the three-point linkage offered obvious advantages, but it was expensive and customers needed special implements to fit the linkage. Sales were disappointing, intensifying friction between the partners when Ferguson refused to make the design changes requested by Brown.

Practical demonstration
A Ferguson-Brown tractor is demonstrated at an agricultural show by Ferguson himself. The experience demonstrating tractors first gave Ferguson the idea for his three-point linkage.

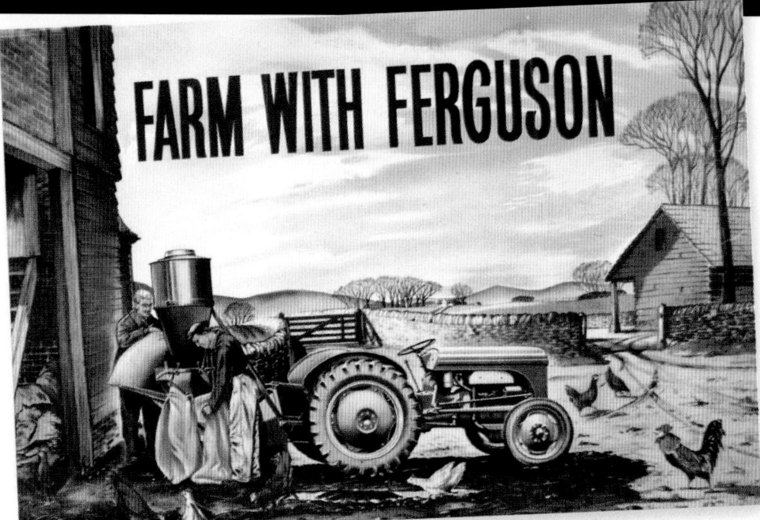

More than just a tractor
A sales poster for Ferguson equipment. The image illustrates one of the many uses to which the Ferguson could be put with the aid of the three-point linkage.

Ferguson parted company with David Brown in 1939. Without telling his partner, Ferguson had arranged to demonstrate his new tractor to Henry Ford. Ford was planning to resume tractor production in the US, and he was impressed by the performance of the Ferguson-Brown. The two men formed a new partnership; Ford agreed to build a new Ferguson System tractor while Ferguson formed a marketing organization to sell the tractor in the US.

Its official name was the "Ford tractor with Ferguson System" or 9N. It was a small machine with a 23-hp Ford engine and Battleship Grey paint, the color Ferguson chose for all his production tractors. The combination of Ford's unrivaled production resources and Ferguson's

Harry Ferguson was the **first** British citizen to **design**, **build**, and **fly** an airplane.

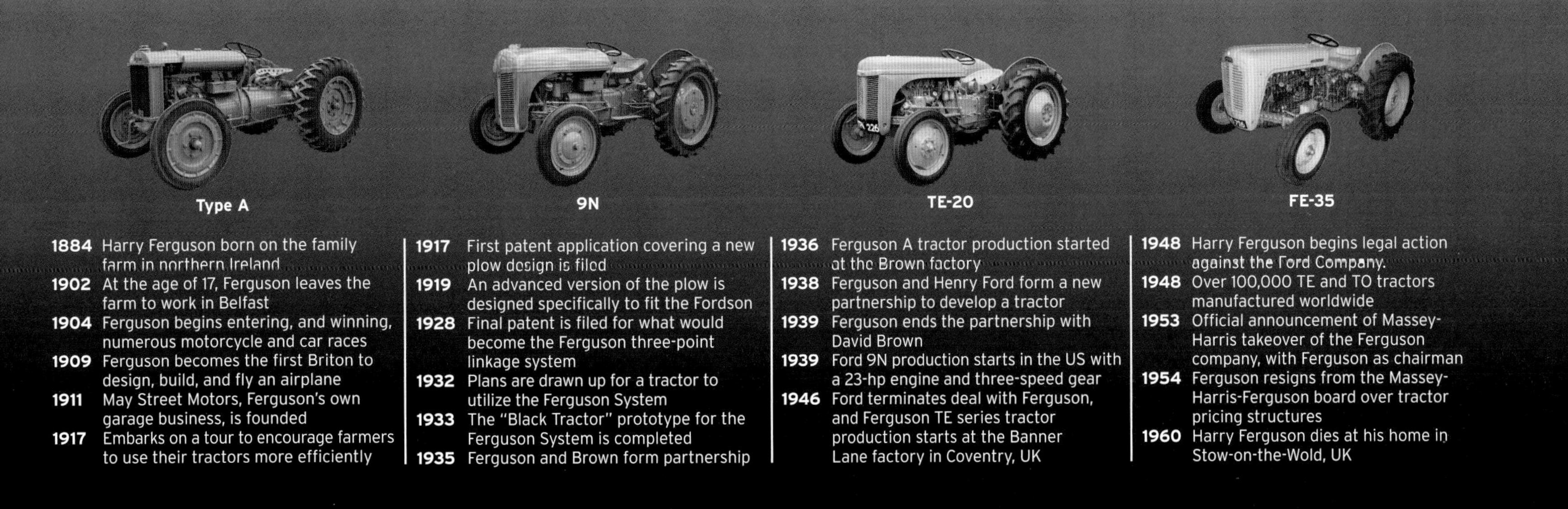

1884 Harry Ferguson born on the family farm in northern Ireland
1902 At the age of 17, Ferguson leaves the farm to work in Belfast
1904 Ferguson begins entering, and winning, numerous motorcycle and car races
1909 Ferguson becomes the first Briton to design, build, and fly an airplane
1911 May Street Motors, Ferguson's own garage business, is founded
1917 Embarks on a tour to encourage farmers to use their tractors more efficiently
1917 First patent application covering a new plow design is filed
1919 An advanced version of the plow is designed specifically to fit the Fordson
1928 Final patent is filed for what would become the Ferguson three-point linkage system
1932 Plans are drawn up for a tractor to utilize the Ferguson System
1933 The "Black Tractor" prototype for the Ferguson System is completed
1935 Ferguson and Brown form partnership
1936 Ferguson A tractor production started at the Brown factory
1938 Ferguson and Henry Ford form a new partnership to develop a tractor
1939 Ferguson ends the partnership with David Brown
1939 Ford 9N production starts in the US with a 23-hp engine and three-speed gear
1946 Ford terminates deal with Ferguson, and Ferguson TE series tractor production starts at the Banner Lane factory in Coventry, UK
1948 Harry Ferguson begins legal action against the Ford Company.
1948 Over 100,000 TE and TO tractors manufactured worldwide
1953 Official announcement of Massey-Harris takeover of the Ferguson company, with Ferguson as chairman
1954 Ferguson resigns from the Massey-Harris-Ferguson board over tractor pricing structures
1960 Harry Ferguson dies at his home in Stow-on-the-Wold, UK

three-point linkage helped ensure the 9N's popularity, and more than 70,000 tractors were built in 1946.

Meanwhile, Ford's plant in the UK was still building outdated tractors based on the original 1917 Fordson. Ferguson had assumed production would switch to the 9N, but when this idea was rejected, he decided to start his own British company. He formed another partnership, this time with the Standard Motor Company, utilizing the spare capacity in their car factory at Banner Lane, Coventry. The agreement was similar to previous partnerships; Standard controlled production while Ferguson handled the engineering and marketing.

Production of the new Ferguson TE series tractors started in 1946. A four-speed gearbox replaced the three speeds of the Ferguson A and 9N, and a 24-hp Continental engine provided the power during the tractor's first two years, to be followed by various Standard engines. The TE series plus the TO version built at a Ferguson factory in the US became one of the tractor industry's biggest successes, with production peaking at more than 100,000 per year.

While his Standard partnership was starting, Harry Ferguson's relationship with Ford ended acrimoniously with a hugely expensive legal action lasting four years. Ferguson eventually accepted a $9.25 million settlement.

In 1952 Ferguson decided to part with his US company, eventually resolving to sell his entire business to Massey-Harris. The North American farm equipment firm wanted to expand in Europe and appointed Ferguson chairman, allowing him to retain some input into the business.

A complete farming system
Flexibility was one of the Ferguson system's greatest strengths. The long list of Ferguson equipment included a circular saw that would probably not meet current safety standards.

Revolutionizing agriculture
A leaflet helped tractor salesmen explain the benefits of the Ferguson system to potential customers.

The UK's Golden Age

The period following World War II was a golden age for the UK's tractor industry. Demand for tractor power grew rapidly around the world and British manufacturers recovered from the war faster than their European rivals. Many North American companies chose to build tractors in UK factories for Britain, Continental Europe, and what was then the British Empire, to circumvent postwar import controls on US tractors introduced to alleviate the dollar shortage. As a result, the UK became the world's biggest tractor exporter and was at the forefront of design improvements.

△ Minneapolis-Moline UDM

Date 1948 **Origin** UK

Engine Meadows 4-cylinder diesel

Horsepower Meadows 65 hp

Transmission 5 forward, 1 reverse

Minneapolis-Moline offered two UK-assembled tractors in the postwar years. The UDS, equipped with a Dorman diesel engine, appeared in 1946, and the Meadows-powered UDM version in 1948. Both tractors were too expensive to have any impact on the market, and production ceased in 1949.

△ Ferguson TE-20

Date 1947 **Origin** UK

Engine Continental 4-cylinder gasoline

Horsepower 24 hp

Transmission 4 forward, 1 reverse

Harry Ferguson's TE series, painted his favorite shade of gray, was one of the tractor industry's biggest successes. It took the benefits of the Ferguson three-point linkage throughout the world. The company attracted a successful takeover from Massey-Harris.

▷ David Brown Cropmaster

Date 1947 **Origin** UK

Engine David Brown 4-cylinder gasoline/kerosene

Horsepower 35 hp

Transmission 6 forward, 2 reverse

The Cropmaster was one of the best-known David Brown tractors. Designated VAK/1C, it was launched in April 1947 to replace the VAK/1A model. It had an integral power lift and a new six-speed transmission. Nearly 60,000 were built by the time its production run ended in 1953.

△ Field Marshall Series 2

Date 1948 **Origin** UK

Engine Marshall single-cylinder horizontal diesel

Horsepower 40 hp

Transmission 3 forward, 1 reverse

The first Field Marshall was a huge success, but continual development led to a number of improvements. The Series 2 featured more power, a better cooling system, a larger-diameter clutch, a more comfortable operator's seat, and larger tires.

▷ Nuffield Universal M3

Date 1951 **Origin** UK

Engine Morris Commercial ETA 4-cylinder gasoline/kerosene

Horsepower 38 hp

Transmission 5 forward, 1 reverse

Forecasts of expanding demand brought a flood of companies into Britain's tractor industry. The Nuffield car and truck group (Nuffield Organization) was one of the successful ones, introducing the Universal range, which included the unusual M3, a three-wheel row-crop model.

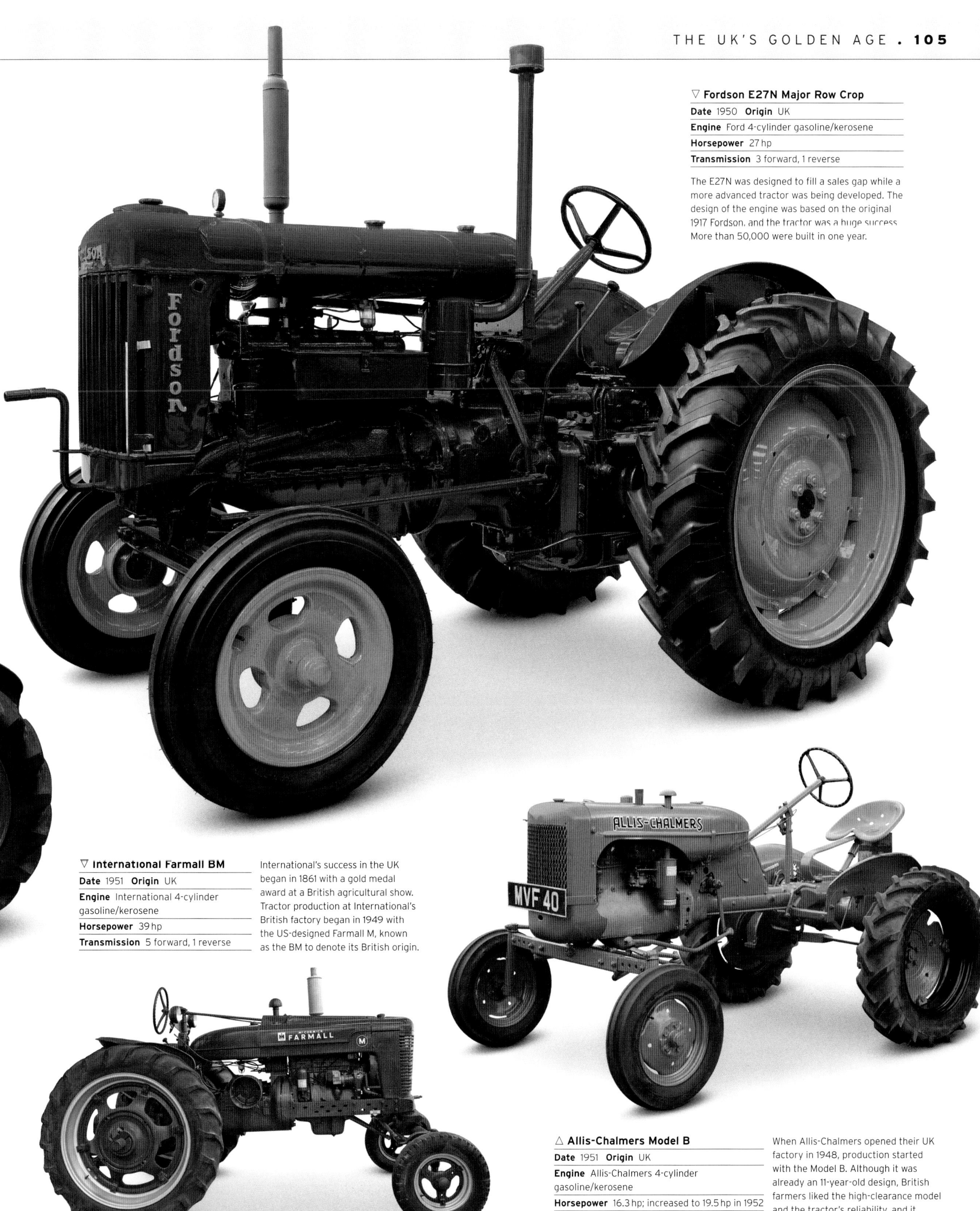

▽ **Fordson E27N Major Row Crop**

Date 1950 **Origin** UK

Engine Ford 4-cylinder gasoline/kerosene

Horsepower 27 hp

Transmission 3 forward, 1 reverse

The E27N was designed to fill a sales gap while a more advanced tractor was being developed. The design of the engine was based on the original 1917 Fordson, and the tractor was a huge success. More than 50,000 were built in one year.

▽ **International Farmall BM**

Date 1951 **Origin** UK

Engine International 4-cylinder gasoline/kerosene

Horsepower 39 hp

Transmission 5 forward, 1 reverse

International's success in the UK began in 1861 with a gold medal award at a British agricultural show. Tractor production at International's British factory began in 1949 with the US-designed Farmall M, known as the BM to denote its British origin.

△ **Allis-Chalmers Model B**

Date 1951 **Origin** UK

Engine Allis-Chalmers 4-cylinder gasoline/kerosene

Horsepower 16.3 hp; increased to 19.5 hp in 1952

Transmission 3 forward, 1 reverse

When Allis-Chalmers opened their UK factory in 1948, production started with the Model B. Although it was already an 11-year-old design, British farmers liked the high-clearance model and the tractor's reliability, and it became a popular choice.

Henry Ford with his experimental tractor in 1907

Great Manufacturers

Fordson

Fordson was the name given to the tractors produced from 1917 by the Ford Motor Company, first in Dearborn, MI, and then in Ireland and the UK, where manufacture continued in Dagenham until 1964. The marque remained a separate entity from the Ford tractor line, established after production was reintroduced to the US in 1939.

HENRY FORD WAS BROUGHT up as a farm boy. He made his fortune in the automobile industry, but never lost his ambition to remove the drudgery from agricultural labor by developing an affordable farm tractor. He built several experimental machines in 1905–08 before announcing plans in 1915 to manufacture a lightweight tractor at a price that would undercut his rivals.

His fellow directors at the Ford Motor Company were opposed to building tractors, so Ford set up a separate venture in partnership with his wife and son. Henry Ford & Son was the name of the business, which operated from a factory in Brady Street, Dearborn, MI, and "Fordson" was its cable address.

Henry Ford
1863-1947

The tractor was eventually rushed into production in 1917 at the request of the British Ministry of Munitions, which ordered 6,000 of the machines for its wartime plowing campaign. These early examples were known simply in the UK as Ministry of Munitions or MOM tractors.

> "I have followed many a **weary mile** behind a plow and I know the **drudgery** of it."
>
> HENRY FORD

During 1918, the tractor was launched in the US market as the Fordson Model F, priced at $750.

In 1921, production of the Model F was transferred to Ford's Rouge River plant in Michigan, where it was built until 1928, when the company exited the tractor market in the US after losing a price war with International Harvester. Satellite production of the Model F was also carried out in Cork, Ireland, from 1919 to 1922. From 1929 Cork became the center of manufacturing for the improved Fordson Model N, which continued until 1932, when production moved to the newly constructed Dagenham plant in Essex, England.

MOM tractors
Tractors imported into the UK by the Ministry of Munitions were leased to farmers, and drivers could be hired from the Army Service Corps. This example was harvesting on a farm in South Lincolnshire.

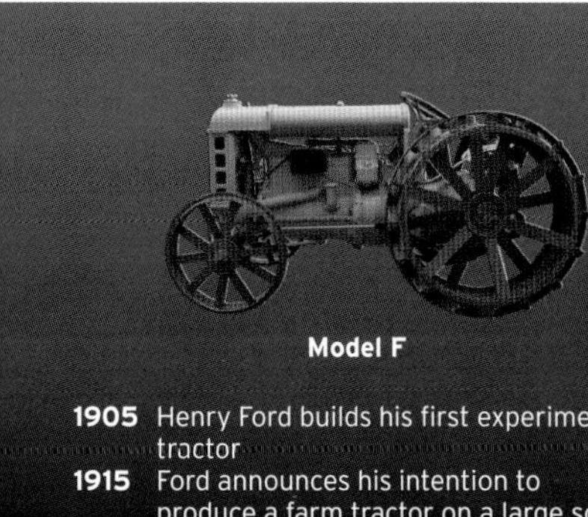

Model F

Model N

E27N Major

"New Performance" Super Major

1905 Henry Ford builds his first experimental tractor
1915 Ford announces his intention to produce a farm tractor on a large scale
1917 First consignment of tractors shipped to the British Ministry of Munitions
1918 Fordson Model F, priced at $750, goes on sale in US
1919 Irish production of the Model F begins in Cork, Ireland
1923 US price of Model F cut to $395 to compete with International Harvester

1928 Model F production ends at factory in Rouge River, MI
1929 Production of Model N begins in Cork, Ireland
1933 Tractor production begins in Dagenham, Essex
1937 Improved Model N launched in "harvest gold" livery
1941 Tractor production in Dagenham disrupted by German bombing raids
1943 100,000th tractor since the outbreak of war built in Dagenham

1944 The British Ministry of Agriculture requests a more efficient tractor
1945 First Fordson E27N Major rolls off assembly line
1950 New vaporizer announced for E27N Major
1951 "New" Major launched in Southend-on-Sea
1954 500,000th Fordson to be built in Dagenham drives off the assembly line.
1955 Dagenham builds its 100,000th Diesel Major

1957 Fordson Dexta launched at Alexandra Palace in London
1958 Power Major replaces the Fordson Diesel Major
1960 Super Major unveiled in Hamburg, Germany
1961 Super Dexta announced at London's Smithfield agricultural show
1963 Fordson "New Performance" range introduced
1964 Production of the Fordson comes to an end

The Model N, now a product of the British Ford Motor Company, was built in Dagenham from 1933 until the end of World War II. Between 1939 and 1945 the factory produced nearly 140,000 tractors—some 95 percent of all the wheeled models made in the UK during the war. The plant operated a seven-day week and, at peak times, one tractor came off the production line every 17 minutes 36 seconds.

In 1944 the Ministry of Agriculture asked Ford to develop a more efficient model with greater horsepower and lower fuel consumption. The result was the Fordson Major (designated E27N), the first of which rolled out of Dagenham on March 19, 1945. The Major inherited the N's outdated gasoline/kerosene side-valve engine, but had a newly designed three-speed gearbox that was quieter and more efficient, without the earlier worm-drive transmission.

The Fordson E27N Major could be temperamental, but it was ruggedly reliable and the basic agricultural model on steel wheels was very affordable. It was the savior of British agriculture in a time of austerity, and more than 230,000 were built. From 1948 the tractor was available with a Perkins P6 diesel engine as a factory-installed option.

The E27N model was replaced in November 1951 by the "New" Major, with a six-speed gearbox and a four-cylinder, overhead-valve engine available as a gasoline, kerosene, or diesel unit. The diesel version was truly outstanding and is still regarded by many as one of the best tractors ever made. The gasoline and kerosene variants were eventually dropped, and the Fordson Diesel Major was sold to almost every part of the globe, including substantial exports to the US.

In 1957 the Fordson Dexta was introduced as Ford's answer to the Ferguson TE-20. It had a three-cylinder diesel engine, jointly developed in conjunction with Perkins, a six-speed transmission, and draft-control hydraulics—a first for a Fordson tractor.

During 1958, the Diesel Major was revamped into the Fordson Power Major with minor changes. More improvements arrived in 1960 with the development of the Super Major, which featured differential lock, independent disc brakes, and a revised hydraulic system incorporating draft control. The Dexta was given a facelift at the same time, continuing in production alongside the more powerful Super Dexta, which was introduced in late 1961 with an enlarged version of the Dexta's three-cylinder engine.

Output at Dagenham had risen to more than 350 tractors per day, but changes were afoot to restructure Ford's worldwide tractor operations and transfer production to a new British plant in Basildon, Essex. The "New Performance" range, launched in 1963 in a new blue and gray livery, marked the swan song of the Fordson era. Production finally ended in 1964.

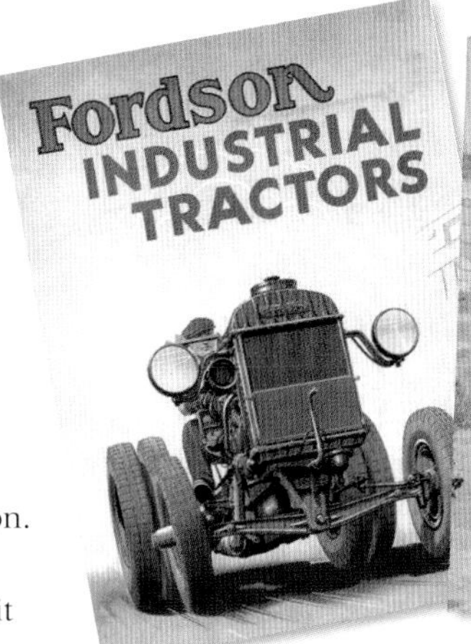

Fordson brochures
The company's sales literature was always colorful. Left to right are a 1937 catalog for the Model N industrial tractor, a 1951 launch brochure for the "New" Major, and literature for the 1961 "Super Class" range.

Dexta workmate
In 1957 the Major got a workmate in the form of the Fordson Dexta, with a three-cylinder diesel engine, a six-speed transmission, and draft-control hydraulics.

Specials and Conversions

The simplicity and compact design of the humble farm tractor made it the basis for an untold number of specialty machines. During and after World War II, there was neither the time nor the money to design and construct dedicated pieces of machinery for every task required. Instead, standard tractors were equipped with more powerful engines. Engines using both gasoline and diesel as fuel, and even "producer gas" conversions, as well as other attachments—by aftermarket suppliers, and not always with the makers' consent—were used.

▽ **Fordson E27N Roadless DG Half-track**

Date 1950 **Origin** UK

Engine Fordson 4-cylinder gasoline/kerosene

Horsepower 27 hp

Transmission 3 forward, 1 reverse (Red Spot)

The Roadless DG conversion of the E27N made this into a useful replacement for full-track crawler tractors, which were in short supply at the end of WWII. Ford offered two gearbox options for the E27N: the low-geared "Red Spot" for heavier work and the "Green Spot" for higher working speeds.

△ **Case-Mercer Crane**

Date 1942 **Origin** US

Engine Case 4-cylinder gasoline

Horsepower 53 hp

Transmission 3 forward, 1 reverse

The Mercer crane was a rough-and-ready machine built around a standard agricultural tractor. It was used for general on-site lifting duties, ranging from assisting in the erection of buildings to engine removal and replacement in army and civilian workshops.

◁ **Ford 2N V8**

Date 1949 **Origin** US

Engine Ford Industrial V8 8-cylinder gasoline

Horsepower 85 hp

Transmission 3 forward, 1 reverse

The Funk Brothers produced aircraft in small numbers during WWII. In 1943 they made a kit available to convert the Ford 2N to six-cylinder power using a Ford side-valve engine. In 1949 an additional kit was offered to allow the 239-cubic-inch (3.9-liter) Ford side-valve V8 engine to be installed.

▷ **Case DEX Roadless**

Date c. 1942 **Origin** UK

Engine Case 4-cylinder gasoline/kerosene

Horsepower 30 hp

Transmission 4 forward, 1 reverse

The Case DEX with a Roadless conversion is a rare machine. Difficult to steer, the operator had to control it using the standard tractor differential with brakes mounted on the inside of the sprockets. Some of these machines were also converted into half-tracks by the provision of a forecarriage and were used on wartime airfields.

▷ Fordson E27N P6 Major

Date 1951 **Origin** UK
Engine Perkins P6 (TA) 6-cylinder diesel
Horsepower 45 hp
Transmission 3 forward, 1 reverse

The Perkins P6 was available on the E27N as a factory-installed option to the Fordson gasoline/kerosene engine from 1948. Perkins offered a P6 conversion kit that included all the parts required to convert tractors to diesel power.

▷ Howard Dungledozer

Date 1944 **Origin** UK
Engine Fordson 4-cylinder gasoline/kerosene
Horsepower 24 hp
Transmission 3 forward, 1 reverse

The Dungledozer was an attempt to mechanize the loading of manure from cattle yards and dung heaps. Technically, it was a great success, but it needed a fleet of tractors and trailers to operate it at its full potential capacity. As a commercial venture it was a failure, since farmers were not ready to invest large sums of money in machinery of this kind. Only two were built.

◁ Aveling Barford 5-Ton Roller

Date c. 1940 **Origin** UK
Engine Fordson 4-cylinder gasoline/kerosene
Horsepower 24 hp
Transmission 3 forward, 1 reverse

Aveling Barford was formed in 1934 from the remnants of Aveling & Porter and the Agricultural and General Engineers (AGE) companies. Their wide product range had motor rollers based on several different power units. The tractor shown here is built around a Fordson Model N, a readily available unit during the war years.

▽ Allis-Chalmers WC-W Patrol

Date 1946 **Origin** US
Engine Allis-Chalmers 4-cylinder gasoline
Horsepower 29 hp
Transmission 4 forward, 1 reverse

Allis-Chalmers produced a wide range of road-building equipment, including the smallest motor grader in their range, the WC-W Patrol. Some 3,000 were built over the production run from 1940 to 1950.

TECHNOLOGY

Producer Gas as Fuel

Vehicles converted to run on "producer gas" were a common sight during World War II, when liquid fuel was often unobtainable. The gas was made by passing air through any heated carbonaceous material—usually wood or coke—via a "producer" unit on the tractor. The unit was unwieldy and cumbersome and the tractor engines could not generate their full horsepower using this fuel.

Gas-powered Fordson This model N, built in 1942, features a gas producer unit on the side of its Fordson four-cylinder 18 hp engine.

Tracks in the UK

The crawler market had until World War II been dominated by North American manufacturers. A few British makers, such as Clayton and Fowler, had had some success with their machines in earlier years. By the late 1940s, due to lack of funds, all nonessential imports to the UK were stopped, including medium-powered crawlers for agricultural use. As a result, numerous British makes and models appeared to satisfy customer demand. Some were unsatisfactory, but the majority were acceptable.

◁ Bristol 10

Date 1945 **Origin** UK
Engine Austin 4-cylinder gasoline/kerosene
Horsepower 18 hp
Transmission 3 forward, 1 reverse

This small tractor was intended for the horticultural and garden sectors of the market, a role that it filled admirably. The early models used Jowett flat four-cylinder engines, which were not ideal for the application. Later machines used Austin engines, which gave the tractor better balance and smoothness.

△ Fowler FD4

Date 1946 **Origin** UK
Engine Fowler 4B 4-cylinder diesel
Horsepower 45 hp
Transmission 6 forward, 2 reverse

The Fowler FD4 was the largest of the four sizes of FD tractors planned by Rotary Hoes Ltd. The FD1 never left the drawing board; the FD3 was the first to be placed in production, followed shortly thereafter by the FD2. Fowler was sold to T.W. Ward of Sheffield in 1947, just after production of the FD4 had begun. Only 14 FD4s were produced.

▷ David Brown Trackmaster

Date 1950 **Origin** UK
Engine David Brown 4-cylinder gasoline/kerosene
Horsepower 34 hp
Transmission 6 forward, 2 reverse

The Trackmaster and later crawler models offered by David Brown were well-engineered, neat, and reliable tractors. They had controlled elliptical differential steering, which made unnecessary the use of numerous potentially troublesome oil seals in the steering gear. These tractors achieved considerable success.

▽ Fowler VF

Date 1948 **Origin** UK
Engine Marshall single-cylinder 2-stroke diesel
Horsepower 40 hp
Transmission 6 forward, 2 reverse

After the FD range was dropped, the only immediate replacement was a combination of the Marshall (also owned by Ward) single-cylinder, two-stroke engine and Fowler's proven running gear. Although a crude and simple machine, it was a huge success—more than 4,000 were built.

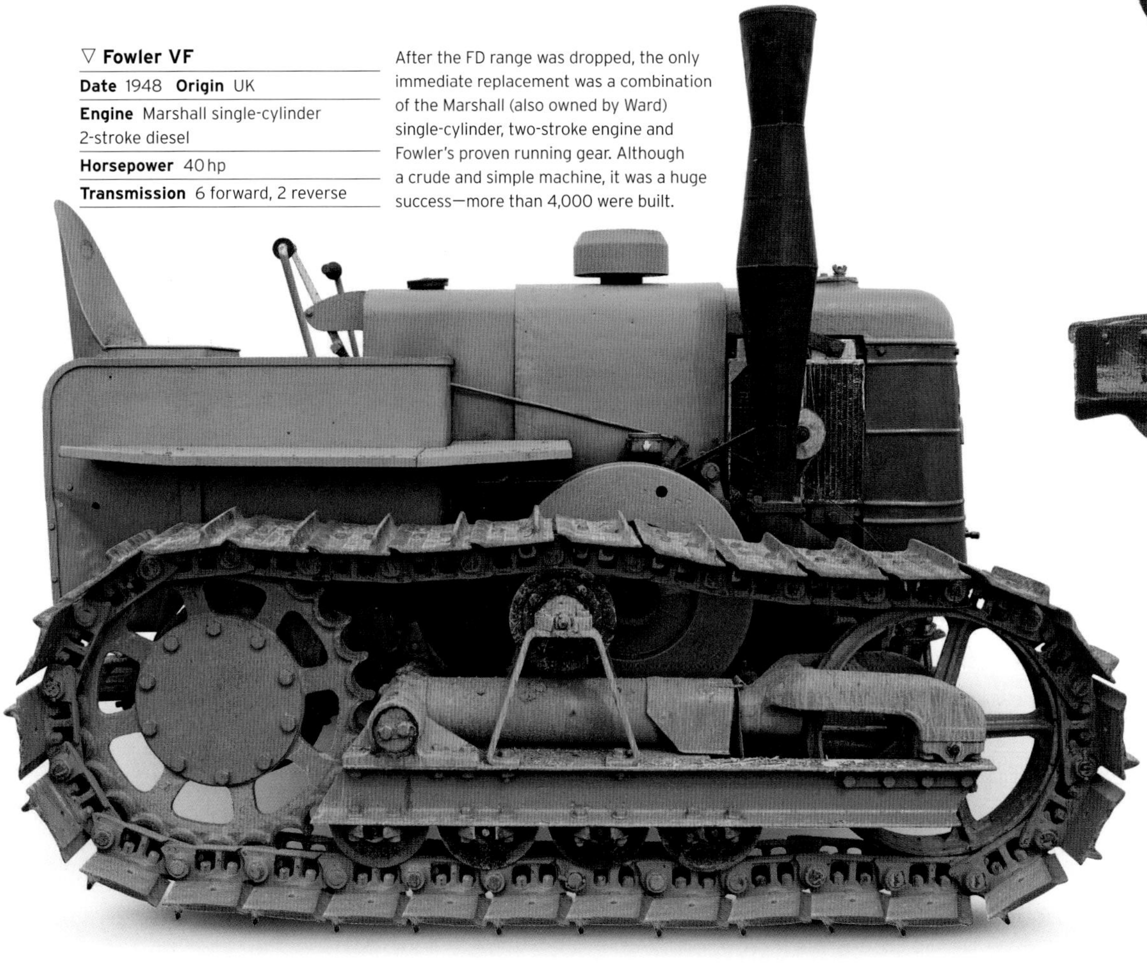

△ **Loyd Dragon**

Date 1951 **Origin** UK

Engine Dorman-Ricardo 4-cylinder diesel

Horsepower 36 hp

Transmission 4 forward, 1 reverse

After WWII, Vivian Loyd started to make tractors from Brengun carrier components. After a number of models, the Dragon was introduced in 1951, and could be supplied with either a Turner V4 or Dorman-Ricardo engine. It did not enjoy great success in the marketplace.

△ **County Full-Track**

Date 1950 **Origin** UK

Engine Perkins P6 6-cylinder diesel

Horsepower 45 hp

Transmission 3 forward, 1 reverse

The Full-Track was the first of a line of crawlers based on Fordson skid units offered by County. The use of large-diameter sprockets and front idlers allowed the skid units to be converted to a crawler with minimum modifications. It also made possible the use of 9-in- (23-cm-) pitch track links, reducing the number of track wearing parts, which helped bring down operating costs.

△ **Roadless Type E**

Date 1950 **Origin** UK

Engine Fordson 4-cylinder gasoline/kerosene

Horsepower 27 hp

Transmission 3 forward, 1 reverse (Red Spot)

The Type E was Roadless Traction Ltd's answer to the demand for medium-powered agricultural crawlers in the early 1950s. Equipped with rubber-jointed tracks, it was offered with either the Fordson gasoline/kerosene engine or the Perkins P6 diesel. Very few of these machines were built.

◁ **Fowler Challenger 3**

Date 1951 **Origin** UK

Engine Meadows 6 DC630 6-cylinder diesel

Horsepower 95 hp

Transmission 6 forward, 4 reverse

The Challenger 3 was one of Fowler's most successful models. It was originally intended to be equipped with the Marshall ED8 two-cylinder, two-stroke diesel engine and sold as the Fowler Challenger Mk2. The engine was a failure, and it was almost immediately replaced by the excellent Henry Meadows engine; the tractor then became the Challenger 3.

War Agricultural Executive Committees

At the outbreak of World War II, War Agricultural Executive Committees (WAEC) were created in most counties in the UK, with powers to supervise food production. The "War Ag," as it was usually known, requisitioned land, controlled cropping, coordinated the use of labor, including the Women's Land Army, and allocated tractors and machinery to farmers.

THE WORK OF THE SURREY "WAR AG"

The WAECs had their own machinery depots, each with its own fleet of tractors and equipment. Every depot had an area foreman, who was responsible for a team of drivers. The work they carried out included cultivating and harvesting requisitioned farms as well as land drainage, clearance, and reclamation operations on a county-wide basis. The Surrey "War Ag" had five machinery depots. The tractors they were allocated were either British Fordsons or US machines brought in under Lend-Lease. Heavy bulldozers were also leased from the military. Surrey WAEC's reclamation work included landscaping the village green in Abinger Hammer by filling in ponds formed by the Tillingbourne River; these so-called "hammer" ponds once powered the forges from which the village took its name.

The Surrey WAEC operating an International TD-18 bulldozer, the largest of IH's new range of diesel "Trac-TracTors," in Abinger Hammer.

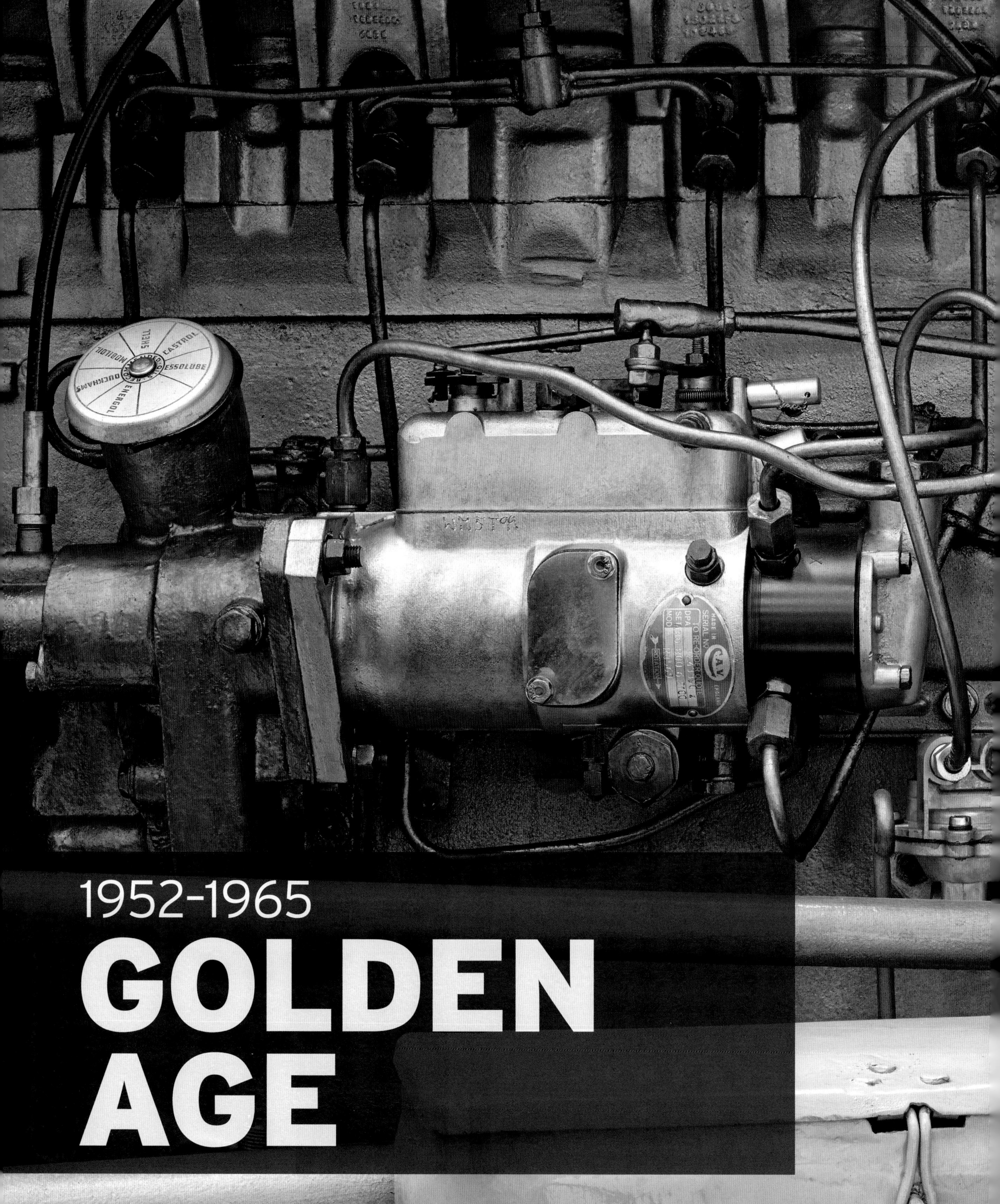

1952–1965

GOLDEN AGE

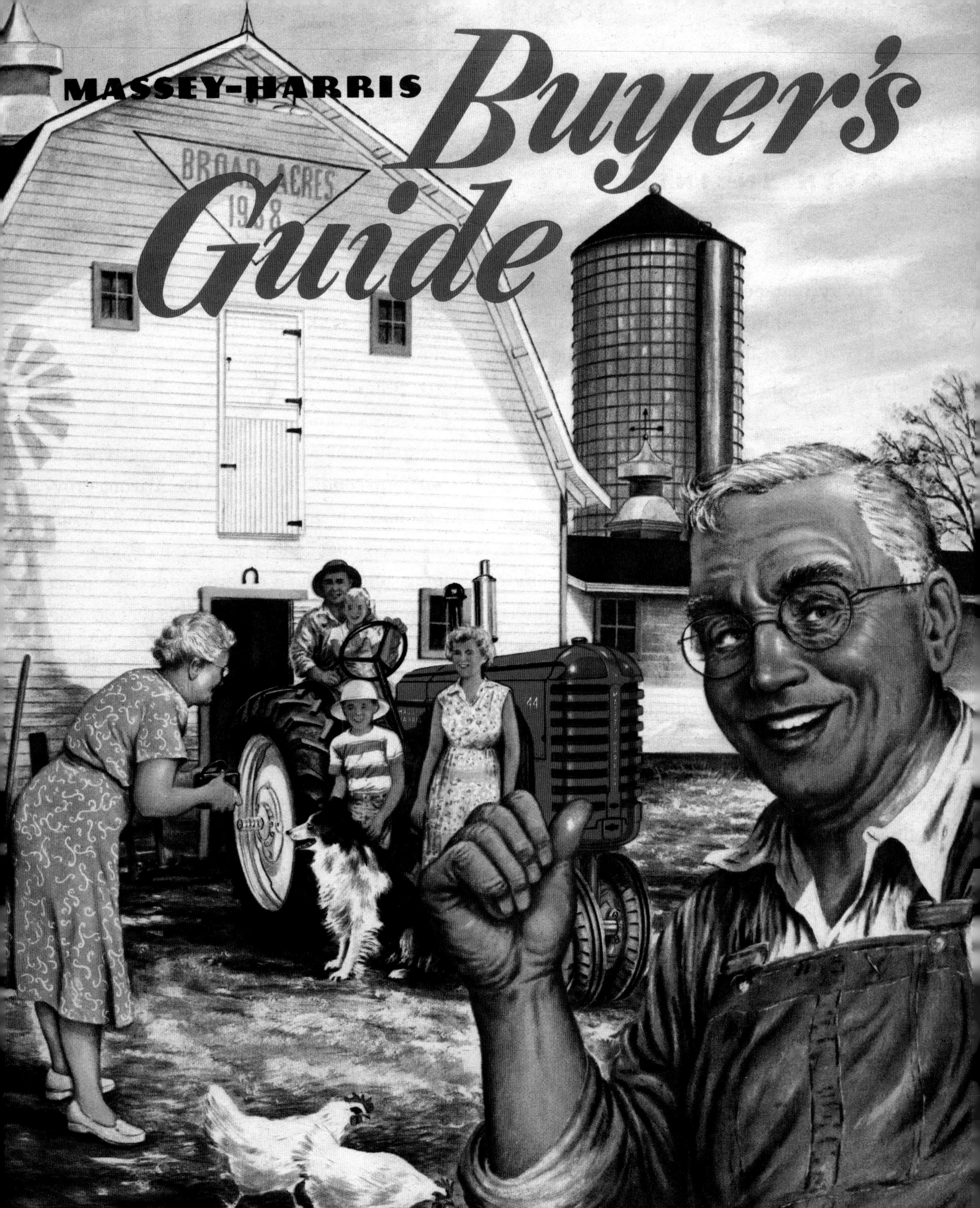
MASSEY-HARRIS
Buyer's
Guide
BROAD ACRES
1968
44

GOLDEN AGE

Once the postwar shortages that had curbed tractor development were resolved, the manufacturers began introducing new models with improved features such as a greater number of transmission speeds, differential lock, and "live" power takeoff.

A hydraulic lift was now standard on many tractors as the market moved from trailed to mounted implements. The lapsing of certain Ferguson patents allowed other manufacturers to offer a similar system of hydraulic draft sensing to control the depth of the implement in work. The high-speed diesel engine was almost universally adopted.

△ **Australian tractors**
The 60D, launched in 1953, was the first Chamberlain tractor to be equipped with a diesel engine.

The drive to mechanize farms and streamline agricultural production led to a golden age for the tractor industry. The world was hungry, and a plethora of new manufacturers and new models came to the market as the industry rose up to meet the challenges of a new era of intensive farming where the timeliness of operations was the key to success.

Farmers demanded more from their machines, and thoughts turned to increasing tractor performance with larger engines and the option of four-wheel drive. The race for power and increased productivity had begun. Customers also expected more from their dealerships, such as on-farm service and rent-to-own. Several manufacturers also offered a full line of matched equipment.

The postwar tractor boom peaked in some of the more industrialized nations during the 1960s as the number of tractors on farms (close to 5 million in the US) reached saturation point. Production in the US had already begun to decline as farmers began buying fewer but larger machines.

△ **Perkins engines**
Perkins's badge symbolized "a square deal all around" and its high-speed diesel engines were supplied to tractor manufacturers worldwide.

"Beauty in **engineering** is that which **is simple**, has **no superfluous** parts and **serves** exactly the **purpose**."

HARRY FERGUSON (1884-1960)

◁ **Massey-Harris buyer's guide** for 1953 offered the farmer a full line of tractors and equipment.

Key events

- ▷ **1953** Harry Ferguson sells his tractor companies to Massey-Harris.
- ▷ **1955** The 100,000th Fordson Diesel Major rolls out of the Dagenham plant.
- ▷ **1955** Fiat enters the four-wheel drive market with its 25R dt model.
- ▷ **1957** Douglas and Maurice Steiger build an articulated, four-wheel-drive tractor in Red Lake Falls, MN.
- ▷ **1958** International's 560 tractor is tainted by failures, allowing John Deere to take the top spot in the US.
- ▷ **1958** Sir Edmund Hillary arrives at the South Pole on Ferguson tractors.
- ▷ **1959** Eicher manufactures India's first indigenous tractor at its factory in Faridabad.
- ▷ **1960** John Deere ditches its two-cylinder line in favor of a new range of multi-cylinder tractors.
- ▷ **1960** Kubota develops the first Japanese tractor to go into full commercial production.
- ▷ **1961** International Harvester unveils its experimental HT-350 tractor, combining a gas-turbine engine with a hydrostatic transmission.
- ▷ **1962** John Deere, Minneapolis-Moline, and Massey Ferguson break the 100 hp barrier for two-wheel-drive tractors.

△ **Air-cooling**
In 1958, German manufacturer Deutz introduced its F4L 514 crawler, powered by an economical, air-cooled diesel engine developing 60 hp.

North American Developments

In the 1950s, the US and Canada saw diesel-engined tractors gain in popularity. The faster travel speeds made possible by the now widespread use of rubber tires meant transmissions with a bigger choice of gear ratios were a popular choice. Driver comfort was making slow headway up the priority list, although weather protection in the form of cabs was still regarded as an unnecessary extravagance on small- and medium-horsepower wheeled tractors. Industrial designers were enjoying greater influence on the appearance of tractors, which can be seen especially in changes in manufacturers' choice of paint colors.

△ International Farmall 100

Date 1954-1956 **Origin** US
Engine International 4-cylinder gasoline
Horsepower 18.3 hp
Transmission 4 forward, 1 reverse

When many farmers were still working with horses, a small tractor such as the Farmall 100 was the first step into power farming. The 100 could pull a single-furrow plow, and a generous underside clearance made it ideal for small-scale production of vegetables and other specialty crops.

◁ Minneapolis-Moline ZBU

Date 1953 **Origin** US
Engine Minneapolis-Moline 4-cylinder gasoline
Horsepower 34.8 hp
Transmission 5 forward, 1 reverse

The ZB was an improved version of the ZA row-crop series, available from 1953 with easier steering and a more comfortable seat. It was offered as the ZBN with a single front wheel. A wide front axle was available in the ZBE version, and a tricycle version with double front wheels was offered in the ZBU version.

▽ Sheppard SD3

Date 1956 **Origin** US
Engine Sheppard 3-cylinder diesel
Horsepower 32 hp
Transmission 4 forward, 1 reverse

The R.H. Sheppard company built diesel-powered tractors during the 1950s. There were three models, all powered by Sheppard engines and with standard, industrial, and orchard versions available. However, sales volumes remained small and production ended in about 1959.

▷ Cockshutt 50

Date 1956 **Origin** Canada
Engine Buda 6-cylinder gasoline
Horsepower 49 hp
Transmission 6 forward, 1 reverse

The Cockshutt 50 from Canada was also sold as the Co-op E5 model. It was available with gasoline or diesel engines. Both units were supplied by Buda and both had the same bore and stroke measurements. Nebraska tests showed almost 20 percent better fuel economy for the diesel version.

▷ International 650

Date 1957 **Origin** US
Engine International 4-cylinder LPG
Horsepower 61.3 hp
Transmission 4 forward, 1 reverse

Although diesel power was becoming popular, there was still a demand for spark-ignition engines. International offered the 650 model in gasoline and LPG (liquefied petroleum gas, a low-cost alternative to gasoline) versions.

△ **John Deere Model 80**

Date 1956 **Origin** US
Engine John Deere 2-cylinder horizontal diesel
Horsepower 65 hp
Transmission 6 forward, 2 reverse

The Model 80 was among the first batch of John Deere tractors to be identified by numbers instead of letters. Driver comforts featured in the new range included better seats, improved controls and, for the first time, power steering as an option.

▽ **John Deere 320**

Date 1957 **Origin** US
Engine John Deere 2-cylinder gasoline
Horsepower 29.2 hp
Transmission 4 forward, 1 reverse

John Deere's original all-green paint finish was replaced by a more eye-catching green and yellow when the new 20 series tractors arrived in 1956. These models offered major hydraulic system improvements, and the new seat suspension system could be adjusted to suit the driver's weight.

△ **J.I. Case 830**

Date 1964 **Origin** US
Engine Case 4-cylinder diesel
Horsepower 64 hp maximum
Transmission 8 forward, 2 reverse; optional torque converter version

The diesel-powered 830 was available in a standard version and with the new Case-O-Matic torque converter transmission, introduced as an option when the 800 series models were announced in 1958. The 830 also featured the recently introduced Desert Storm paint finish, which replaced the previous red color.

British Diesel Takes Supremacy

During the 1950s and early 1960s, diesel power gained increasing importance. Apart from a few tracklayers and the unconventional, single-cylinder Marshall, diesel-powered tractors had been virtually unknown, especially on British farms, before the late 1940s. Companies such as Ford, Perkins, and David Brown helped the UK become the world leader in developing easy-starting, smooth-running multi-cylinder diesels for tractors. Engine power was also on the increase, and the arrival of transmissions with more gears made it easier for operators to use the increased engine power efficiently.

△ **Turner Yeoman of England**

Date 1953 **Origin** UK
Engine Turner V4 diesel
Horsepower 40 hp
Transmission 4 forward, 1 reverse

The Turner V4 engine looked big and powerful, an impression supported by advertisements claiming unbelievable work rates and "astonishing lugging power." Unfortunately, the engine's reputation also included starting problems and poor reliability.

△ **Marshall MP6**

Date 1954 **Origin** UK
Engine Leyland 6-cylinder diesel
Horsepower 70 hp
Transmission 6 forward, 2 reverse

Belatedly, perhaps, dwindling sales of their aging single-cylinder models forced Marshall to follow the rest of the industry and use multi-cylinder diesel power. The result was the big, powerful, and expensive MP6, which sold in small numbers.

▷ **Field Marshall Series 3A**

Date 1955 **Origin** UK
Engine Marshall single-cylinder horizontal diesel
Horsepower 40 hp
Transmission 6 forward, 2 reverse

This was the final version of Marshall's single-cylinder diesel tractor. Unveiled in 1952, the Series 3A featured a number of small improvements to enhance engine performance. A pressurized cooling system was introduced to allow the tractor to warm up more quickly.

◁ **Ferguson TE-F20**

Date 1955 **Origin** UK
Engine Standard 4-cylinder diesel
Horsepower 26 hp
Transmission 4 forward, 1 reverse

Harry Ferguson took a keen interest in most technical developments affecting his tractors. However, he disliked diesel engines. He appears to have accepted the addition of a diesel-powered model to his TE tractor range as a necessary evil.

Narrow Tractors

Special situations sometimes need special tractors, and narrow tractors are slimmed-down standard models. Conversions are usually handled by specialty companies using skid units from a major manufacturer. The biggest demand for them is in vineyards, but they are also needed for hops production, some types of fruit growing, or simply for traditional farm buildings that have narrow passageways and doors.

◁ **Ferguson TE-L20**

Date 1954 **Origin** UK
Engine Standard 4-cylinder gasoline/kerosene
Horsepower 26 hp
Transmission 4 forward, 1 reverse

The system for labeling TE-20 tractors can be confusing. TE means built in the UK; TE-D20 is a kerosene version. One of the rarer models is the TE-L20 Vineyard, an extra-narrow, low-profile tractor for wine-growing and sugarcane production.

△ BMB President

Date 1954 **Origin** UK

Engine Morris 4-cylinder gasoline/kerosene

Horsepower 10 hp

Transmission 3 forward, 1 reverse

Smaller models arriving in the post-war tractor boom were often poorly designed and underpowered, but the President was better than some. It was equipped with wheel-track adjustment, and a hydraulic lift and belt pulley options were available.

△ Fordson Diesel Major

Date 1957 **Origin** UK

Engine Ford 4-cylinder diesel

Horsepower 52 hp

Transmission 6 forward, 2 reverse

Often called the New Major to distinguish it from the previous Fordson model, the E1A is one of the tractor industry's all-time classics. Production peaked at more than 350 tractors per day, helped by Ford's easy-starting, reliable diesel engine.

◁ International B275

Date 1967 **Origin** UK

Engine International 4-cylinder diesel

Horsepower 35 hp

Transmission 8 forward, 2 reverse

Announced in 1956, the International B250 established International's new range of small British tractors, which were built at a subsidiary plant near Bradford, Yorkshire. The improved B275 model arrived in 1958 and remained in production until 1968. This is a late model.

▷ Allis-Chalmers ED-40 Depthomatic

Date 1967 **Origin** UK

Engine Standard 4-cylinder diesel

Horsepower 41 hp

Transmission 8 forward, 1 reverse

The ED-40 was Allis-Chalmers' final new British model. The Depthomatic version, introduced in 1963, offered extra power, but the improved hydraulics turned out to be overcomplicated. Production ended in 1968.

△ Fordson Super Major KFD 68

Date 1964 **Origin** UK

Engine Ford 4-cylinder diesel

Horsepower 52 hp

Transmission 6 forward, 2 reverse

With parts available worldwide and an excellent reputation for reliability, Fordson's Super Major was popular with companies producing tractor conversions for special purposes. They included Kent Ford Dealers or KFD narrow tractors for vineyard, fruit, and industrial work.

◁ David Brown 850 Implematic

Date 1963 **Origin** UK

Engine David Brown 4-cylinder diesel

Horsepower 35 hp

Transmission 6 forward, 2 reverse

In 1961 David Brown introduced narrow versions of its 850 and 880 Implematic models to meet demand for tractors for hops, grapevines, and fruit cultivation. The overall width was reduced to 48 in (122 cm) on the narrowest wheel-track.

Renault N73 Junior

Model numbering was turned on its head for Renault's N-series tractors, which made their debut in 1960. Replacing the D series, they comprised the N70, N71, N72, and N73, but it was the last that had the smallest power output. The 20 hp tractor was the foundation of a line whose larger members—with respective horsepowers of 40 (N70), 35 (N71), and 25 (N72)—all had German MWM (Motoren-Werke Mannheim) diesel engines.

IN THE DECADE following World War II, Renault established itself as a strong presence in the tractor, car, and truck sectors in Europe. After the war the firm was nationalized by the French government, and by the time the N series was introduced, tractor production had been relocated to a new factory in Le Mans. With the N series came a new level of styling, with hood and radiator grille taking on a new shape for the new era. Typical of many small tractors at the time, it featured a twin-cylinder diesel engine and an underslung exhaust. The N-series tractors were popular for light field tasks, such as spraying and haymaking, but were also capable of taking on tillage tasks on smaller farms and handling equipment, such as single-furrow plows.

FRONT VIEW

REAR VIEW

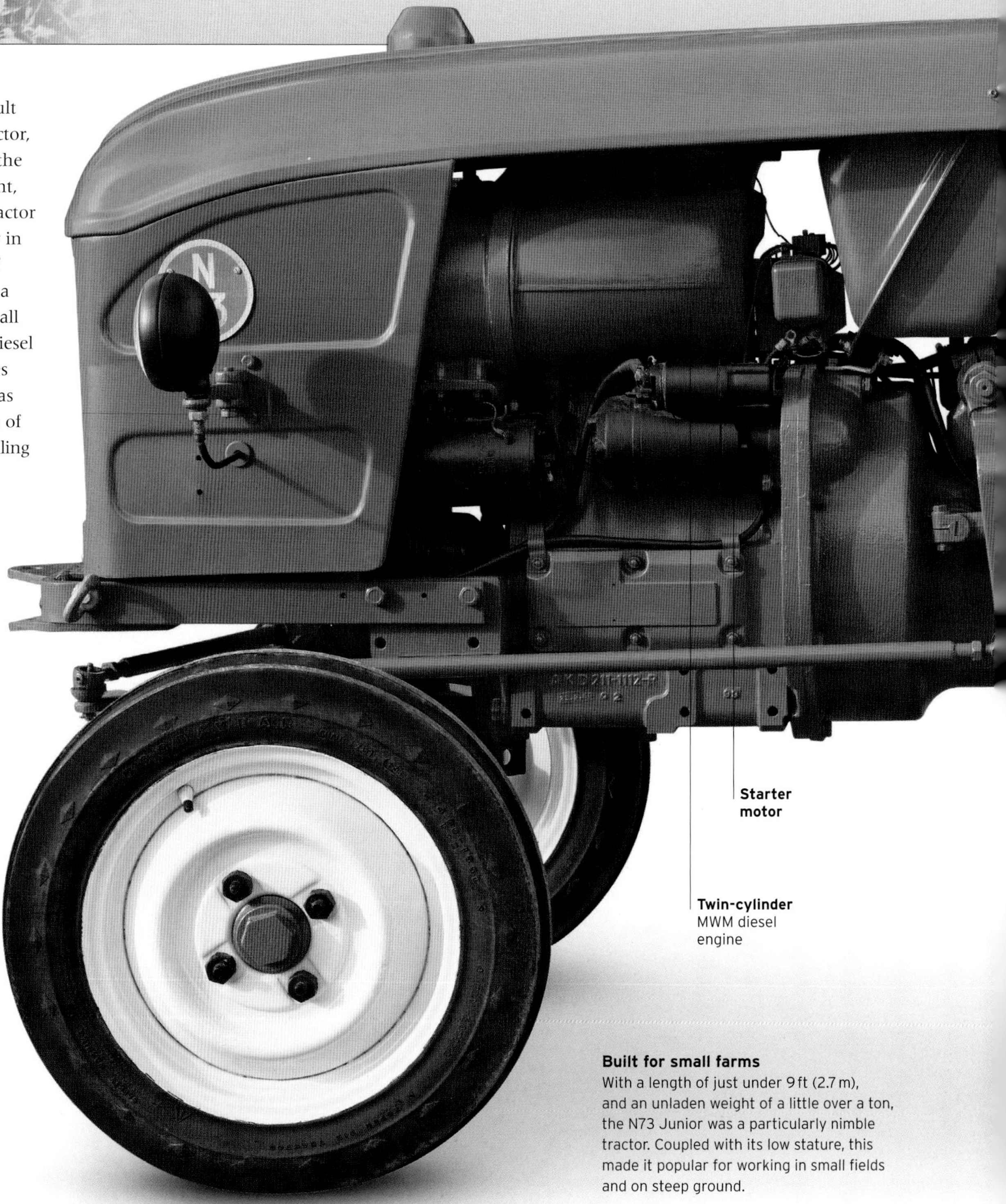

Starter motor

Twin-cylinder MWM diesel engine

Built for small farms
With a length of just under 9 ft (2.7 m), and an unladen weight of a little over a ton, the N73 Junior was a particularly nimble tractor. Coupled with its low stature, this made it popular for working in small fields and on steep ground.

SPECIFICATIONS			
Model	Renault N73 Junior	**Capacity**	85.4 cu in (1,400 cc)
Built	1961	**Transmission**	6 forward, 1 reverse
Origin	France	**Top speed**	Not known
Production	9,020	**Length**	8 ft 11 in (2.71 m)
Engine	20 hp MWM 2-cylinder diesel	**Weight**	1.1 tons (1 metric ton)

THE DETAILS

1. 1960s Renault badge on front of tractor **2.** Restyled headlights on the N series **3.** Driver's seat and the gear lever **4.** Underslung exhaust **5.** Rear wheel hub

Changing Markets

In the 1950s and 1960s, many independent tractor makers were producing machines of individual design, incorporating features relevant to their local markets. For instance, Hürlimann was one of the first to offer a four-wheel-drive tractor specially designed for the hilly terrain in Switzerland. Slowly, however, most of these companies disappeared. Some were taken over by larger firms; others did not survive because their products were inferior or too expensive.

▷ Massey-Harris Pony 820 V

Date 1952 **Origin** France
Engine Simca 4-cylinder gasoline
Horsepower 18 hp
Transmission 3 forward, 1 reverse

The Pony was Massey-Harris's answer to the demand for a small tractor, and competed with models from International and John Deere. As with the other manufacturers of this size of tractor, the Pony could be equipped with plows, cultivators, ridgers, and other specialty attachments.

◁ International Farmall Cub

Date 1956 **Origin** France
Engine International C-60 4-cylinder side-valve gasoline
Horsepower 9.25 hp
Transmission 3 forward, 1 reverse

This was the smallest tractor of the International Farmall range. The Cub was designed as a row-crop tractor and featured "culti-vision"—the engine and gearbox were offset from the centerline of the tractor to give the driver an almost uninterrupted front view. Proving very popular with customers, the Cub had a production run of more than 245,000 units in 1947-81.

▽ Hürlimann D200S

Date 1958 **Origin** Switzerland
Engine Hürlimann 4-cylinder diesel
Horsepower 65 hp
Transmission 10 forward, 2 reverse full sychromesh

Hans Hürlimann founded his tractor manufacturing company in 1929. The D200S came with a sychromesh gearbox, a very advanced feature for a farm tractor at the time. It was aimed at both agricultural and forestry markets. Hürlimann tractors are still being built under the parentage of SAME Deutz-Fahr (SDF).

▷ Vierzon 201

Date 1957 **Origin** France
Engine Vierzon single-cylinder 2-stroke hot-bulb
Horsepower 25 hp
Transmission 3 forward, 1 reverse

The Vierzon 201 was built by the French company Société Française Vierzon (SFV). As with the other tractors in the Vierzon range, the 201 was a licensed copy of the Lanz Bulldog. Its modern appearance hid the fact that under the sheet metal was a simple and reliable, if relatively old and outdated, design.

△ Kubota L13G

Date 1960 **Origin** Japan
Engine Kubota single-cylinder diesel
Horsepower 13 hp
Transmission 6 forward, 2 reverse

The L13G was one of Kubota's first tractors and was built to the high specifications and quality that are a feature of all their products. Although only a small machine, the L13G had features not available on similar-sized competitor machines. A range of matched equipment including front-end loaders became available.

△ Volvo-BM 470

Date 1961 **Origin** Sweden
Engine Bolinder 4-cylinder diesel
Horsepower 73 hp
Transmission 5 forward, 1 reverse

At 73 hp, the Volvo-BM 470 was considered a large tractor in its day—in 1961, farm tractors averaged 45 hp. A strong and reliable machine, the 470 was expensive and was not produced in large numbers as 75 hp tractors were yet to become popular. The 470 remained a rare machine outside its native market. Volvo-BM tractor production finally ceased in 1984.

▽ Steyr 185

Date 1962 **Origin** Austria
Engine Steyr WD 313 3-cylinder diesel
Horsepower 45 hp
Transmission 6 forward, 1 reverse

The Steyr 185 had a large following in Austria, but it was never exported in large numbers because competition from makers such as Ford and Massey Ferguson was too strong in most European countries. With its engine produced in house, the Steyr 185 was a high-quality machine. Steyr tractors are still produced as part of the Case-New-Holland (CNH) tractor range.

▷ Renault N73 Junior

Date 1961 **Origin** France
Engine Motor Werke Mannheim (MWM) 2-cylinder diesel
Horsepower 20 hp
Transmission 6 forward, 1 reverse

The N73 was a small, lightweight machine designed for the French market, which was mainly based on the small family farm. Renault bought most of its engines from outside suppliers and the N73 was no exception. Economical to operate and maintain, the N73 had power takeoff and hydraulics—features that had only recently become standard in this class of machine.

Germany Moves On

Production of agricultural machinery in Germany quickly resumed after 1945. Established companies such as Mercedes-Benz, Lanz, and Hanomag soon began to turn out products in quantity. There was a market for smaller tractors, with many small farms to be restored and returned to production of their traditional crops. The acquisition of Lanz by John Deere was particularly important—Deere had the resources to launch multi-cylinder models, and this gave the company a manufacturing base in Europe, which today is a vast and expanding concern.

△ Lanz Bulldog D2206

Date 1952 **Origin** Germany
Engine Lanz single-cylinder semi-diesel, gasoline-assisted starting
Horsepower 22 hp
Transmission 6 forward, 2 reverse

The 150,000th Bulldog tractor produced was a D2206 that left the factory in February 1953. These tractors finally did away with the need to heat the hot bulb with a blowtorch before the engine could be started. The cylinder had to be warmed up with gasoline ignition, and then the engine would run as a semi-diesel.

▽ Unimog 401

Date 1952 **Origin** Germany
Engine Daimler-Benz OM636 4-cylinder diesel
Horsepower 25 hp
Transmission 6 forward, 2 reverse; optional creeper box

Albert Friedrich designed the first Unimog soon after WWII as an agricultural multipurpose vehicle. Mercedes-Benz took over the manufacture of the Unimog in 1951 and introduced a new model, the Series 401. A long wheel-base model, the 402, was also made available, and a closed cab was offered from 1953.

▽ Lanz 6017

Date 1957 **Origin** Germany
Engine Lanz single-cylinder diesel
Horsepower 60 hp
Transmission 9 forward, 3 reverse

Lanz had a long tradition of offering a specially equipped tractor for general road use, and the largest in their range was no exception. The D6017 was offered with special gear ratios to give road speeds up to 19 mph (31 km/h). It was not John Deere's policy to implement abrupt changes in Mannheim, and these tractors continued to appear in Lanz colors for some time after the takeover.

▷ Normag Zorge C10

Date 1952 **Origin** Germany
Engine Farymann DL2 single-cylinder horizontal diesel
Horsepower 10 hp
Transmission 5 forward, 2 reverse

The C10 was the smallest tractor offered by Normag. The drive from the hopper-cooled, single-cylinder engine to the gearbox was transmitted by twin belts. The gear ratios were much higher than usual for an agricultural tractor. Although a small tractor, it was very well-equipped with steering brakes, lights, pulley, and differential lock.

◁ Porsche Coffee Train P312

Date 1954 **Origin** Germany
Engine Porsche 2-cylinder gasoline
Horsepower 24 hp
Transmission Direct drive

Porsche produced about 300 of these tractors. The streamlined bodywork was designed so as not to damage the bushes set out in rows in the coffee plantations of Brazil—the destination for the entire production run. Although all the other Porsche tractors had diesel engines, the P312 was built with a gasoline engine, since coffee growers did not want to subject their crops to diesel engine fumes.

△ **Fahr D130**

Date 1956 **Origin** Germany

Engine Guldner 2LD 2-cylinder diesel

Horsepower 17 hp

Transmission 5 forward, 1 reverse; optional low creeper gearbox

Johann Georg Fahr founded his company in 1870 to produce agricultural equipment. This model came in four versions: the regular D130, the D130H high-clearance model, the D130A with longer wheelbase, and the D130AH, which was a combination of A and H. Later, in a series of mergers, Fahr became part of SAME Deutz-Fahr (SDF).

△ **John Deere-Lanz 2416**

Date 1961 **Origin** Germany

Engine John Deere-Lanz single-cylinder full diesel

Horsepower 25 hp

Transmission 9 forward, 3 reverse

John Deere acquired Heinrich Lanz AG in 1956, at which time there were 19 Lanz models in production. John Deere continued with the single-cylinder tractors, but slowly introduced updates and modifications to the range. This included replacing the traditional Lanz colors with their own green and yellow livery.

▷ **Schlüter AS503**

Date 1961 **Origin** Germany

Engine Schlüter 3-cylinder diesel

Horsepower 50 hp

Transmission Spur gear

The company was founded in 1898 and was run by three generations of Schlüters until the 1980s. Tractor production started in 1937 but was disrupted by WWII. Known for its high-quality products, Schlüter produced the first 100 hp tractor built by a German manufacturer in 1964 and went on to build the first German 500 hp tractor in 1978.

△ **Hanomag R460 ATK**

Date 1962 **Origin** Germany

Engine Hanomag D57 4-cylinder diesel

Horsepower 60 hp

Transmission 5 forward, 1 reverse

The R460 ATK (A for pneumatic tires, and TK for the Voith fluid coupling between the engine and gearbox) was a robust, high-quality tractor that was specially adapted for road work. A wide range of attachments was available for this machine: the model shown here is equipped with a front pusher plate and a full driver's canopy.

Lanz Bulldog

In 1921, the simple, reliable Bulldog, the most famous of all the tractors to carry the Lanz name, began what would be a 35-year production run. The idea of developing the German firm's first true tractor was the brainchild of Dr. Karl Lanz, son of the company's founder, Heinrich Lanz, but it was his decision to employ Dr. Fritz Huber as his chief engineer that brought the concept to fruition.

BY THE EARLY 1950s Lanz was offering two distinct ranges of "semi-diesel" Bulldog tractors, based on the 06 machines launched just before World War II. The D2206 was the largest in the smaller line of models, all of which used the same engine with a 130 mm bore and 170 mm stroke. The first two numbers of the model designation signified maximum horsepower output, which could be varied via the use of different fixed engine speeds. The smaller range also consisted of 17 hp and 19 hp models.

The larger line's 28-, 32-, and 36 hp machines shared a bigger engine with a 150 mm bore and 210 mm stroke. Lanz kept pace with advancing technology, and later tractors were equipped with a transmission offering six forward and two reverse speeds and equipped with pneumatic tires.

Hydraulic lift lever for implements

FRONT VIEW

REAR VIEW

Rear tires larger than on previous models

SPECIFICATIONS	
Model	Lanz Bulldog D2206
Built	c. 1952–53
Origin	Germany
Production	6,298
Engine	22 hp Lanz horizontal single-cylinder semi-diesel, gasoline-assisted starting
Capacity	138 cu in (2,260 cc)
Transmission	6 forward, 2 reverse
Top speed	12.4 mph (20 km/h)
Length	9 ft 1 in (2.76 m)
Weight	1.5 tons (1.4 metric tons)

THE DETAILS

1. Bulldog badge with horsepower rating on the front of the tractor
2. Front-wheel hub 3. Fuel tap
4. Pressure gauge 5. Transmission levers 6. Implement lift lever

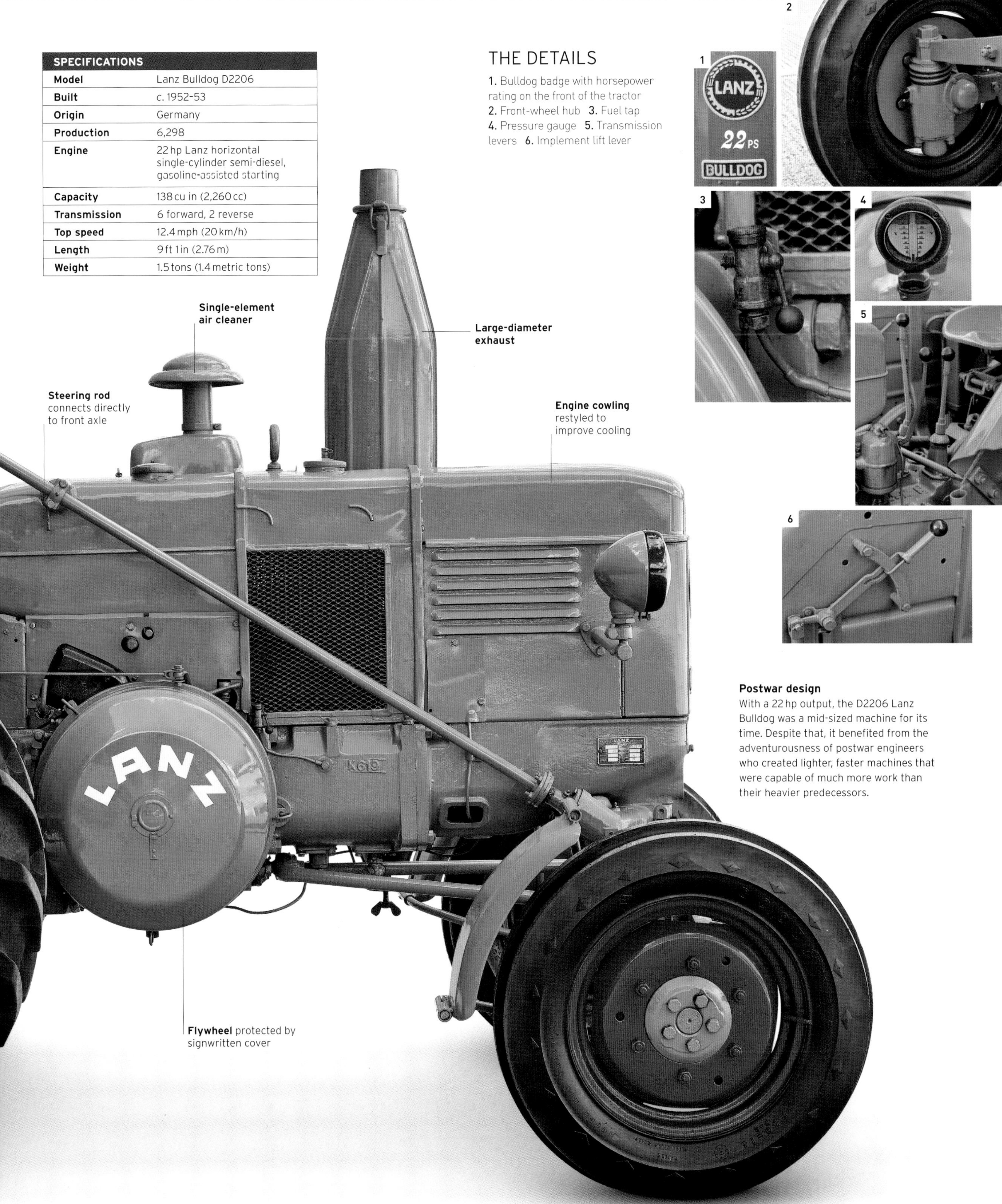

Single-element air cleaner

Large-diameter exhaust

Steering rod connects directly to front axle

Engine cowling restyled to improve cooling

Flywheel protected by signwritten cover

Postwar design
With a 22 hp output, the D2206 Lanz Bulldog was a mid-sized machine for its time. Despite that, it benefited from the adventurousness of postwar engineers who created lighter, faster machines that were capable of much more work than their heavier predecessors.

Tracks Across Antarctica

In 1957, New Zealand explorer Sir Edmund Hillary was tasked with establishing a supply chain for the Commonwealth Trans-Antarctic Expedition. Hillary's team would set out from Scott Base on the Ross Ice Shelf to support the main party, led by British scientist Dr. Vivian Fuchs, which was crossing in the opposite direction—from Shackleton Base on the Weddell Sea.

DASH TO THE SOUTH POLE

Hillary's team had three Ferguson TE-A20 tractors, which ran on gasoline and were modified for antarctic conditions with flexible tracks around the front and rear wheels. Fuchs's party was even better equipped, with the latest US Tucker Sno-Cats.

Leaving Scott Base on October 14, 1957, Hillary and his team made excellent progress despite having to cope with blizzards, ice crevasses, and temperatures down to –33°F (–36°C). Leaving the final drop on December 20, and hearing that Fuchs was delayed, Hillary decided to push his little tractors another 500 miles (800 km) to the South Pole. The Fergusons, pressed to the limit, swept into the US's South Pole Station just 17 days later, having crossed 1,250 miles (2,000 km) of some of the most inhospitable terrain on earth, beating Fuchs by 16 days.

Sir Edmund Hillary (left) and his team make their triumphant arrival at the South Pole at 12:30 pm on January 4, 1958.

Crawlers Around the World

During the 1950s, mid-horsepower crawler tractors were the main source of power on farms all over the world. Manufacturers from several countries competed to satisfy the demand. There were two basic versions: the multi-cylinder machine and the single-cylinder type. The US and UK manufacturers favored almost exclusively the multi-cylinder type, the exception being the single-cylinder British Fowler VF/VFA from the Marshall Organization. Builders in Continental Europe produced both types, but preferred the single-cylinder type.

◁ Ursus OMP 55C

Date 1952 **Origin** Poland

Engine Ursus single-cylinder lamp-start semi-diesel

Horsepower 50 hp

Transmission 4 forward, 1 reverse

The OMP 55 used a single-cylinder, lamp-start, semi-diesel engine. These engines were capable of producing their full rated horsepower on a variety of low-grade liquid fuels. The inconvenience of having to heat the hot bulb with a blowtorch was outweighed by the economy that was returned.

△ Fiat 55

Date 1951 **Origin** Italy

Engine Fiat 4-cylinder diesel

Horsepower 50 hp

Transmission 5 forward, 1 reverse

Fiat crawlers had a reputation for being among the best, and the 55 was no exception. The design was practically a copy of the Caterpillar D4, even down to the horizontal, two-cylinder gasoline engine for starting.

▽ Breda 50TCR

Date 1952 **Origin** Italy

Engine Breda single-cylinder hot bulb lamp start diesel

Horsepower 50 hp

Transmission 4 forward, 2 reverse

The 50TCR had a massive 14-liter, single-cylinder engine and was in production from 1947 to 1953. The Italian market for crawlers was the largest in the world outside the US. Breda's crawlers were popular in Italy and certain parts of Europe, but never made any inroads into the US market.

▽ Bubba Ariete

Date 1953 **Origin** Italy

Engine Bubba single-cylinder lamp start semi-diesel

Horsepower 40 hp

Transmission 6 forward, 1 reverse

The Bubba Ariete was in production from 1938 to 1954. The tractor followed the basic design used by Lanz of Germany, the popular features of the design being mechanical simplicity, reliability, and the ability to run on low-grade fuel.

▷ David Brown Trackmaster Diesel 50

Date 1953 **Origin** UK

Engine David Brown 6-cylinder diesel

Horsepower 50 hp

Transmission 6 forward, 2 reverse

David Brown made a serious attempt to enter the crawler market and was comparatively successful with agricultural sales. The Diesel 50, later called the 50TD, was DB's first six-cylinder model and it shared its engine with the 50D wheeled tractor. The crawlers were produced entirely in house.

△ Howard Platypus PD4 (R6)

Date 1955 **Origin** UK

Engine Perkins R6 6-cylinder diesel

Horsepower 70 hp

Transmission 6 forward, 2 reverse

The Platypus Tractor Co. was a subsidiary of Rotary Hoes Ltd., Horndon, Essex. The PD4 (R6) was their most powerful offering, but it had limited success. Only 15 were recorded as being built; a Leyland-engined version is also listed, but it did not sell.

△ Track-Marshall

Date 1958 **Origin** UK

Engine Perkins L4 4-cylinder diesel

Horsepower 48 hp

Transmission 6 forward, 2 reverse

The Track-Marshall was one of the most successful British-built crawlers. It satisfied the need for more power than was available from the earlier VF machine with its single-cylinder Marshall engine. It had a Perkins L4 in an upgraded VF chassis, and came with a four- or five-roller-track frame and a full range of attachments.

▷ Motomeccanica CP3C

Date 1960 **Origin** Italy

Engine Perkins P3 3-cylinder diesel

Horsepower 24 hp

Transmission 4 forward, 1 reverse

This small CP3C crawler was built for use on small hillside farms and vineyards. The company started business as Pavesi & Tolotti, but when Tolotti retired in 1919 it was renamed Motomeccanica. The first tractors were sold under the name Balilla.

△ Minneapolis-Moline Motrac

Date 1960 **Origin** US

Engine Minneapolis-Moline D206A-4 4-cylinder diesel

Horsepower 59 hp

Transmission Direct drive or torque converter with shuttle forward and reverse

Minneapolis-Moline (MM) was never a serious crawler manufacturer. The Motrac appeared in 1960 with 160 units being produced; one further unit was built in 1961. The diesel version used the MM D206A-4 engine, and the gasoline version used the MM 206M-4 engine. Only 39 of the gasoline tractors were built.

Straddling Crops

High-clearance and row-crop tractors have much in common. Both are specially adapted versions of standard models designed mainly for working between rows of crop plants. The special feature of high-clearance models is that they provide extra height for straddling taller plants, which complicates the final drive to the wheels. As sales volumes were small, the conversion work was often left to specialty companies that used skid units provided by the leading manufacturers. Some high-clearance models had special row-crop driving wheels and tires, resulting in a much slimmer alternative that reduced the risk of crop damage.

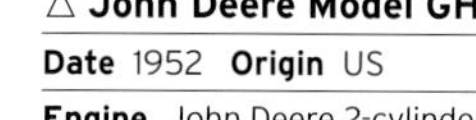

△ John Deere Model GH

Date 1952 **Origin** US
Engine John Deere 2-cylinder horizontal kerosene
Horsepower 36 hp
Transmission 6 forward, 2 reverse

John Deere engineers produced numerous conversions to meet special crop requirements, including the Hi-Crop, or GH, version of their Model G. The unstyled Model G arrived in 1938, followed in 1941 by a styled version with an uprated specification.

△ International Farmall MDV

Date 1954 **Origin** US
Engine International 4-cylinder diesel
Horsepower 38 hp
Transmission 5 forward, 1 reverse

The Farmall M was essentially the row-crop version of the McCormick-Deering W-6. Versions included the diesel-powered MD and the high-clearance MV—the V stands for "vegetable." The tractor here is a diesel-engined, high-clearance model.

◁ David Brown 850 "Super" High Clearance

Date 1962 **Origin** UK
Engine David Brown 4-cylinder diesel
Horsepower 35 hp
Transmission 6 forward, 2 reverse

David Brown's experimental department built this tractor for a black-currant grower. It was a special order and only one was built. The reduction units incorporated a vertical train of gears to raise the rear axle more than 3 ft (1 m) off the ground.

◁ Massey Ferguson 35 Hi-Clear

Date 1963 **Origin** UK
Engine Perkins A3-152 3-cylinder diesel
Horsepower 30 hp
Transmission 6 forward, 2 reverse

This high-clearance version of the MF35 was one of several adaptations made by Standen of St. Ives, Cambridgeshire, and Lenfield Engineering, Ashford, Kent. The conversion included a modified front axle and row crop rear wheels and tires. This tractor was used by vegetable and produce growers in Lincolnshire, UK.

△ **Ford Powermaster 951**

Date 1959 **Origin** US
Engine Ford 4-cylinder gasoline
Horsepower 44 hp
Transmission 5 forward, 1 reverse

The 900 series tractors were row-crop models in the Powermaster range. They were equipped with a 2.8-liter uprated version of the Ford Red Tiger engine, first developed in 1953. The 951 was offered in LPG (liquefied petroleum gas) and gasoline versions plus a new diesel model.

△ **County Hi-Drive**

Date 1962 or 1963 **Origin** UK
Engine Ford 4-cylinder diesel
Horsepower 52 hp
Transmission 6 forward, 2 reverse

Most County tractors were tracklaying or four-wheel-drive conversions, but the Hi-Drive was a two-wheel drive with its ground clearance increased to 30 in (76 cm). Built mainly for export, particularly for sugarcane work, it was based on the Fordson Super Major.

TALKING POINT

Extreme Hi-Clear

Nicknamed "Daddy Long-Legs," this special high-clearance tractor, based on an ex-military Ford WOT6 truck, was designed and manufactured by County for Pest Control of Cambridge in 1946. Used to spray cordon apple orchards in the UK and coffee plantations overseas, it had a ground clearance of 7 ft (2.1 m).

Tallest tractor Seen outside County's works at Fleet, Hampshire, the tractor's front forks were mounted on telescopic dampers.

△ **Oliver 1650**

Date 1964 **Origin** US
Engine Oliver 6-cylinder diesel
Horsepower 66 hp
Transmission 6 forward, 2 reverse

These distinctively styled 50 series tractors were available from the mid-1960s, with six models for sale. The Oliver 1650 was offered with either gasoline or diesel engines, and it was made with tricycle and wide front axle models as well as high clearance.

Smaller Tractors for Special Tasks

The rise in average horsepower had been a tractor marketing feature since the 1920s and continued as farmers demanded greater output and efficiency, but not everyone needed a big tractor. In some situations, compact size, reduced weight, and lower running costs can be more useful than massive pulling power—examples include market gardens, intensive livestock farms, specialty fruit growers, and vineyards. Small size need not mean a basic specification: features such as articulated steering, four-wheel drive, and versatility were available on some low-horsepower tractors in the 1950s and 1960s.

△ **International Super H**
Date 1958 **Origin** US
Engine International 4-cylinder gasoline
Horsepower 30 hp
Transmission 5 forward, 1 reverse

Good maneuverability and generous ground clearance made the Super H ideal for working in vegetables and other specialty crops. The compact size and light weight made it a popular choice for small farms and market gardens.

△ **Lanz Alldog A1806**
Date 1956 **Origin** Germany
Engine MWM 2-cylinder air-cooled diesel
Horsepower 18 hp
Transmission 6 forward, 1 reverse

Few small tractors match the Alldog's versatility. Removing the dumping container converts it into a tool carrier for mid-mounted implements—the list of options included haymaking equipment and a portable milking machine.

▷ **Kiva**
Date 1955 **Origin** France
Engine Bernard W112 single-cylinder gasoline
Horsepower 8 hp
Transmission Belt and pulley drive

The Kiva tractor was designed in the 1930s for mountainous areas. Most had a mowing attachment for making hay; the plow on this tractor is rare. Early versions used Chaise engines, later replaced by Bernard, and production ended with a VM diesel version.

△ **David Brown 2D**
Date 1957 **Origin** UK
Engine David Brown 2-cylinder air-cooled diesel
Horsepower 14 hp
Transmission 4 forward, 1 reverse

David Brown's tool carrier was designed with a rear engine, an open frame, and space for mid-mounted implements to give the driver a clear view for accurate row-crop steering. The implement lift mechanism used compressed air.

▷ **Holder A10**
Date 1957 **Origin** Germany
Engine Holder air-cooled diesel
Horsepower 10 hp
Transmission 3 forward, 1 reverse

Holder began building pedestrian-controlled, two-wheeled tractors in the 1930s. Its first four-wheel-drive tractor with articulated steering, the A10, was designed in 1954 for working in vineyards; it was also a popular market garden tractor.

△ **Oliver 660**

Date 1959 **Origin** US
Engine Oliver 6-cylinder diesel
Horsepower 30 hp
Transmission 6 forward, 2 reverse

The Oliver 60 model became the 66 in 1948. It was updated to become the Super 66 in 1954, and reemerged as the 660 in 1959. Row-crop and regular wheel layouts were available, plus diesel and gasoline engine options.

◁ **John Deere 720**

Date 1957 **Origin** US
Engine John Deere 2-cylinder LPG or gasoline/kerosene
Horsepower 55 hp
Transmission 6 forward, 2 reverse

The 20 series added yellow to John Deere's green color and introduced an improved hydraulic system. The cylindrical tank on this tractor holds liquefied petroleum gas (LPG) fuel, a low-cost option in some areas in the US.

▽ **SAME SameCar SA42T**

Date 1961 **Origin** Italy
Engine SAME 4-cylinder air-cooled diesel
Horsepower 42 hp
Transmission 6 forward, 1 reverse

Francesco Cassani, who owned the SAME company, sketched the initial SameCar design during a family vacation. His idea was a 25-mph (40-km/h) load-carrying tractor with a comfortable cab, but it was costly to produce and did not sell well.

△ **OTO Vigneron R3**

Date 1960 **Origin** Italy
Engine OTO single-cylinder diesel
Horsepower 17 hp
Transmission 6 forward, 2 reverse

The OTO Melara company started making military equipment in 1905 but turned briefly to building tractors after WWII. They started with the R3 vineyard tractor, powered by an air-cooled engine, and also available as a highly maneuverable three-wheeler and in a four-wheeled version.

David Brown at the wheel of a DB 25 tractor in 1953

Great Manufacturers
David Brown

David Brown exuded that great British tradition of putting engineering excellence before monetary considerations. It did not have the resources of the global manufacturers, but it took them on, and often beat them, on the world stage. The little company with big ideas was responsible for many industry firsts.

DAVID BROWN FOUNDED HIS business in the Yorkshire town of Huddersfield in the mid-19th century, and its roots were in gear manufacturing. The man with the vision to propel the organization into tractor building was the founder's grandson, also named David Brown, who entered into an agreement to manufacture Harry Ferguson's Type A model in 1936.

David Brown emblem showing the roses of Lancashire and Yorkshire.

The venture with Ferguson was short-lived and fractious; it ended in 1939 after Ferguson formed a new partnership with Ford in the US. However, young (later Sir) David Brown was eager to continue tractor production, forming David Brown Tractors Ltd. to build a new model in a complex of empty cotton mills in nearby Meltham.

Painted in David Brown's trademark livery of "Hunting Pink," the new David Brown tractor (later designated VAK1) was modern in both appearance and specification, but its production was limited by the outbreak of World War II. During the war, the Meltham Mills factory built transmissions for the British Royal Air Force's Spitfire aircraft's Merlin engines, which put the tractor division on a firm financial footing.

Postwar tractors included the famous Cropmaster, which was built from 1947 to 1953 and enjoyed a phenomenal production run of nearly 60,000 units sold worldwide. Crawler, narrow, and industrial tractors were soon added to the range as David Brown tried to live up to its slogan of "Mechanizing the World's Farms."

The early 1950s lineup consisted of the 25, 25D, 30C, 30D, and 50D wheeled tractors, the 30T, 30TD, and 50TD crawlers, and a whole host of industrial models. David Brown had built aircraft-towing tractors for the Royal Air Force during World War II, and this association continued into peacetime with a number of specialty models built for both military and civilian use.

David Brown's small but skilled engineering team was headed by technical director Herbert Ashfield and chief engineer Charles Hull. Both were truly inventive and made their budget go a long way as the company became one of the most innovative in the entire industry, eclipsing the developments of many larger manufacturers.

Many of the features taken for granted on tractors by the 1950s, such as the turnbuckle top-link and dished wheel centers for altering track width, were David Brown firsts. The company also pioneered high-speed, direct-injection diesel engines, six-speed gearboxes,

Six-cylinder model
The largest wheeled tractor in David Brown's new range of 1953 was the six-cylinder 50D model, which was a powerful contender, but suffered without a hydraulic linkage. Sales were disappointing and most were exported.

Record sales
The introduction of a new range of Implematic models in 1961 broke the record for the number of tractors dispatched from the Meltham factory, with 70 percent going for export.

> "Tractor production ... not only **survived**, it **flourished**."
>
> SIR DAVID BROWN

1860 David Brown, the founder of the organization, begins trading as a pattern maker in Huddersfield
1931 The grandson of the company's founder, also named David Brown, becomes managing director
1936 Agreement to build Harry Ferguson's Type A tractor at Park Works
1939 David Brown Tractor Ltd. is formed and a New David Brown tractor is exhibited at the Royal Show in Windsor Great Park

1940 Contract awarded to build towing tractors for the Air Ministry
1947 First Cropmaster tractor rolls off the assembly line
1949 David Brown builds its first diesel tractor
1953 New tractor range launched in Harrogate
1955 Unique 2D toolbar tractor released; David Brown is awarded the Royal Warrant for agricultural machinery; Harrison, McGregor & Guest acquired

1959 Implematic hydraulic system introduced
1960 Meltham plant supplies tractors to the Oliver Corporation in the US
1964 Selectamatic hydraulic system launched
1965 Livery of David Brown tractors changes from red to white
1971 Hydra-Shift semiautomatic transmission goes into production
1972 David Brown Tractors Ltd. acquired by Tenneco and merged with Case

1973 Unified color scheme adopted for David Brown and Case tractors
1977 Half-millionth tractor is auctioned for the Queen's Silver Jubilee Appeal
1979 New David Brown 90 series launched in Monaco
1983 David Brown name dropped from hood with launch of Case 94 series
1985 Tenneco merges International Harvester with Case to form Case IH
1988 Final David Brown tractor, a Case IH 1594 model, built in Meltham

and the two-speed power takeoff. The first David Brown diesel tractor, called the Cropmaster Diesel, was launched in 1949.

Perhaps the most exciting development of this period was the introduction of the innovative 2D row-crop tractor with its ground-breaking air-operated lift. The company also diversified into farm machinery after buying Harrison, McGregor & Guest Ltd. and its Albion range of implements in 1955.

David Brown now had the largest tractor line in the UK, but was also in danger of becoming overstretched. Some rationalization was needed, and the range was slimmed down during the late 1950s. The new 900 tractor, launched in 1956, suffered from reliability problems and failed to live up to expectations, but the Implematic models that followed heralded a new era of success for the organization.

By 1964, David Brown was exporting to 95 countries, and overseas sales accounted for four-fifths of production. The following year saw the launch of the new Selectamatic range of tractors with a more sophisticated hydraulic system and a change of livery from red to white.

With sales buoyant, David Brown invested heavily in new production facilities at Meltham, but their completion coincided with a downturn in world markets. At the time, the tractor business was also parent to David Brown's Aston Martin division, which was another drain on resources. It was rumored that the sports car company was draining the tractor company to the tune of £1 million per year. The banks stepped in and David Brown was forced to divest both tractor and car operations.

The David Brown tractor division was acquired by US industrial giant Tenneco, and merged with its J.I. Case organization in 1972. A unified, red and white color scheme was adopted for David Brown and Case tractors the following year. Under Case's ownership, it looked as if Meltham's future was assured, and a new range of DB 90 series tractors was launched in a fanfare of publicity at Monte Carlo in 1979, but a volatile world economy meant more changes were on the horizon. An improved range of Case 94 series tractors, but without the David Brown name on the hood, was introduced at Meltham in 1983. Two years later, Tenneco acquired International Harvester and merged it with Case to form Case IH. The inevitable rationalization of models and production facilities saw the former David Brown factory under threat when Doncaster was chosen as the main UK production center for the merged organization.

The announcement that Meltham was to stop tractor production was made in 1986. The last machine, a red Case IH 1594 model, rolled off the plant's assembly line at 10:43 am on March 11, 1988—and the David Brown era was over. The factory continued to operate as a much-reduced facility, supplying components to Case's worldwide operations. The ax finally fell in June 1993 when Meltham's closure was announced.

Queen's tractor
Included on David Brown's stand at the 1977 Smithfield Show was its 500,000th tractor, a special DB 1412 model, auctioned on behalf of the Queen's Silver Jubilee Appeal.

Greater Horsepower, More Cylinders

Choosing a bigger tractor with more engine power can be a cost-effective way to boost output and efficiency, and tractor horsepower in the 1950s and 1960s had already started on the upward trend that still continues some 50 years later. The engine details show diesel power was becoming a popular choice for US-built tractors in the medium- to high-horsepower sector, often in a six-cylinder format. Some of the manufacturers were also offering better transmissions with a more generous choice of gear ratios for converting the engine power into drawbar pull.

△ Chamberlain 60DA

Date 1953 **Origin** Australia
Engine General Motors 3-71 3-cylinder diesel
Horsepower 66 hp
Transmission 9 forward, 3 reverse

The Chamberlain family designed tractors for Australian farms, and the 60DA model was their first diesel machine. At 66 hp it was said to be the most powerful tractor available in Australia when production started in 1953.

◁ Oliver Super 99

Date 1954 **Origin** US
Engine Oliver-Waukesha 6-cylinder diesel
Horsepower 65 hp
Transmission 6 forward, 2 reverse

The Oliver Corp. was probably the first wheeled tractor manufacturer in the US to fully embrace diesel technology. The Super 99, the most powerful model in the Super range, introduced in 1954, was equipped with a six-cylinder diesel engine that was developed in collaboration with Waukesha and featured the Lanova combustion system.

▽ John Deere 830

Date 1960 **Origin** US
Engine John Deere 2-cylinder horizontal diesel
Horsepower 75.6 hp
Transmission 6 forward, 2 reverse

Mechanically, the 830 was similar to the previous 820 model, but there were new driver safety features. A significant change was adding electric starting to the options list, an indication that the big diesel engine was becoming easier to start.

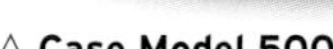

△ Case Model 500

Date 1956 **Origin** US
Engine Case 6-cylinder indirect injection diesel
Horsepower 64 hp
Transmission 4 forward, 1 reverse

Some features on the 500 were far from new, including the roller-chain final drive Case had featured in 1929, but this was its first diesel-powered tractor. The 500 was also the last new model launched before Case switched from its familiar red paint color.

△ **Massey Ferguson 98**

Date 1960 **Origin** US
Engine General Motors 3-71 3-cylinder diesel
Horsepower 79 hp
Transmission 6 forward, 2 reverse

The 90 series models brought in by Massey Ferguson to expand their big tractor range included the 98 model built by Oliver and supplied as an MF lookalike. Oliver supplied 500 of these tractors, which were powered by an unusual "soft-blown" engine.

△ **Massey Ferguson 97**

Date 1962 **Origin** US
Engine Minneapolis-Moline 6-cylinder diesel
Horsepower 102 hp
Transmission 5 forward, 1 reverse

The 97 was one of several models acquired from rival companies to fill a power gap at the top of the Massey Ferguson range. The MF97, a Minneapolis-Moline G705 tractor in disguise, made history as the first MF tractor to exceed 100 hp.

△ **International Farmall 560**

Date 1961 **Origin** US
Engine International 6-cylinder diesel
Horsepower 65 hp
Transmission 5 forward, 1 reverse; optional torque amplifier

The 560 was available with gasoline and LPG (liquefied petroleum gas) engines as well as diesel, and it was the diesel version that was gaining customers. Design problems led to a massive recall, allowing John Deere to steal market leadership from International.

◁ **John Deere 5010**

Date 1963 **Origin** US
Engine John Deere 6-cylinder diesel
Horsepower 121 hp
Transmission 8 forward, 2 reverse

The 5010 was part of the "New Generation of Power" range of John Deere tractors announced in the early 1960s to replace more than 40 years of two-cylinder tractor engines. The new models used four- and six-cylinder power with the emphasis on diesel.

Doe Triple-D

Originating from the US and Australia, the idea of linking two tractors in tandem to multiply power and traction was introduced to the British farming fraternity by the Essex agricultural engineering firm of Ernest Doe & Sons. Doe, a Ford dealership, adopted the idea of one of its customers, George Pryor, who began experimenting in 1957 with an articulated machine based on two Fordson tractors to plow his heavy clay soil.

PRYOR'S DESIGN used a turntable arrangement to join the two tractors together, with the vehicle steered by two pairs of two hydraulic rams. Doe refined the design and put it into production in 1958 as the Doe Dual Power, based on two Fordson Power Majors. After changes were made to satisfy the requirements of vehicle licensing authorities, the tractor reappeared the following year as the Doe Dual Drive or "Triple-D" for short.

The driver sat on the rear tractor and could select the gears on the front unit using a manual linkage that ran across the hood. Concerns over the safety of this arrangement saw it replaced in May 1960 with a system of hydraulic master cylinder and slave assemblies to operate the various controls remotely. Later machines were based on Super Major tractors, and more than 280 were built before the model was replaced by the Doe 130 in 1964.

New performance
The final version of the Triple-D, based on the "New Performance" Super Major, was introduced in 1963 with stronger bearings and bushes for the turntable. Doe offered an optional heavy-duty linkage assembly.

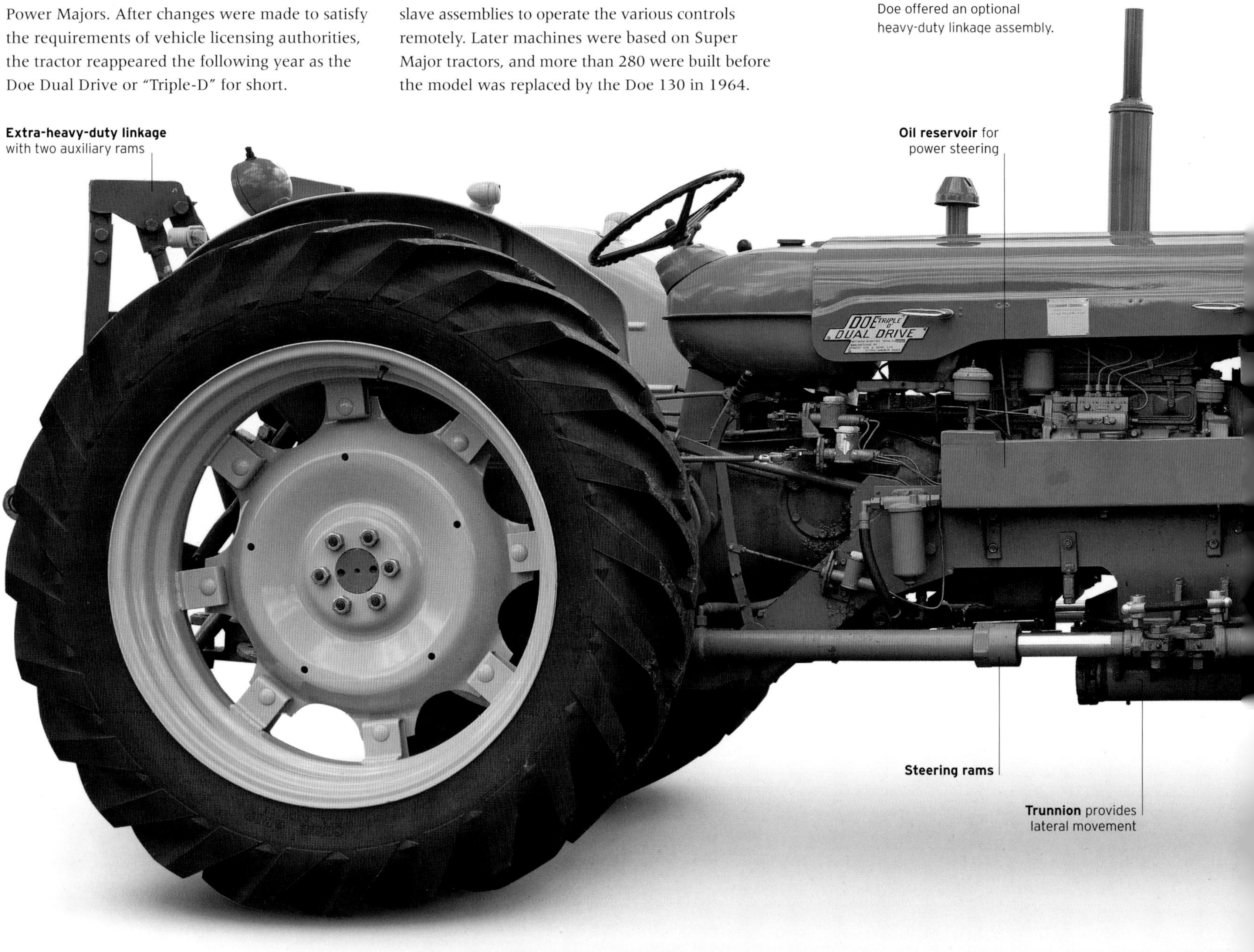

Extra-heavy-duty linkage with two auxiliary rams

Oil reservoir for power steering

Steering rams

Trunnion provides lateral movement

THE DETAILS

1. Serial number of each Triple-D stamped on manufacturer's plate affixed to the hood **2.** Control box on the front tractor contains the slave rams that operate gear selection **3.** Turntable links the two tractor units with two steering rams mounted on either side **4.** Remote master cylinder arrangement on the rear tractor controls the gear-change on the front unit **5.** Driver can control the functions of both tractor units from the rear seat

FRONT VIEW

REAR VIEW

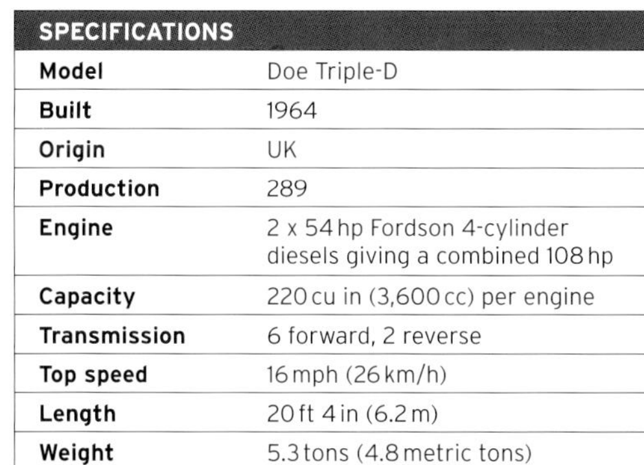

SPECIFICATIONS	
Model	Doe Triple-D
Built	1964
Origin	UK
Production	289
Engine	2 x 54 hp Fordson 4-cylinder diesels giving a combined 108 hp
Capacity	220 cu in (3,600 cc) per engine
Transmission	6 forward, 2 reverse
Top speed	16 mph (26 km/h)
Length	20 ft 4 in (6.2 m)
Weight	5.3 tons (4.8 metric tons)

Fuel tank holds 17.5 gallons (66 liters)

Toolbox

Four-cylinder Fordson diesel engine delivers 54 hp

Increasing Power

There were several reasons for the increase in four-wheel-drive (4WD) tractor sales during the 1950s and 1960s, especially in the UK and other European countries. One factor was evidence that 4WD can increase pulling power by 10 percent or more in difficult conditions, boosting work rates and making more efficient use of labor and fuel. There were also safety benefits resulting from increased stability, particularly when handling a heavy load in the mud or on a steep downhill slope.

◁ Steiger No.1

Date 1958 **Origin** US

Engine Detroit Diesel V-6 diesel

Horsepower 238 hp

Transmission N/A

When the Steiger brothers were unable to buy a big tractor for their 4,000-acre (1,600-hectare) farm in Minnesota, they built their own in 1957–58, and this was it. The result attracted requests for a similar tractor and was the start of the Steiger tractor business.

◁ Fiat 25R 4RM

Date c. 1960 **Origin** Italy

Engine Fiat 4-cylinder gasoline/kerosene

Horsepower 27 hp

Transmission 4 forward, 1 reverse or 10 forward, 2 reverse

Available from 1951, the 25 was the first tractor with Fiat's new orange paint finish. It was produced in various versions: with diesel and kerosene engines, two- and four-wheel drive, vineyard and forestry models, plus the 25C tracklayer model.

▽ County Super-4

Date 1961 **Origin** UK

Engine Ford 4-cylinder diesel

Horsepower 52 hp

Transmission 6 forward, 2 reverse

The Super-4 started County's four-wheel-drive success story. Previously, the company built tracklayers plus the skid-steer Four-Drive, but the Super-4 was based on a Fordson Super Major skid unit with two propeller shafts driving the front wheels.

△ Wagner TR-9

Date 1955 **Origin** US

Engine Cummins 4-cylinder diesel

Horsepower 120 hp

Transmission 10 forward, 2 reverse

Wagner four-wheel-drive tractors were sold under their own name through the 1950s, and they also made some for John Deere. Production began with the TR series models with outputs up to 165 hp, but they were followed by updated WA models.

◁ Doe Triple-D (Dual Drive)

Date 1964 **Origin** UK

Engine 2 x 54 hp Fordson 4-cylinder diesels

Horsepower 108 hp

Transmission 6 forward, 2 reverse

Four-wheel-drive traction and around 100 hp output made Doe tractors a popular choice, while the big companies were concentrating on smaller models and two-wheel drive. The downside was having two engines to service and two fuel tanks to fill.

△ **Matbro Mastiff**

Date 1962 **Origin** UK
Engine Ford 6-cylinder industrial diesel
Horsepower 100 hp
Transmission 6 forward, 2 reverse

The Mathews brothers, after whom the Matbro company was named, developed an improved pivot-steering system used on Matbro four-wheel-drive tractors and, with more success, on its agricultural and industrial loaders. Matbro later licensed its use by Caterpillar.

TALKING POINT

TOYS

Diecast toy tractors by makers such as Britains, Dinky Toys, and Siku were made to be played with, and used ones have often lost some paint and a tire or two. Good examples are highly collectible, and there is also a growing range of tractor and machinery models from Ertl and others to meet the increasing demand from collectors.

Boxed-up Early examples of diecast model tractors are popular with collectors, especially those with the original box, such as this Matchbox series Fordson.

△ **Roadless 6/4**

Date 1963 **Origin** UK
Engine Ford 6-cylinder commercial diesel
Horsepower 76 hp
Transmission 6 forward, 2 reverse

The first of Roadless Traction's Ploughmaster models was the 6/4, with the figures indicating a six-cylinder engine and four-wheel drive. Roadless conversions were based on various tractor makes and models, with Fordson and Ford at the top of the list.

▷ **Dutra UE-28**

Date 1965 **Origin** Hungary
Engine Csepel 2-cylinder diesel
Horsepower 28 hp
Transmission 6 forward, 2 reverse

Dutra is best-known for high-horsepower four-wheel drives, but the 28 series was sold mainly with two-wheel drive, and this UE four-wheel version is unusual. The large engine compartment is deceptive as it houses a small, two-cylinder diesel.

Amphibious Tractor

In 1963, British manufacturer County Commercial Cars Ltd. developed an amphibious tractor that could be used for offshore survey work. Based on County's four-wheel-drive Super-4 tractor, the machine was marketed as the Sea Horse. The tractor was equipped with flotation tanks front and rear, and had watertight compartments in each wheel for extra buoyancy. Special oversized tires were put on with the tread facing to the rear so that the deep lugs provided forward propulsion.

CROSSING THE CHANNEL

In an unrivaled publicity stunt to promote the Sea Horse, County's engineering director, David Tapp, drove the machine across the English Channel on July 30, 1963. While making its crossing from Cap Gris Nez on the French coast to Kingsdown near Dover, Kent, the County Sea Horse became the only tractor ever logged by the UK's South Goodwin lightship.

On its arrival in England, the Sea Horse was immediately put to work cultivating a field to show that it could perform equally well on land as on water. After a couple of minutes, it ran out of diesel, demonstrating that the sea crossing had taken it very close to the limit of its fuel capacity.

The crossing of 28 nautical miles (52 km) took 7 hours 50 minutes at an average speed of 3.5 knots; David Tapp's only problem was boredom.

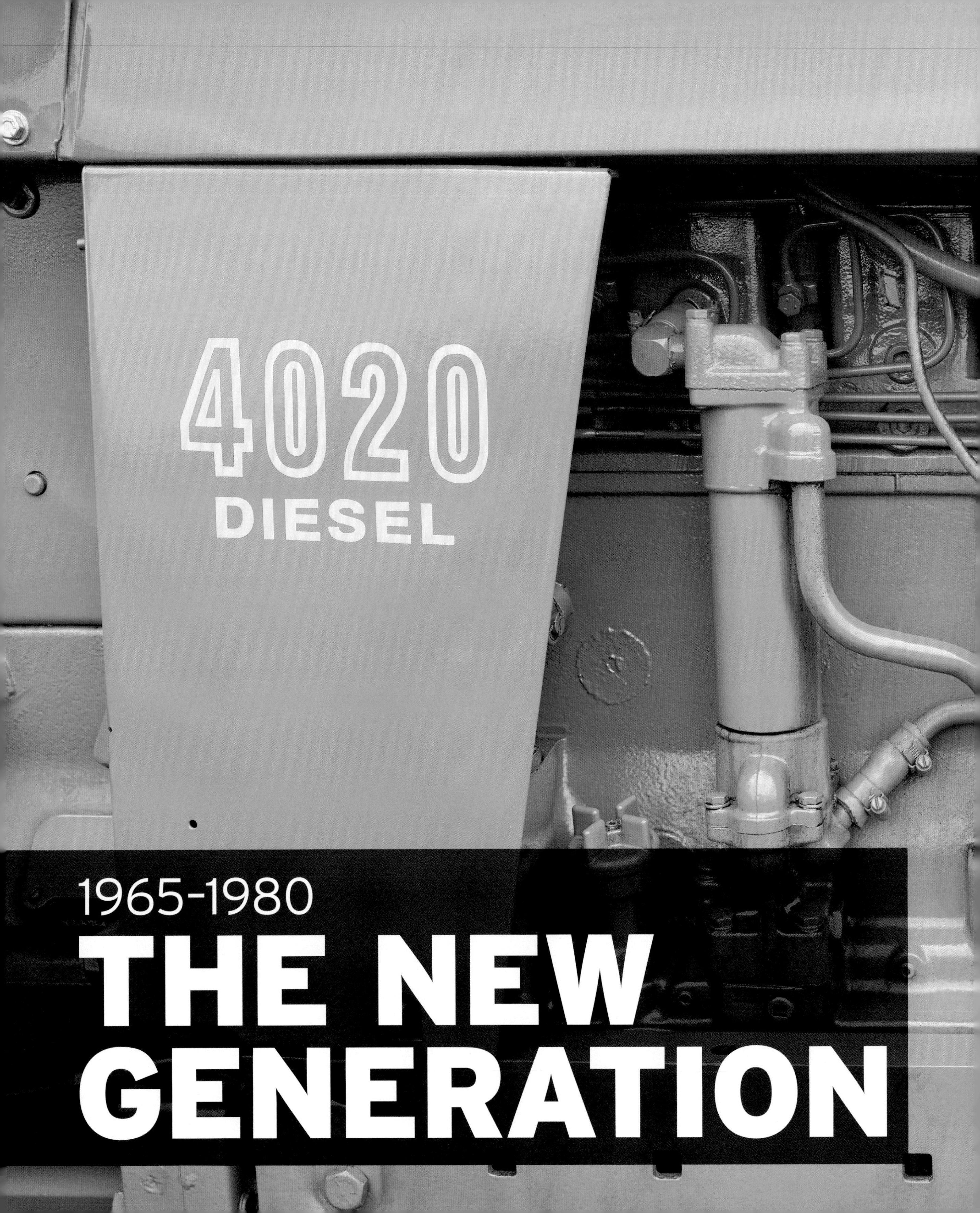

1965-1980

THE NEW GENERATION

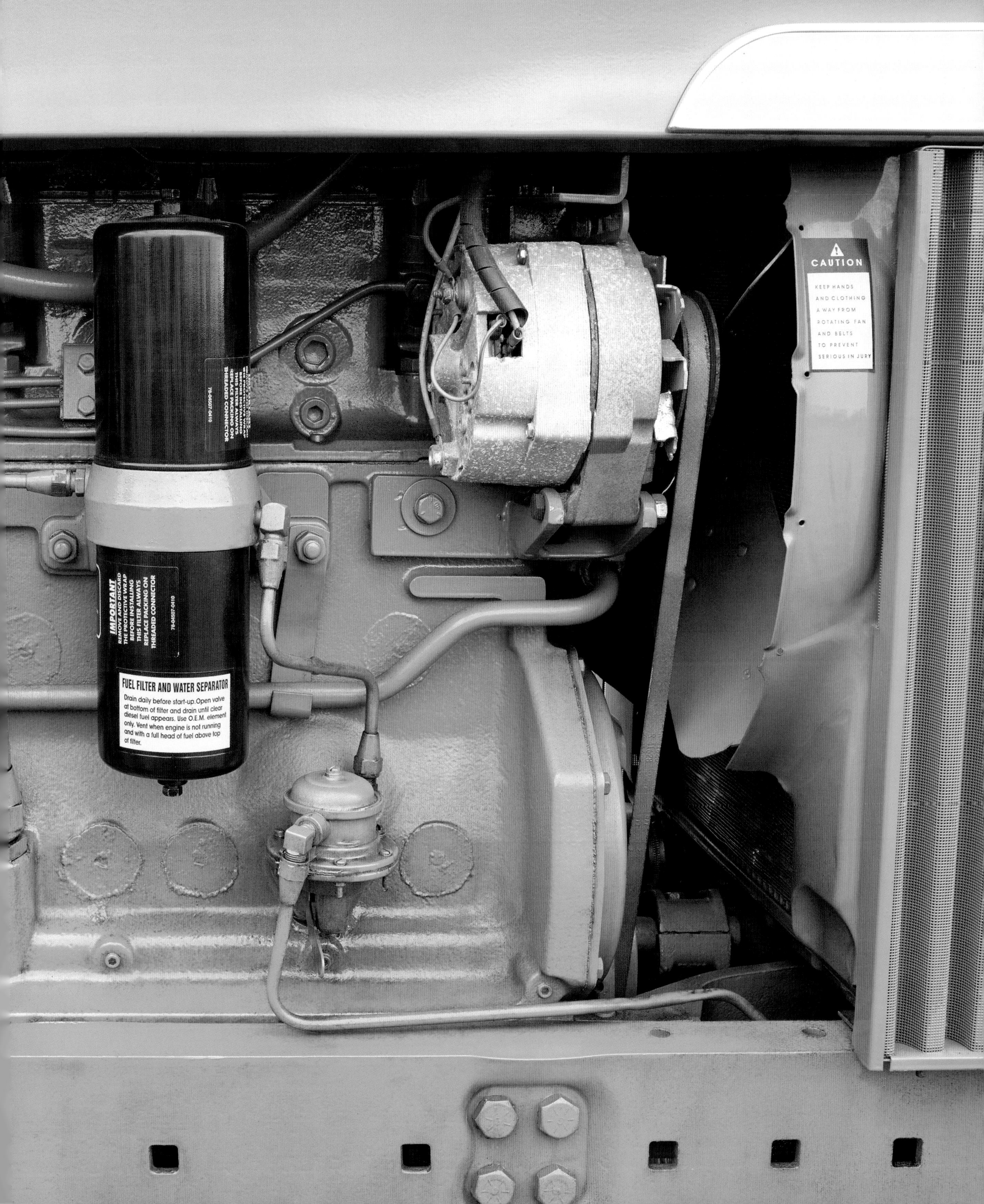

IMPORTANT
REMOVE AND DISCARD THE PROTECTIVE WRAP BEFORE INSTALLING THIS FILTER ALWAYS REPLACE PACKING ON THREADED CONNECTOR
78-04507-0410
FUEL FILTER AND WATER SEPARATOR
Drain daily before start-up.Open valve at bottom of filter and drain until clear diesel fuel appears. Use O.E.M. element only. Vent when engine is not running and with a full head of fuel above top of filter.
CAUTION
KEEP HANDS AND CLOTHING A WAY FROM ROTATING FAN AND BELTS TO PREVENT SERIOUS IN JURY

MINNEAPOLIS-MOLINE® MM

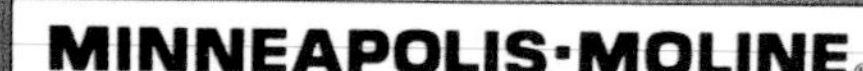

Built to make your farm future more productive

Models G850/G940

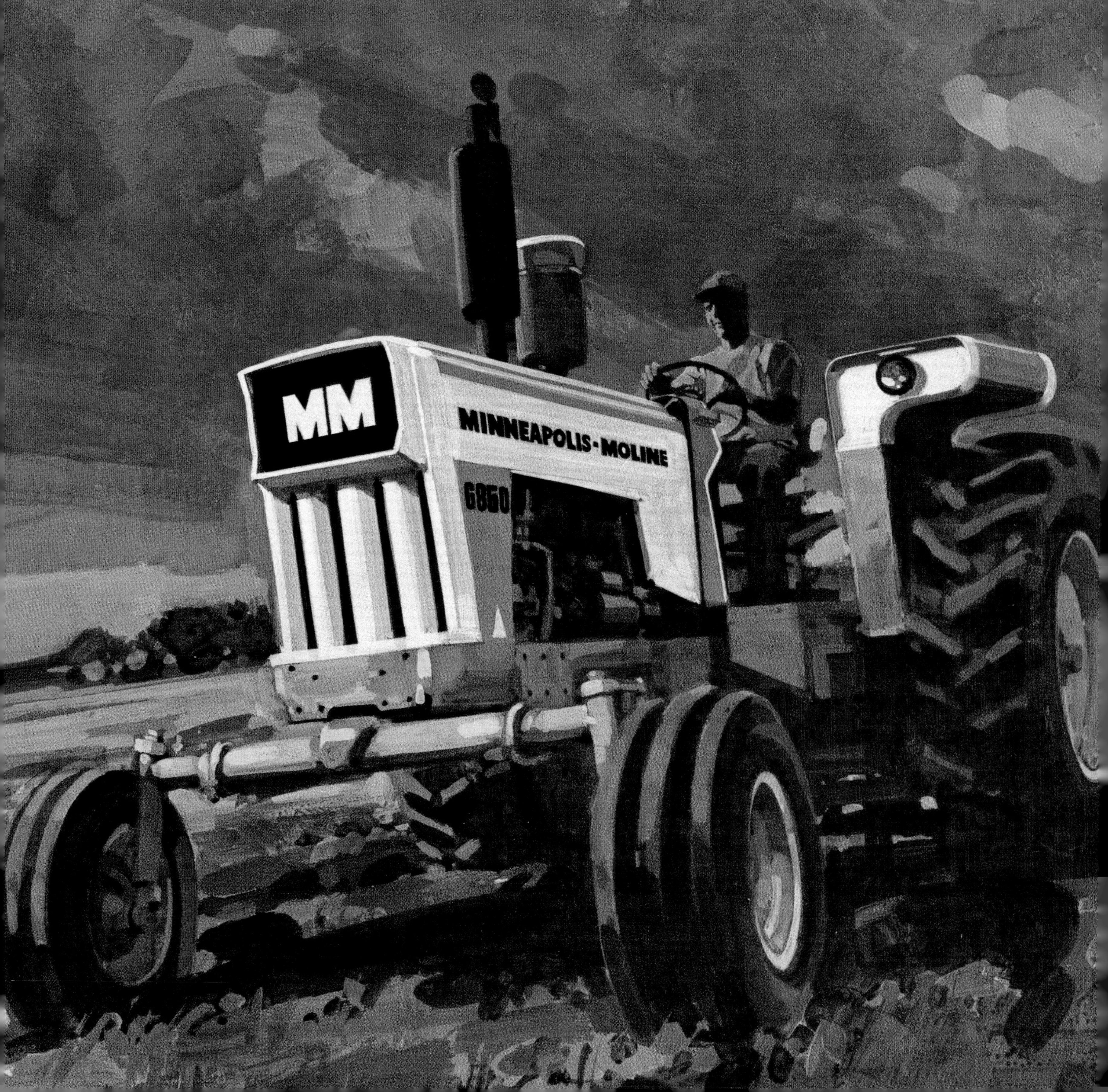

THE NEW GENERATION

The mid-1960s saw the introduction of what industry commentators referred to as the second generation of postwar tractors, with sleeker styling, ergonomic controls, multiple-speed transmissions, and more sophisticated hydraulic systems. Reliability was improved, along with better accessibility for routine servicing from dealers.

Typical of the new introductions were Massey Ferguson's "Red Giants" and Ford's 6X range, which were both launched at the end of 1964. Significantly, these new ranges were homogenous models aimed at global markets, and designed to go into simultaneous production in factories around the world. Sharing common components allowed for enormous savings in development and production costs. More importantly, it gave the tractor manufacturers a worldwide corporate identity as the industry embraced globalization.

△ **Quiet cabs**
Ford introduced sound-insulated cabs to meet UK noise level regulations for agricultural tractor cabs that came into force in 1976.

With drivers spending more hours in the seat, there were concerns about operator comfort and well-being, which lead to the introduction of rollover protection structures (ROPs), safety cabs, and quiet cabs, or "Q-cabs." Heaters, radios, air-conditioning, and power steering became the norm. The continuing race for power and productivity saw the rise of the articulated giants, particularly in North America and Australia, as the prairie tractors returned.

Overall tractor sales increased during the 1970s on the back of rising commodity prices, but by the end of the decade the outlook was gloomy. Mechanization had caused a seismic shift in agriculture and had depopulated the land. The improved performance of the latest machines meant greater acreages could be farmed with fewer tractors, and sales began to tumble in the face of growing competition from Eastern Europe, Japan, and India in many of the global markets.

"... the tractor did not **create more jobs** than it **abolished.**"

ROBERT C. WILLIAMS, US HISTORIAN (1931-)

◁ **Futuristic artwork** on this 1971 Minneapolis-Moline brochure emphasizes the arrival of a new generation of tractors.

Key events

- ▷ **1964** Sweden makes rollover protection mandatory on all tractors driven by employed labor. Safety cabs become compulsory in the UK in 1970.
- ▷ **1965** China introduces its first mass-produced tractor—the Dongfanghong 28, built by Changchun Tractor Group.
- ▷ **1966** David Brown is granted a Queen's Award to Industry for export achievement. It also received the same award in 1968, 1971, and 1978.
- ▷ **1970** Lely of the Netherlands launches its 87 hp Hydro 90 tractor with a hydrostatic transmission.
- ▷ **1971** The US Department of Transportation prepares a report on Federal Safety Standards for Tractors, but concludes that the industry should set its own voluntary code.
- ▷ **1972** John Deere unveils Generation II tractors with Sound-Gard cabs.
- ▷ **1976** Versatile of Canada introduces "Big Roy"—a 600 hp articulated tractor with four axles and eight-wheel drive.
- ▷ **1976** UK noise level regulations for agricultural tractor cabs implemented.
- ▷ **1977** Ford celebrates 60 years of tractor production.
- ▷ **1977** Big Bud of Montana builds the world's largest agricultural tractor—the 900 hp 16V-747.
- ▷ **1978** Fiat becomes the world's fifth-largest tractor manufacturer.

△ **Four-wheel drive**
Rugged four-wheel-drive system and luxury Spacecab were the selling points on Muir-Hill's 121 series III tractor in 1978.

Progress in the US

Major changes in US midrange tractors during the late 1960s and early 1970s included a rapid increase in the number of diesel models, encouraged partly by the success of the diesel tractors arriving from the UK. For driver comfort and convenience, features such as power steering, improved seat designs, and occasionally even canopies and cabs were added. Transmission improvements included a greater choice of gears plus easier shifting, although four-wheel-drive progress remained slow.

▷ **Case 1200 Traction King**

Date 1964 **Origin** US

Engine Case 6-cylinder turbocharged diesel

Horsepower 120 hp

Transmission 6 forward, 6 reverse

Case entered the four-wheel drive arena in 1964 with its 1200 Traction King, a rigid-frame machine with crab steering and a heavy-duty industrial transmission. It was hurried onto the market using a turbocharged version of an existing Case engine.

△ **Case 1030 Comfort King**

Date 1967 **Origin** US

Engine Case 6-cylinder diesel

Horsepower 102 hp

Transmission 8 forward, 2 reverse

The 1030 diesel arrived in 1966 to take Case into what was then the exclusive 100 hp plus sector of the market. It was available in two versions: the general-purpose 1031 seen here, plus a Western Special model known as the 1032.

▽ **John Deere 4020**

Date 1963 **Origin** US

Engine John Deere 6-cylinder diesel

Horsepower 91 hp

Transmission 8 forward, 2 reverse powershift

Developments featured on the 20 series included a powershift transmission enabling the driver to shift on the go. The improved diesel engines had become an increasingly popular option, and a four-wheel drive 4020 was introduced in 1966.

△ **John Deere 6030**

Date 1972 **Origin** US

Engine John Deere 6-cylinder diesel with turbocharger and intercooling

Horsepower 176 hp

Transmission 8 forward, 2 reverse

The 6030 was the most powerful two-wheel-drive tractor on the market at the time, with the option of either a 141 hp naturally-aspirated or a 176 hp turbocharged engine. This tractor was available with a canopy, but there was no full cab option.

◁ **Minneapolis-Moline G1350**

Date 1969 **Origin** US
Engine Minneapolis-Moline 6-cylinder diesel
Horsepower 141 hp
Transmission 10 forward, 2 reverse

In 1963 Minneapolis-Moline was bought by the White Motor Co., which also owned the Oliver brand. The Minneapolis-Moline G1350, which was also sold in Oliver colors as the 2155, was built for only two years before being replaced by the G1355.

▷ **International "Gold Demonstrator" 1456**

Date 1970 **Origin** US
Engine International 6-cylinder turbocharged diesel
Horsepower 131 hp
Transmission 8 forward, 4 reverse

This was the row-crop 1456 carrying the Farmall name, which was absent from the general-purpose model. International painted its demonstrators gold for the 1970 "Gold Demo" program. The 1456's options included a full cab plus, significantly, four-wheel drive.

◁ **International Hydro 100 Diesel**

Date 1974 **Origin** US
Engine International 6-cylinder diesel
Horsepower 104 hp
Transmission 2-range hydrostatic

International made a big effort to promote hydrostatic drive systems using an oil flow instead of gears to transfer power to the driving wheels. The Hydro 100 was one of these special machines, but the idea was not widely adopted for tractors due to increased power losses.

▽ **Ford Commander 6000**

Date 1967 **Origin** US
Engine Ford 6-cylinder diesel
Horsepower 66 hp
Transmission 10 forward, 2 reverse

Introduced in 1961 with gasoline, diesel, and LPG (liquefied petroleum gas) engine options, the Ford 6000 was plagued by reliability issues. The problems were addressed in the improved Commander 6000, launched in 1964 as part of Ford's new Worldwide tractor program.

△ **Oliver 2255 Diesel**

Date 1974 **Origin** US
Engine Caterpillar V8 diesel
Horsepower 147 hp
Transmission 18 forward, 2 reverse

The 2255 was the Oliver entry in the tractor power contest that was developing in the US. It was one of the most powerful two-wheel-drive models available, and was unusual in having a V8 engine. Dual driving wheels were standard equipment.

Demonstration of a Case 20-40 tractor at a country fair

Great Manufacturers
Case

Case IH is a global leader in agricultural equipment, with over 4,900 dealers operating in more than 160 countries. Today, it is part of the CNH organization, which incorporates the New Holland brand. It has a complicated legacy that can be traced back more than 380 years.

THE FOUNDER OF THE MARQUE, Jerome Increase Case, was descended from English settler John Case, who left Kent, England, in 1633 to seek a new life in the Massachusetts Bay Colony. Jerome began experimenting with threshing machines in Rochester, WI, in 1842. He eventually had holdings in two separate concerns: the J.I. Case Threshing Machine Company and the J.I. Case Plow Works; both were located in Racine, WI.

In 1869 the J.I. Case Threshing Machine Company built its first steam engine. This portable engine, which had to be moved by horses, had a locomotive-style boiler. Case's first self-moving engine appeared in 1877.

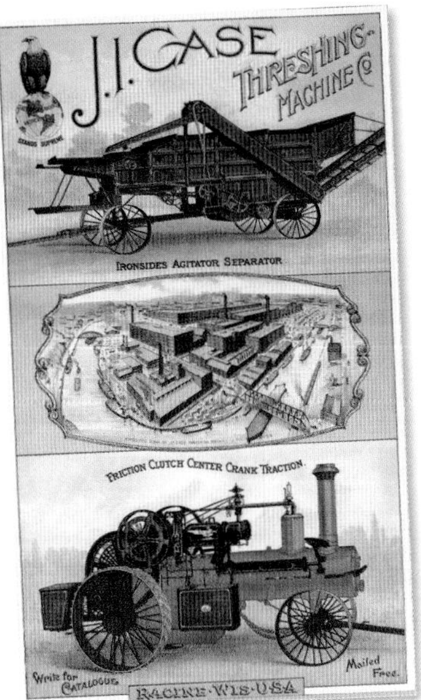

"Old Abe" is the famous Case trademark.

Following Jerome's death in 1891, control of the J.I. Case Threshing Machine Company passed to his brother-in-law, Stephen Bull.

The Bull family were astute businessmen and the company continued to prosper despite tough competition. The same could not be said for the J.I. Case Plow Works, which floundered financially.

Jerome's son, Jackson Case, had desperately thrown in his hat with the latter business, using both his and his sisters' stock in the other company as collateral for loans from the family trust to try to keep the J.I. Case Plow Works afloat.

By doing so, they lost all of the Case family's interests in the Threshing Machine Company.

Henry M. Wallis, Jackson Case's brother-in-law, eventually gained control of the Plow Works, leading to a period of intense rivalry between the two Case factions. The J.I. Case Threshing Machine Company had experimented with tractors as early as 1892, but was a latecomer to the market, and it was 1912 before it began full-scale production with the 20-40 and 30-60 models. Not to be outdone, Henry M. Wallis established the Wallis Tractor Company the same year.

The acrimony continued until 1928 when Massey-Harris purchased both the Plow Works and the rights to Wallis tractors. The J.I. Case Threshing Machine Company was then renamed the J.I. Case Company to reflect its emergence as a manufacturer with a full product line.

Case's tractor line in the 1920s consisted of a range of rugged machines with transverse engines. These "cross-motor" models had an enviable reputation for quality engineering, but were too heavy and too expensive to compete with the new generation of tractors coming on the market.

The Case Model C and L tractors that appeared in 1929 had their engines mounted longitudinally in the conventional fashion, but the high standard of engineering remained the same. These tractors were lighter and more powerful than their predecessors, setting the design for Case tractors for years to come. In

Early Case products
In 1887 the J.I. Case Threshing Machine Co. added a center-crank steam traction engine with a friction clutch to the line of products built in Racine.

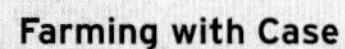

Farming with Case
The Case tractor range for 1965 included the 1200 Traction King with four-wheel steering, the 30 Series models in six different power sizes, and the Colt compact tractor.

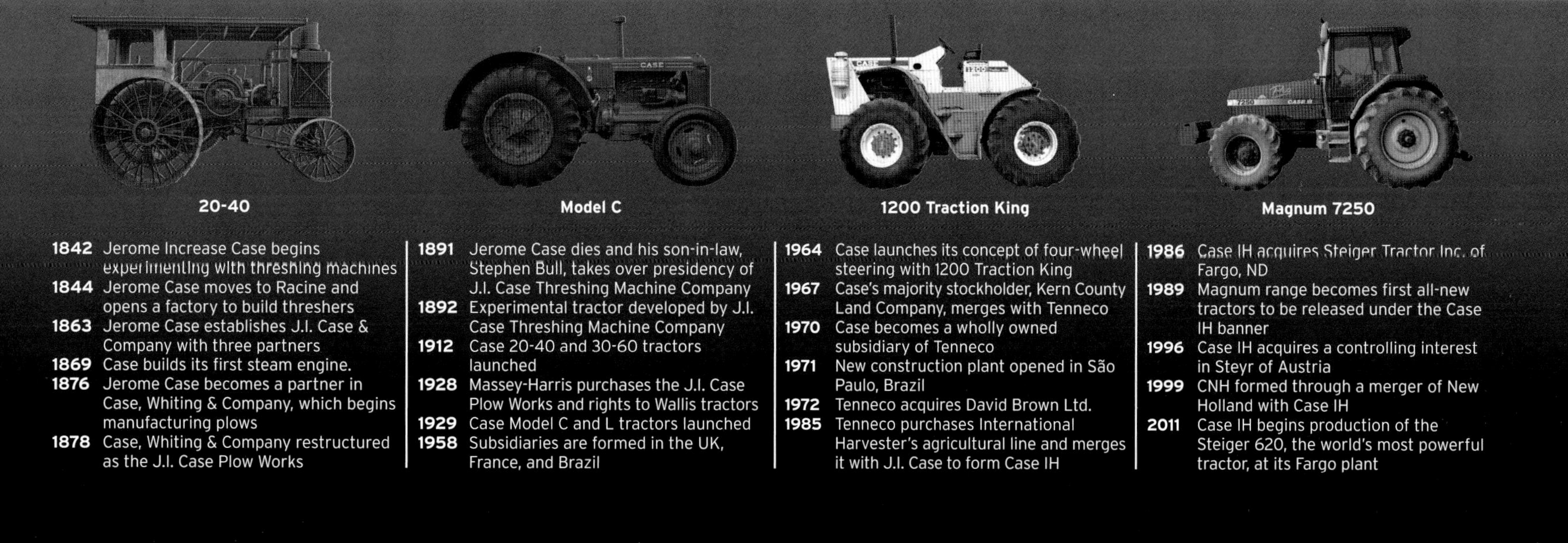

1842 Jerome Increase Case begins experimenting with threshing machines
1844 Jerome Case moves to Racine and opens a factory to build threshers
1863 Jerome Case establishes J.I. Case & Company with three partners
1869 Case builds its first steam engine.
1876 Jerome Case becomes a partner in Case, Whiting & Company, which begins manufacturing plows
1878 Case, Whiting & Company restructured as the J.I. Case Plow Works
1891 Jerome Case dies and his son-in-law, Stephen Bull, takes over presidency of J.I. Case Threshing Machine Company
1892 Experimental tractor developed by J.I. Case Threshing Machine Company
1912 Case 20-40 and 30-60 tractors launched
1928 Massey-Harris purchases the J.I. Case Plow Works and rights to Wallis tractors
1929 Case Model C and L tractors launched
1958 Subsidiaries are formed in the UK, France, and Brazil
1964 Case launches its concept of four-wheel steering with 1200 Traction King
1967 Case's majority stockholder, Kern County Land Company, merges with Tenneco
1970 Case becomes a wholly owned subsidiary of Tenneco
1971 New construction plant opened in São Paulo, Brazil
1972 Tenneco acquires David Brown Ltd.
1985 Tenneco purchases International Harvester's agricultural line and merges it with J.I. Case to form Case IH
1986 Case IH acquires Steiger Tractor Inc. of Fargo, ND
1989 Magnum range becomes first all-new tractors to be released under the Case IH banner
1996 Case IH acquires a controlling interest in Steyr of Austria
1999 CNH formed through a merger of New Holland with Case IH
2011 Case IH begins production of the Steiger 620, the world's most powerful tractor, at its Fargo plant

1939 the range received a fresh identity with a change of color from gray to "Flambeau Red."

In the decades that followed, Case went from strength to strength, by expanding and consolidating its lines of agricultural and construction machinery. The company entered the four-wheel-drive arena in 1964 with its 1200 Traction King. The 1200 model, a rigid-frame machine with steering on all four wheels, had a heavy-duty industrial transmission and marked Case's entry into the market for large agricultural tractors.

In 1967 the company's majority stockholder, Kern County Land Company, merged with the US company Tenneco, the world's largest distributor of natural gas, with interests in oil, chemicals, packaging, and other commodities. Tenneco steadily increased its holdings in Case, with the agricultural company eventually becoming a wholly owned subsidiary of the industrial conglomerate in 1970.

> "It is a **major achievement** for a company to have **endured successfully** for 150 years."
>
> EDWARD J. CAMPBELL, PRESIDENT, CASE CORPORATION 1992–94

Case's turnover reached an all-time high of $919 million with profits of $65.8 million in 1973. Some of the previous year's profits had been used to purchase David Brown Tractors Ltd., which became an operating division of its US parent. For a time, the David Brown and Case concerns operated independently, but a new "Power Red" and "Orchid White" color scheme was introduced for both tractor ranges to show the family connection. The merger also gave Case a valuable foothold in the lucrative European markets.

In 1984 Tenneco made a bid for International Harvester's agricultural division, which was on the brink of bankruptcy. The $475 million deal was approved by the US Justice Department the following year, and J.I. Case and International Harvester were merged into Case IH. The US manufacturer Steiger was brought into the company's fold in 1986.

In 1994, Case IH was floated on the New York stock exchange as Tenneco sought to decrease its ownership. Five years later, the agricultural world was rocked by the announcement that New Holland and Case were to merge to form a massive global agricultural and construction equipment business. The merger, ratified in late 1999, led to the inevitable rationalization of production facilities, but the two lines retained their separate identities.

The innovation continues: Case IH's Steiger 620 model, the world's most powerful tractor, has set new industry records for fuel efficiency and was awarded the Machine of the Year 2014 trophy at Agritechnica in Hanover, Germany.

Tracks and wheels
Case introduced the latest concept in tractors in 2014 with its Magnum Rowtrac, which combined individual track units on the rear with a driven front axle.

British Progress

The UK's tractor success continued through the late 1960s and early 1970s, with subsidiaries of North American companies leading the way. Some of the growth was due to the increasing worldwide popularity of diesel power, which remained a mainly British success story in the tractor market. There was modest progress in driver comfort and convenience with improved seats and more user-friendly controls and instruments, but the high-volume tractor companies continued to neglect the growing demand for four-wheel drive.

▷ **David Brown 880 Selectamatic**

Date 1966 **Origin** UK
Engine David Brown 3-cylinder diesel
Horsepower 46 hp
Transmission 12 forward, 4 reverse

The 880 was an example of David Brown's eye-catching new paint finish and of the simple but efficient Selectamatic hydraulic system that was available throughout the range from 1965. Customers were given a choice of 6- and 12-speed transmissions; the latter offered more speed than any other British manufacturer.

▷ **International B614**

Date 1966 **Origin** UK
Engine International 4-cylinder diesel
Horsepower 60 hp
Transmission 8 forward, 2 reverse

Introduced at London's Smithfield Show in 1963, the new B614 model offered the International customer greater performance and a larger number of gears. Another feature was an independent power takeoff, which was operated via a separate, multi-plate clutch.

▽ **Ford 5000**

Date 1968 **Origin** UK
Engine Ford 4-cylinder diesel
Horsepower 65 hp
Transmission 8 forward, 2 reverse plus optional 10 forward, 2 reverse

Ford's new 6X models arrived in 1964, the first tractors built at the new factory in Basildon, Essex. The 5000 was the biggest of four models, and this tractor was built in 1968, just before the updated Ford Force 5000 version was announced.

◁ **International 434**

Date 1968 **Origin** UK
Engine International 4-cylinder diesel
Horsepower 43 hp
Transmission 8 forward, 2 reverse

In 1966 International's 434 replaced the B-414, which had been the company's first tractor to feature a draft-control hydraulic system. Both were built at the Idle factory near Bradford, Yorkshire. The 434's larger brother, the International 634 from Doncaster, was launched at the same time.

◁ Nuffield 4/65

Date 1969 **Origin** UK

Engine BMC 4-cylinder diesel

Horsepower 65 hp

Transmission 10 forward, 2 reverse

Nuffield's success faltered during the 1960s, mainly due to lack of investment. Its Universal series was 19 years old when the new-look 3/45 and 4/65 models arrived in 1967 with significant design improvements. However, they were soon replaced by new Leyland-badged models.

△ Massey Ferguson 135

Date 1969 **Origin** UK

Engine Perkins 3.152 3-cylinder diesel

Horsepower 45.5 hp

Transmission 6 forward, 2 reverse

This was the smallest of the DX project tractors, and it was immensely successful. It inherited many of the specification features from the previous MF35X model, including the engine, which was modified to give a 12.4 percent power increase.

◁ Massey Ferguson 165

Date 1970 **Origin** UK

Engine Perkins 4-cylinder diesel

Horsepower 58 hp, increased to 60 hp in 1968

Transmission 12 forward, 4 reverse Multi-Power

The MF165 was part of the DX series tractor project that Massey Ferguson began developing in 1962. Driver comfort featured on the options list, with a suspension seat, a weather-protection cab, and power steering all available at extra cost.

◁ Leyland 154

Date 1971 **Origin** UK

Engine BMC 4-cylinder diesel

Horsepower 27 hp

Transmission 9 forward, 3 reverse

The 15 hp Mini tractor joined the Nuffield range in 1965, but sales were poor, partly due to competition from Japanese compact tractors. Boosting the power to 27 hp and, in 1969, changing the name to Leyland and the color to blue did little to improve demand.

PIONEERS

Global Tractor Ranges

Ford and Massey Ferguson both introduced all-new tractor ranges for global markets in 1965. Both already had a long history of designing tractors for world farming, but introducing complete ranges was more ambitious and needed big resources. Ford's 6X program was spread over their British, Belgian, and US factories, and Massey Ferguson invested more than a million hours in design and development work for their DX tractor project.

Sales material Ford 1000 series tractors for global markets included the 5000 model built at their new British factory in Basildon, Essex. Massey Ferguson's DX project produced the Red Giant tractor range, with six totally new or substantially updated models.

Driver Protection

The tractor industry and its customers must share the blame for the slow arrival of safety frames and quiet or "Q-cabs" to protect drivers. Tractors had arrived in the 1890s, and by the early 1920s there was evidence that driver injuries and deaths caused by tractors overturning had become a serious problem. Virtually no action was taken until 1959, when Sweden became the first country to introduce legislation requiring new tractors to be built with an approved safety cab or frame; other countries followed suit. As tractors with safety cabs were introduced, there was a dramatic reduction in the number of driver fatalities caused by overturning tractors.

◁ Ford 7000

Date 1975 **Origin** UK
Engine Ford 4-cylinder turbocharged diesel
Horsepower 94 hp
Transmission 8 forward, 2 reverse

Although it was a big tractor by early 1970s standards, the 7000's power unit was developed from the engine used in the smaller 5000 model. To boost the power output, the 7000 became the first Ford tractor engine to be equipped with a turbocharger.

△ David Brown 1210

Date 1976 **Origin** UK
Engine David Brown 4-cylinder diesel
Horsepower 72 hp
Transmission 12 forward, 4 reverse synchromesh

Launched in 1971, the DB1210 was fitted with an easy-to-use synchromesh gearbox. This is a four-wheel drive model with a German Kramer axle and the DB "Q-cabs" with raised floor level and sound-insulating cladding, supported on rubber mountings.

△ Massey Ferguson 1080

Date 1974 **Origin** France
Engine Perkins 6-cylinder diesel
Horsepower 92 hp
Transmission 12 forward, 4 reverse Multi-Power

The design of the first safety cabs often magnified noise levels, leading to a new generation of "Q-cabs". The MF1080 is an example; the design changes on the updated Mark II version included an extra 2 hp and much better noise insulation in the cab.

▷ Massey Ferguson 1155

Date 1974 **Origin** US
Engine Perkins V8 diesel
Horsepower 155 hp
Transmission 12 forward, 4 reverse Multi-Power

Massey Ferguson's response to the demand for more power included the MF1155, equipped with a big V8 that filled the engine compartment. Surprisingly for a 155 hp tractor in the mid-1970s, there was no four-wheel-drive version.

◁ **Deutz D10006**

Date 1972 **Origin** Germany
Engine Deutz 6-cylinder air-cooled diesel
Horsepower 100 hp
Transmission 12 forward, 6 reverse

Deutz was one of the few tractor companies to use air cooled engines, competing with the rest of the industry's water-cooled diesels. The 100 hp D10006 model was available in both two- and four-wheel-drive versions.

▷ **Lamborghini R1056**

Date 1977 **Origin** Italy
Engine Lamborghini 6-cylinder diesel
Horsepower 105 hp
Transmission 12 forward, 3 reverse

The Italian who started the Lamborghini tractor business also made some of the world's most exotic sports cars. He decided to sell the tractor business but continued making cars. The R1056 tractor was available in two- and four-wheel-drive versions.

△ **Case 1570**

Date 1976 **Origin** US
Engine Case 6-cylinder turbo diesel
Horsepower 180 hp
Transmission 12 forward, 4 reverse

The US celebrated its bicentennial in 1976, and to mark the occasion Case produced the "Spirit of '76," a limited edition of the big 1570 model. The standard 1570 specification included a well-equipped cab and dual rear wheels, but no four-wheel drive.

▷ **Fiat 680H**

Date 1977 **Origin** Italy
Engine Fiat 4-cylinder diesel
Horsepower 68 hp
Transmission 12 forward, 3 reverse

Some of the early cabs were add-ons for existing tractor models, but Fiat's Comfort cab was designed by Pininfarina, the Italian sports car design studio. The 680 was offered in two- and four-wheel-drive versions, plus a tracklayer model called the 665C.

▽ **Leyland 272 Synchro**

Date 1978 **Origin** UK
Engine Leyland 4-cylinder diesel
Horsepower 72 hp
Transmission 9 forward, 3 reverse synchromesh

The Nuffield range built at the Bathgate factory in Scotland changed its name and color to Leyland and two-tone blue in 1969. The new 272 model arrived in 1976, complete with a "Q-cab;" it was also available as a 472 model with four-wheel drive.

A demonstration of Fiat's first tractor, the 702

Great Manufacturers
Fiat

Fiat is the last firm still building tractors, cars, and trucks. Major expansion in the 1990s saw the acquisition of key rivals like Ford New Holland and the Case Corporation, and the adoption of their brands. Although new tractors no longer carry the Fiat name, the company remains one of the world's largest manufacturers.

FOUNDED IN TURIN, Italy, in 1899 by a group of engineers and investors including Giovanni Agnelli, who soon became the company's first managing director, Fabbrica Italiana di Automobili Torino soon became known for simplicity's sake by the acronym of its initials: Fiat. The firm first ventured into car construction, but soon diversified into heavier products and developed its own trucks and buses.

Giovanni Agnelli (1866-1945)

Expanding into tractor production seemed a natural next step, and the heat and rugged landscape of much of Italy demanded durable machines. Fiat's first design used their military truck engine, a 1.5-gallon (5.6-liter) gasoline unit with four cylinders producing 20 hp, driving through a three-speed gearbox. Labeled the 702, the tractor was first demonstrated in 1918, just a year after Fordson's Model F went into production.

Expansion for Fiat
Introduced in 1926, the 35 hp 700 weighed more than a ton less than the 702, with a more compact, higher-revving engine incorporating overhead valve technology.

The 702 model was expanded into a number of varieties, including the 702 A, B, and BN options. It remained in production until 1925, by which time 2,000 tractors had been built. In 1926 Fiat introduced the lighter 35 hp 700 model. By 1932 Fiat launched its first crawler version, the 700C, and, as the market in mountainous Italy began to boom, the firm became a leader in this sector. Among the most famous of its early crawlers was the Model 40, which featured a Boghetto engine notable for its elongated combustion chamber that featured a "biconical" venturi-type top and was capable of running on a range of different fuels.

In 1944, tractor operations at Fiat's Modena plant came to a halt. Limited raw materials and Nazi occupation meant production was diverted to the repair of military vehicles. In secret, however, the innovation continued, and Fiat's technical department built a prototype for a new crawler—the Series 50.

Following the end of World War II and the return to full tractor production in Fiat's factories, the firm could once more focus on product development. The 18 hp 600 tractor, introduced in 1949, was the first to feature a power takeoff and run on pneumatic tires. By the time the Model 55 was launched just a year later, available power had jumped to 55 hp. Small tractors remained popular, however, and the 19–20 hp Piccola, introduced in 1957, marked out Fiat's commitment to miniature tractors.

Fiat's first series
One of the tractors in Fiat's 1965 Diamante range. This series was the first to feature synchronized speeds and differential lock. It firmly established Fiat's position in Europe.

But it was arguably the 411, in both wheeled and crawler guises, that was Fiat's most significant tractor in the immediate postwar period. Launched at the 1958 Verona Agricultural Fair, it used a four-cylinder engine to develop 41 hp at 2,300 rpm. A cast-iron crankcase enabled the engine to be set within a stressed chassis.

Power outputs continued to rise over the next decade. The first full series of Fiat tractors, the Diamante, was launched in 1965. Comprising four models, the 215, 315, 415, and 615, the new line spanned 22–70 hp, and featured new styling by the Pininfarina vehicle design company. A key feature was an innovative, new transmission with an "Amplicouple" shift-on-the-move splitter.

As the 50th anniversary of Fiat tractor production approached, the company introduced the Nastro d'Oro (Golden Ribbon) range, a development of the Diamante line, in 1968. The tractors' bodywork remained orange but the engine area received deep blue paintwork. Under the hood there were new direct-injection engines with rotary injection pumps, gear synchronization, and a draft/position-control rear linkage.

In 1971 the firm's tractors breached the 100 hp barrier with the 1300, and three years later Fiat manufactured its one millionth tractor. By 1975, the firm had launched a completely new range of tractors, the 80 series. Its distinctive styling, again provided by Pininfarina, featured sharply angled cab lines.

A distinct Fiat Trattori tractor group was formed within Fiat, and in the same decade the group expanded into new areas by acquiring the harvester

702

505C

680

Winner 110

1899 Fabbrica Italiana di Automobili Torino (Fiat) founded in Turin by a group of engineers and investors including Giovanni Agnelli
1918 The firm demonstrates its first tractor, the 702, which goes on sale the following year
1926 Introduction of 35 hp 700, lighter than the 702, with a more compact engine and overhead valve technology
1932 Launch of the first Fiat crawler tractor, badged as the 700C or Type 30
1933 Fiat takes over Società Anonima Officine Meccaniche (OM)
1939 Multi-fuel 40 Boghetto crawler unveiled at the Agricultural Fair in Verona, Italy
1949 Introduction of the 18 hp 600 tractor, the first Fiat to feature a power takeoff and pneumatic tires
1957 The 19–20 hp Piccola, aimed at smaller farmers, is introduced to the market
1965 First full series of Fiat tractors, the Diamante, is launched, with four models spanning a 22–70 hp power band
1968 Introduction of Nastro d'Oro (Golden Ribbon) range, a development of the Diamante line. The tractors' bodywork remains orange but the engine area is re-liveried in a new deep blue
1971 The first 100 hp Fiat tractor built
1974 Fiat celebrates the production of its one millionth tractor
1975 Introduction of the 80 series, with a new cab design by Pininfarina
1975 Fiat acquires Italian combine maker Laverda and its Breganze factory
1977 Fiat buys Hesston, the US forage equipment and baler specialist
1983 Fiatagri is formed to bring all of Fiat's agricultural interests together
1991 Fiat buys 80 percent of Ford New Holland, later acquiring the rest
1997 The Fiatagri terra-cotta color scheme is phased out on New Holland tractors
1999 Fiatagri purchases Case Corp. and merges it with New Holland to form CNH

In **1977**, to mark the change of brand from **Fiat Trattori** to **Fiatagri**, the company's livery was **changed** to a new **orange** color.

interests of fellow Italian firm Laverda, followed by the US forage equipment company Hesston.

By the mid-1980s, Fiat Trattori was expanding again, purchasing Braud, a French company manufacturing grape harvesters. Fiat's 80 series design was carried through to the 90 series, but by 1990 a whole new concept was introduced with the square-cabbed, 100–140 hp Winner series.

Soon afterward, Fiat reached a deal with Ford to acquire 80 percent of the latter's Ford New Holland business. Fiat brought the two agricultural divisions together under a whole new brand that combined the Fiatagri emblem and the New Holland name.

Radical departure
In 1975 Fiat launched the completely restyled 80 series of tractors, built in the company's Modena factory.

Initially, both marques kept their own liveries, with smaller tractors made in Italy and larger machines constructed in the former Ford factories in the UK and North America. Eventually, the Ford and Fiatagri "sub-badging" was dropped, as was the terra-cotta Fiatagri livery, in favor of a focus on blue paintwork. Behind the branding, Fiat had hugely increased its presence in the world market, broadening it again in 1999 with its acquisition of the Case Corporation and subsequent creation of Case New Holland.

Final Fiat
The 100–140 hp Winner series were the last tractors designed exclusively for Fiat, and the last to bear its name.

Europe's Industry

The tractor industry made a slow start in mainland Europe, with less technical innovation compared to the US and UK. One of the few advances was using diesel power, which started in Germany, with Italy close behind, although the technology came from the UK. The 1950s and 1960s saw increased production in much of Europe—by the mid-1960s, tractors from both Eastern and Western Europe were competing successfully in world markets.

◁ **SAME Sametto V**

Date 1965 **Origin** Italy

Engine SAME twin-cylinder air-cooled vertical diesel

Horsepower 25 hp

Transmission 5 forward, 1 reverse

The success of Fiat's little Piccola tractor persuaded the Cassani company to add smaller models to its SAME range. The result was the Sametto in 1961, available with two- or four-wheel drive and as the Sametto V vineyard and orchard model.

◁ **Fiat 411R**

Date c. 1965 **Origin** Italy

Engine Fiat 4-cylinder diesel

Horsepower 37 hp

Transmission 6 forward, 2 reverse

Fiat's tractor production expanded rapidly during the 1950s and 1960s with a comprehensive range of wheeled and crawler models. The 411 and this four-wheel-drive 411R version arrived in 1958 to become Fiat's top-selling midrange tractors.

△ **Deutz A110**

Date 1965 **Origin** Germany

Engine Deutz 6-cylinder air-cooled diesel

Horsepower 110 hp

Transmission 5 forward, 1 reverse

The unusual feature of the A110 was having two vertical exhausts side by side, each serving three of the engine's cylinders. Air-cooled engines were a Deutz specialty, which avoided the risk of frost damage. This tractor was a German design built for assembly in Argentina.

▷ **Fendt Favorit 3 FWA**

Date 1965 **Origin** Germany

Engine MWM 4-stroke diesel

Horsepower 52 hp

Transmission 16 forward, 4 reverse

This is the FWA version with four-wheel drive; there was also a two-wheel-drive FW model. The structure on each mudguard is a passenger seat, a relic of a less safety-conscious age when this type of seat was popular in some European countries.

▽ John Deere 1120

Date 1967 **Origin** Germany
Engine John Deere 3-cylinder diesel
Horsepower 50 hp
Transmission 8 forward, 2 reverse

John Deere's original plan was to site their European factory in the UK, but a policy change had resulted in the purchase of the Heinrich Lanz company in Germany. This is where the mid-range 1120 model was built, with production starting in 1967.

△ Zetor 3045

Date 1966–68 **Origin** Czechoslovakia
Engine Zetor 3-cylinder diesel
Horsepower 39 hp
Transmission 10 forward, 2 reverse

Tractors from Eastern Europe still had a reputation for outdated design in the 1960s, but some Zetor models had more advanced features. The 3045 was one of the models that helped to establish Zetor as a major tractor exporter.

△ International McCormick 523

Date 1967 **Origin** Germany
Engine International 3-cylinder diesel
Horsepower 52 hp
Transmission 8 forward, 2 reverse

Contributions to Germany's expanding tractor production came from the John Deere factory in Mannheim plus International's production in Neuss. The 523 was one of the big-selling International models, available as a two-wheel-drive (shown) and in a four-wheel-drive version.

△ Fendt F231 GTS

Date 1967 **Origin** Germany
Engine MWM air-cooled diesel
Horsepower 32 hp
Transmission 8 forward, 4 reverse

The versatile tool-carrier tractors were a success for Fendt. They were used as ordinary tractors for pulling implements: the dumping container at the front was handy for carrying loads and, with the container removed, there was a clear view for row-crop work.

▷ Steyr 540

Date 1974–77 **Origin** Austria
Engine Steyr 3-cylinder diesel
Horsepower 40 hp
Transmission 8 forward, 6 reverse

Steyr was an important manufacturer of military and sporting guns before adding cars, trucks, and, later, tractors to their product range. Tractor specifications were usually high, with front axle suspension offered on some models from the early 1950s.

Global Expansion

European and North American agriculture had become heavily mechanized during World War II and the decades that followed, but globally there were many other parts of the world where tractors and associated equipment were still small in number. To satisfy the growing demand to increase food production, tractor manufacturers began to set up factories in "new" parts of the world, such as South America. Meanwhile, makers in areas including Eastern Europe, Japan, and Australia began to step up production and, in some cases, look for new markets.

▽ **Valmet 360D**

Date c. 1965 **Origin** Brazil
Engine MWM 3-cylinder diesel
Horsepower 40 hp
Transmission 6 forward, 2 reverse

The 360D was one of the first products to come from Valmet's expansion into building tractors outside its Finnish homeland when, in 1960, it set up a new plant in Mogi das Cruzes, Brazil. It used a German-made engine, but otherwise was similar to the Finnish-made 359D.

△ **Kubota L245 FP**

Date 1979 **Origin** Japan
Engine Kubota 3-cylinder diesel
Horsepower 25 hp
Transmission 8 forward, 2 reverse

While more recently it has started to expand its product offering farther up the power scale, Kubota's traditional tractor market has been in the sub-50 hp sector. The L245 was at the heart of the Japanese firm's range in 1976–85.

◁ **Belarus MTZ-50**

Date 1975 **Origin** Belarus
Engine Belarus 4-cylinder diesel
Horsepower 70 hp
Transmission 9 forward, 2 reverse

The Minsk Tractor Works was founded in 1946 in the town of Minsk, in the Soviet republic of Belarus, and from 1949 the tractors were branded with the name of the state. MTZ, or Minsk Tractor Zavod (Works), preceded the model number on all of the maker's early tractors. The MTZ-50 was also available as a four-wheel-drive MTZ-52.

▽ **Ford 8 BR**

Date 1967 **Origin** Brazil
Engine Perkins 4-cylinder diesel
Horsepower 35 hp
Transmission 6 forward, 2 reverse

Like Valmet and Massey Ferguson, in the 1960s Ford developed a new tractor factory in Brazil in response to the government's insistence that tractors sold in the country had to be made there with a level of locally sourced components. The 8 BR used a Brazilian-built Perkins engine.

△ **Satoh D-650G**

Date 1971 **Origin** Japan
Engine Mitsubishi 4-cylinder gasoline
Horsepower 25 hp
Transmission 6 forward, 2 reverse

Founded in Japan in 1914, the Satoh Agricultural Machinery Manufacturing Co. merged in 1980 with fellow Japanese manufacturer Mitsubishi Machinery Co. After the merger, the combined tractor line was rebranded under the Mitsubishi banner.

▷ **Ursus 1204**

Date 1978 **Origin** Poland
Engine Zetor 6-cylinder diesel
Horsepower 110 hp
Transmission 8 forward, 2 reverse

In Communist-era Eastern Europe, Polish state tractor maker Ursus manufactured high-horsepower tractors for sale under its own name, as well as that of Czechoslovakia's Zetor, for many years. In the latter's range, the 1204 was badged as the Zetor Crystal 12011.

▽ **Upton HT-14 350**

Date 1978 **Origin** Australia
Engine Cummins 6-cylinder diesel
Horsepower 350 hp
Transmission 14 forward, 2 reverse

Built for the flat, dry terrain of Australia's arable heartland, the HT-14 350 was broadly conventional in configuration, with smaller front wheels through which the machine was steered. The tractor's 350 hp was put to the ground through the rear wheels only.

△ **Shandong TS-25**

Date 1979 **Origin** China
Engine Shandong 3-cylinder diesel
Horsepower 25 hp
Transmission 6 forward, 2 reverse

Chinese manufacturer Shandong made a vast number of different versions of its TS-25 tractor, including specialty builds for rice fieldwork, transportation, and narrow-tread jobs. Available in both two- and four-wheel-drive versions, this is one of China's top selling tractors.

△ **UTB Universal 530**

Date 1979 **Origin** Romania
Engine Universal 3-cylinder diesel
Horsepower 53 hp
Transmission 12 forward, 3 reverse

Romanian tractor maker Universal developed a close relationship with Italian manufacturer Fiat, and for many years used, under license, a number of the latter's mechanical components. Later models of the 530 were available with a shuttle reverse gearbox.

Northrop 5004T

Like several British four-wheel-drive conversions of the time, the Northrop was based on the easily adaptable Ford 5000 skid unit. This saved on development costs and allowed manufacturers to take advantage of Ford's global sales, spares, and service network. Developed by David J.B. Brown, one of the UK's leading developers of off-road vehicles, the Northrop had a unique layout with a raised power-train to optimize ground clearance.

AFTER DEVELOPING MINING VEHICLES in Africa, David J.B. Brown was appointed chief executive of British Northrop's Chaseside division in 1964. He began developing a four-wheel-drive agricultural tractor with good maneuverability and excellent ground clearance to meet the needs of the forestry and sugarcane industries. With the drive layout on the underside of the engine and transmission, the machine had a raised center of gravity, but its performance was remarkable.

The tractor was launched as the Northrop 5004, but there were concerns that the 67 hp four-cylinder Ford engine lacked power. To overcome this, Northrop developed the 5004T model with a CAV turbocharger, which boosted output to 80 hp. In 1967, a six-cylinder version was introduced, but just three were built before production ended.

FRONT VIEW

REAR VIEW

Contour-shaped seat sprung for maximum comfort

Lever to engage drive to front axle

Rear tire same size as front wheel's for maximum traction

Turbocharged tractor
Dating from 1965, this is the 5004T version of the Northrop tractor, which had a CAV turbocharger as well as a new manifold, air breather, and uprated injectors. The power of the four-cylinder Ford 5000 engine was increased by approximately 25 percent from 67 hp to just over 80 hp.

Straight-through exhaust pipe part of turbocharging kit

Air pre-cleaner bowl

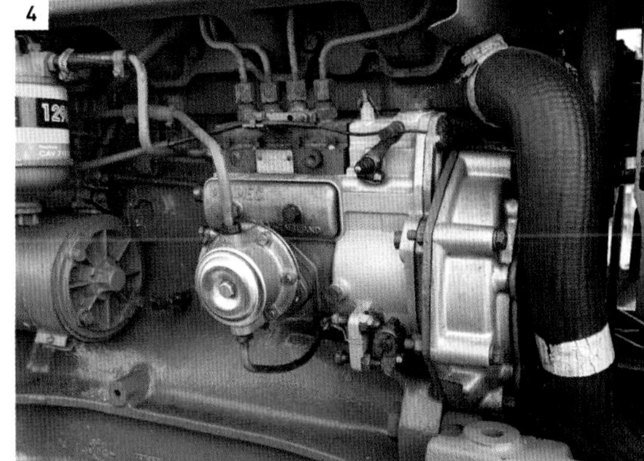

THE DETAILS

1. Front drive-steer axle based on Ford components **2.** Ford instrumentation with a proofmeter registering engine speed **3.** Eight-speed Ford gearbox mated to a transfer box with a hand lever to engage drive to the front axle **4.** Four-cylinder Ford engine with a Simms injector pump **5.** Drive to the front axle taken from the transfer box via a propeller shaft

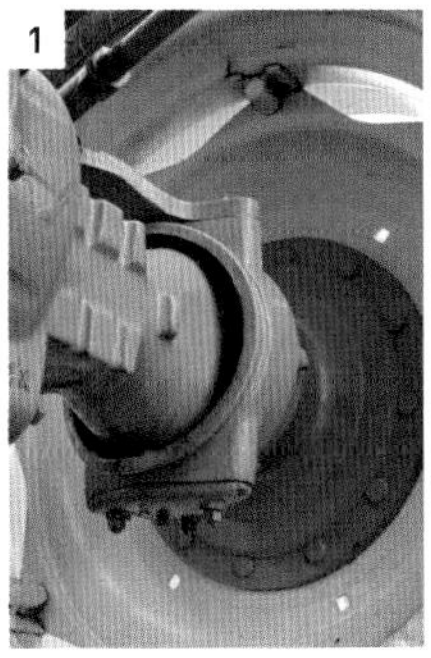

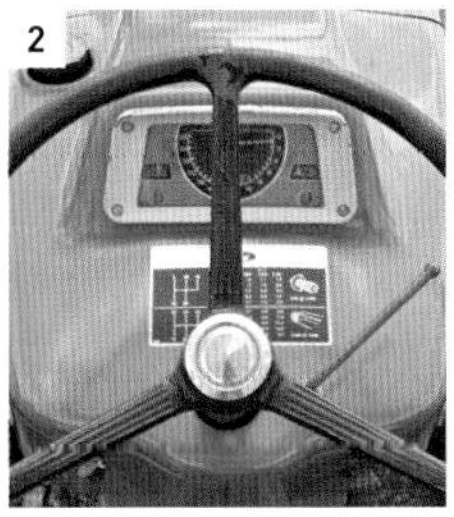

Cast front ballast weight

Front drive-steer axle has planetary hub-reductions

SPECIFICATIONS	
Model	Northrop 5004T
Built	1965
Origin	UK
Production	100–150
Engine	80 hp Ford 4-cylinder turbocharged diesel
Capacity	233 cu in (3,818 cc)
Transmission	8 forward, 2 reverse
Top speed	19.6 mph (32 km/h)
Length	12 ft 3 in (3.7 m)
Weight	4 tons (3.6 metric tons)

All-wheel Drive

There was little interest in four-wheel drive during the first 60 years or so of tractor history, with tracklayers remaining the obvious choice for farmers needing extra pulling efficiency. The breakthrough that turned four-wheel drive into a success during the 1950s and 1960s was mainly because specialized companies in the UK demonstrated the increased traction available, particularly in difficult working conditions. The most effective traction comes from equal-sized front and rear wheels, often called all-wheel drive.

▷ Northrop 5004T

Date 1965 **Origin** UK
Engine Ford 4-cylinder turbocharged diesel
Horsepower 80 hp
Transmission 8 forward, 2 reverse

Ford skid units were popular for four-wheel-drive conversions, and Northrop based their 67 hp 5004 tractor on the Ford 5000. This was followed by the 5004T, a turbocharged version of the Ford engine, boosting the output to 85 hp.

△ Doe 130

Date 1967 **Origin** UK
Engine 2 x Ford 5000 4-cylinder diesels
Horsepower 130 hp
Transmission 8 forward, 2 reverse

A farmer's idea to link two tractor units nose-to-tail to provide increased pulling power attracted the interest of Ernest Doe, a Ford tractor dealer. He developed it commercially as the Doe Triple-D, which was followed by the Doe 130, a later, more powerful version.

△ International 634 Four-Wheel Drive

Date 1968 **Origin** UK
Engine International 4-cylinder diesel
Horsepower 66 hp
Transmission 8 forward, 2 reverse

International offered customers a choice of two- and four-wheel-drive versions of the 634 model. The 4WD tractors were available in an all-wheel-drive version with big front wheels, and in this four-wheel assist type using Roadless drive equipment.

◁ International 634 All-Wheel Drive

Date 1971 **Origin** UK
Engine International 4-cylinder diesel
Horsepower 66 hp
Transmission 8 forward, 2 reverse

The fact that big tractor companies such as International were moving into four-wheel drive brought increasing competition for specialists such as County, which makes it surprising that County agreed to sell their equal-size wheel technology to companies such as International.

△ County 754

Date 1968 **Origin** UK
Engine Ford 4-cylinder diesel
Horsepower 75 hp
Transmission 8 forward, 2 reverse

County made tracklaying conversions of Fordson and Ford tractors before becoming the UK's leading four-wheel-drive specialist. The County 754 with power delivered through equal-sized front and rear wheels was based on the skid unit for the new Ford Force 5000.

◁ Roadless 120

Date 1972 **Origin** UK
Engine Ford 2715E 6-cylinder diesel
Horsepower 120 hp
Transmission 8 forward, 2 reverse

Roadless had a long history as a tracklayer specialist before switching to four-wheel drive in the 1950s. It also produced all-wheel-drive tractors with equal-sized wheels, including this 120, an uprated version of the 115.

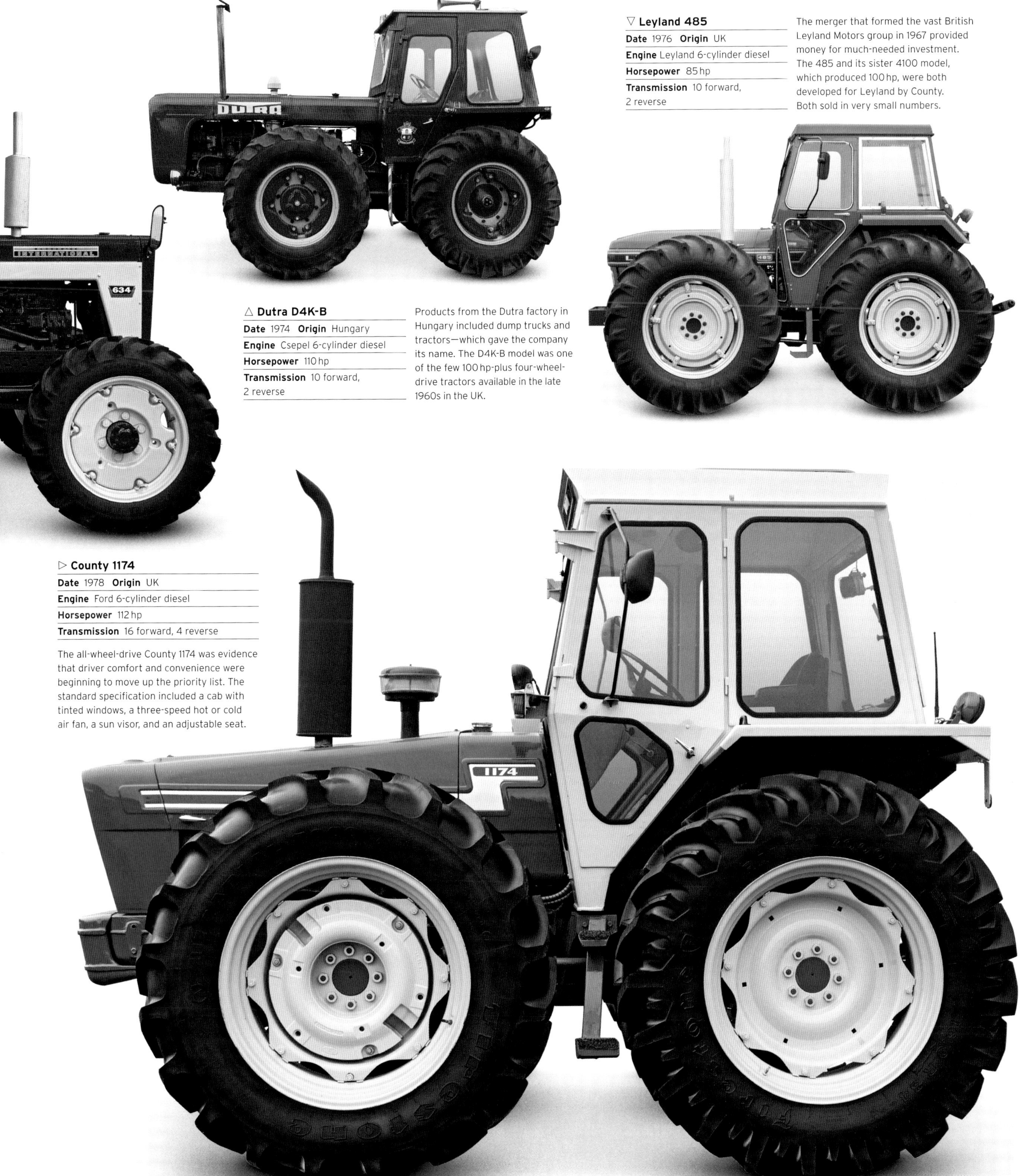

▽ **Leyland 485**

Date 1976 **Origin** UK
Engine Leyland 6-cylinder diesel
Horsepower 85 hp
Transmission 10 forward, 2 reverse

The merger that formed the vast British Leyland Motors group in 1967 provided money for much-needed investment. The 485 and its sister 4100 model, which produced 100 hp, were both developed for Leyland by County. Both sold in very small numbers.

△ **Dutra D4K-B**

Date 1974 **Origin** Hungary
Engine Csepel 6-cylinder diesel
Horsepower 110 hp
Transmission 10 forward, 2 reverse

Products from the Dutra factory in Hungary included dump trucks and tractors—which gave the company its name. The D4K-B model was one of the few 100 hp-plus four-wheel-drive tractors available in the late 1960s in the UK.

▷ **County 1174**

Date 1978 **Origin** UK
Engine Ford 6-cylinder diesel
Horsepower 112 hp
Transmission 16 forward, 4 reverse

The all-wheel-drive County 1174 was evidence that driver comfort and convenience were beginning to move up the priority list. The standard specification included a cab with tinted windows, a three-speed hot or cold air fan, a sun visor, and an adjustable seat.

High Horsepower

Using large amounts of engine power efficiently requires a special type of tractor. The biggest problem to overcome is wheel slip. As power output increases, the tires on a traditional two-wheel-drive tractor lose their grip, which wastes time and fuel. Four-wheel drive is more efficient, ideally with equal-sized front and rear wheels. To achieve good maneuverability with large front wheels, the biggest tractors need articulated, or "bend-in-the-middle," steering, with a hinge point in the center. This became the standard layout for high-horsepower wheeled tractors built since the late 1960s, when engine outputs passed the 200 hp barrier—they are now heading toward 1,000 hp.

▷ **Oliver 2655**

Date 1969 **Origin** US

Engine Minneapolis-Moline 6-cylinder diesel

Horsepower 143 hp

Transmission 10 forward, 2 reverse

Oliver and Minneapolis-Moline were both part of the White company in the early 1970s, and some models were sold under both brand names. The 2655 was Oliver's biggest tractor—the same model was also available as the A4T-1600 in MM colors.

△ **Massey Ferguson 1200**

Date 1967 **Origin** US

Engine Perkins 6-cylinder diesel

Horsepower 105 hp

Transmission 12 forward, 4 reverse

The appearance of the 1200 could be misleading. It looked like a typical high-horsepower, four-wheel-drive tractor with articulated steering and hydraulic linkage, but the 105 hp engine lacked the power of most of its rivals.

△ **Steiger Panther ST350 Series III**

Date 1977-81 **Origin** US

Engine Cummins V8 diesel

Horsepower 350 hp

Transmission 10 forward, 2 reverse

The Series III range of Steiger tractors, introduced in 1976, had four basic models—Wildcat, Bearcat, Cougar, and Panther—all with 10-speed transmissions and either Caterpillar or Cummins engines. ST denoted standard frame, as opposed to the row-crop RC models.

▷ **Waltanna 4-250**

Date 1977 **Origin** Australia

Engine Caterpillar 6-cylinder turbo diesel

Horsepower 250 hp

Transmission 14 forward, 2 reverse

The first Waltanna tractor was made in 1975, but by the early 1980s the company built at least 12 different four-wheel-drive models up to about 400 hp—a large range for a company with only 12 employees. The 4-250 was one of the mid-range models, producing 250 hp.

◁ **Versatile 1080 "Big Roy"**

Date 1976 **Origin** Canada

Engine Cummins 6-cylinder diesel

Horsepower 600 hp

Transmission 6 forward, 1 reverse

As competition increased at the top end of the market, Versatile took the lead by building the 600 hp, eight-wheel-drive "Big Roy," the world's most powerful tractor until it was overtaken by the "Big Bud." Design problems and a lack of implements meant it never went into production.

▽ Big Bud 16V-747

Date 1977 **Origin** US
Engine Detroit Diesel 16-cylinder 2-stroke
Horsepower 900 hp
Transmission 6 forward, 1 reverse

Described as the world's biggest farm tractor, the 900 hp Big Bud had an impressive work rate, covering 1 acre (0.4 hectare) per minute when pulling a 79-ft (24-m) wide cultivator at up to 8 mph (13 km/h). With 1,000 gal (3,785 liters) in the fuel tank, the 16V-747 weighed more than 40 tons.

◁ Schlüter 5000 TVL

Date 1978 **Origin** Germany
Engine MAN 12-cylinder turbo diesel
Horsepower 500 hp
Transmission 8 forward, 1 reverse

Schlüter specialized in tractors with high power and style. The 5000 TVL was the top model in its Profi Trac series. Unusually for such a big four-wheel-drive tractor, it was built on a rigid frame with no articulated steering.

▷ Fiat 44-28

Date 1979 **Origin** Canada
Engine Cummins NT855 6-cylinder turbocharged diesel
Horsepower 280 hp
Transmission 12 forward, 4 reverse

Fiat had close links in Eastern Europe during the 1970s, and the Italian company realized that there was a potential market for high-horsepower tractors in the region. Versatile of Canada agreed to supply some models in Fiat colors, including the Fiat 44-28, available from 1979.

16V
BIG

World's Biggest Tractor

The Big Bud 16V-747, built in 1977, was the largest agricultural tractor of its day, and the most powerful tractor ever built. Equipped with a 747 hp V16 Detroit Diesel engine, the tractor was sold to the Rossi brothers of Bakersfield, CA. Replacing two Caterpillar D9 crawlers, Big Bud was used for deep tillage.

While in California, the 16V-747's engine was opened up to deliver 900 hp. In the 1980s the tractor was sold to a farm in Florida, where it worked for 20 years before going to the Williams brothers of Big Sandy, MT. The new owners used the tractor to pull an 80-ft (24-m) cultivator, covering 1 acre (0.4 hectare) per minute.

ORIGINS OF BIG BUD

"Big Bud" Nelson was the workshop foreman of a tractor dealership in Montana. In 1969 Nelson went into partnership with Willie Hensler, the dealership owner. They formed the Northern Manufacturing Company and set out to build heavy-duty, high-horsepower, articulated tractors—and so the Big Bud line was born.

Beginning with a 280-hp machine, the company introduced increasingly larger and more powerful models. Plans to put the huge 16V-747 into full production never reached fruition, and the tractor shown was the only one of its type ever made. Some 516 Big Bud tractors were built during a 22-year period, and production ended in 1991.

The 900 hp Big Bud 16V-747 stands 14 ft (4.3 m) high and is 28 ft 6 in (8.7 m) long. It has an operating weight of 65 tons (59 metric tons).

Steel-track Sunset

The steel-tracked tractor remained popular in the agricultural industry until the arrival of the rubber-tracked tractor, and to a lesser extent the reliable four-wheel-drive models. Four-wheel drives worked well on lighter soils, but on heavy clay they caused soil compaction and subsequent water retention. A problem that only the light tread and sure traction of a track-type tractor could help prevent. As farms increased in size and the movement of machinery on the road became a necessity, the popularity of the steel-track waned. This, and the slow working speeds, saw the type almost completely disappear from the modern farming scene.

△ Track Marshall 55

Date 1968 **Origin** UK
Engine Perkins 4.270 4-cylinder diesel
Horsepower 55 hp
Transmission 6 forward, 2 reverse

More Track Marshall 55s were produced than any other model built by the company. The 55 was a very reliable and popular machine with both agricultural and industrial customers. It came with a full range of attachments, including driver's cabins, angle dozers and bulldozers, and a rear-mounted toolbar.

△ Fiat 505C

Date 1976 **Origin** Italy
Engine Fiat 3-cylinder diesel
Horsepower 54 hp
Transmission 6 forward, 2 reverse

Fiat produced a range of small crawlers, the main markets for which were vineyards and small farming operations. These tractors were true crawlers in every respect. Other manufacturers had produced small track-type tractors over the years, but they were either conversions or did not use the best features found in the large-sized tractors.

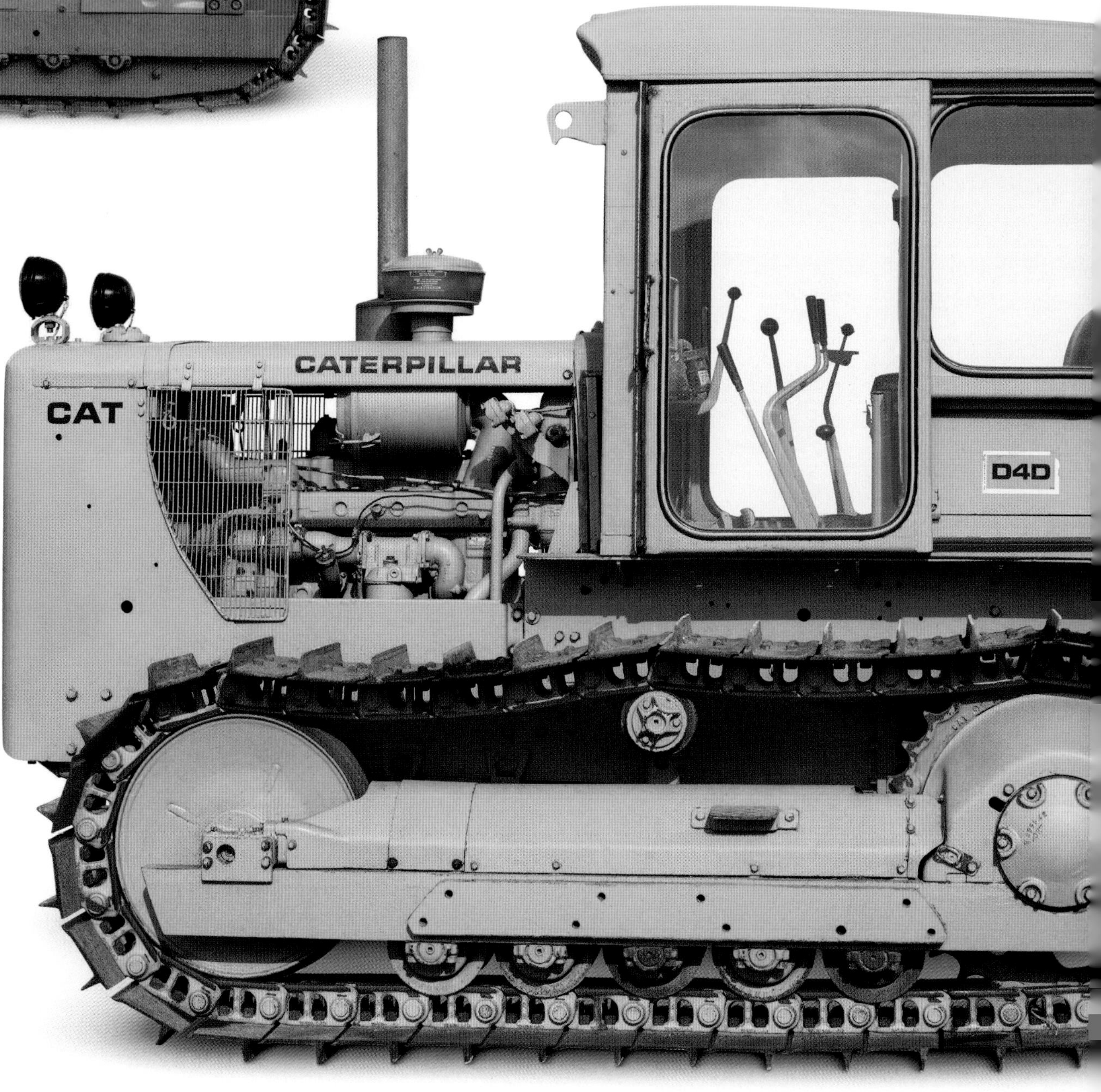

▷ Caterpillar D4D

Date 1970 **Origin** US
Engine Caterpillar 4-cylinder diesel
Horsepower 75 hp
Transmission 5 forward, 5 reverse

The early D4D tractors were equipped with the D330 engine, producing 65 hp. This engine was replaced by a 3304 unit with a 75 hp rating; later modifications to the fuel injection system of the engine increased the output to 90 hp. The D4D was produced in several Caterpillar factories around the world: the tractors were identical, but the product of each factory was identified by a separate serial number prefix.

△ **Mailam 5001**

Date 1967 **Origin** Italy
Engine Ford 4-cylinder diesel
Horsepower 65 hp
Transmission 8 forward, 2 reverse

The early 5001 crawler used a 65 hp 6X Ford 5000 engine; later models used the improved Ford 75 hp 6Y engine. A six-cylinder option was also offered using the Ford 2703 engine. The chassis was closely based on the 92 series International TD9 crawler. Three were imported into the UK in 1970.

△ **Rixmann Knapp 4000**

Date 1973 **Origin** US
Engine Caterpillar 3306 6-cylinder diesel
Horsepower 180 hp
Transmission 6 forward, 3 reverse

This Rix tractor was a re-engined Vickers VR180, built in very small numbers. Its running gear was based on Vickers WWII tank designs, which allowed much higher operating speeds. Engine issues were reduced, but it had transmission and track problems.

△ **International TD8 CA**

Date 1979 **Origin** UK
Engine International 4-cylinder diesel
Horsepower 83 hp
Transmission 5 forward, 1 reverse

The TD8 CA was the last agricultural crawler built at the Doncaster Works. When it was introduced, the power requirements of the British farmer had moved on, the minimum requirement for a steel-tracked crawler then being 125–150 hp: it had become uneconomical to employ a driver to operate a tractor with any lower power output.

△ **Belarus DT75**

Date 1977 **Origin** USSR
Engine Belarus 4-cylinder diesel
Horsepower 101 hp
Transmission 7 forward, 1 reverse

The DT75 was produced in vast numbers, to service the enormous state farms of the USSR. The track technology was derived from Soviet military tanks—made of cast links, these required very little machining, and were cheap to produce and replace.

▷ **Caterpillar D4E Special Application**

Date 1980 **Origin** US
Engine Caterpillar 4-cylinder diesel
Horsepower 97 hp
Transmission 5 forward, 5 reverse

The D4E was manufactured with full agricultural specification. It was fitted with a three-point linkage, an air-conditioned cab, and a special close-ratio gearbox. The tractor was available with a longer, optional six-roller track frame; later models could be supplied with variable horsepower (vhp).

1981-2000

THE NEW TECHNOLOGY

THE NEW TECHNOLOGY

Electronic and computer technology came to the fore during the 1980s as tractors became more sophisticated to meet the challenging needs of a scientific age of agriculture. However, the technical wizardry also pushed up tractor prices, which was not good news as the market went into recession.

The inevitable slump was worsened by a global downturn in commodity prices arising from political tensions and overproduction. The 1979 fuel crisis had created raging inflation, and the tractor firms were faced with the difficult combination of rising inputs, falling exports, and stagnant sales. Several leading manufacturers were saved from bankruptcy only by a massive injection of capital.

The market continued to deteriorate during the 1990s, and the tractor industry had to change to survive. It became a time of takeovers and mergers as many of the independent manufacturers were merged into global organizations.

Despite the economic upheaval, this era stands out as a time of technical brilliance in terms of tractor development. New concepts were introduced, including the rubber-tracked crawler and the high-speed tractor, cab suspension, powershift transmissions, and programmable controls. Environmental concerns led to more fuel-efficient engines with lower emissions, as well as the adoption of global satellite positioning systems to place fertilizer and chemicals accurately.

△ **Silver Jubilee celebration**
Ford's Basildon plant in Essex, UK, commemorated its 25th anniversary with the launch of a limited-edition 7810 "Silver Jubilee" tractor in 1989.

" Though we are **mindful** of the **challenges** awaiting us, we are greatly **encouraged** by **our progress.**"

UMBERTO QUADRINO, CHIEF EXECUTIVE OFFICER OF NEW HOLLAND, 1996-2000

◁ **JCB's Fastrac** introduced the concept of a high-speed tractor that was equally at home on field or road.

Key events

- ▷ **1984** Minsk Tractor Works, Belarus, builds its two-millionth tractor.
- ▷ **1985** Tenneco, owners of Case, acquires International Harvester and merges both under the Case IH brand.
- ▷ **1986** Massey Ferguson's Autotronic and Datatronic models are the first tractors to incorporate computerized electronic systems.
- ▷ **1987** Caterpillar launches the rubber-track Challenger 65 tractor.
- ▷ **1990** AGCO is formed from a management buyout of the Allis-Gleaner Corp.; AGCO takes over Massey Ferguson four years later.
- ▷ **1990** JCB launches the Fastrac, which has a top speed of 40 mph (64 km/h).
- ▷ **1991** Fiat acquires Ford New Holland.
- ▷ **1994** John Deere's 8000 series are the first tractors to have their design concept patented.
- ▷ **1995** Italian manufacturer SAME adds Deutz-Fahr to its portfolio.
- ▷ **1996** Tier 1 engine emissions proposals for off-road diesels come into force in the US.
- ▷ **1997** German manufacturer Fendt becomes part of AGCO.
- ▷ **1999** New Holland and Case IH merge into a global equipment company with a turnover of close to $12 billion.

△ **Ready for delivery**
The Mahindra brand, which began as a joint venture with International in 1962, had become India's best-selling tractor by 1983.

The Big Three

Many small and medium tractor makers disappeared in the 1980s and 1990s through takeovers and mergers, which left control to just a few giants. Ford, John Deere, and Massey Ferguson remained the most familiar global brand names in an industry based mainly in North America and Europe. The success of new arrivals such as JCB and rapid tractor production in Japan, India, and China would ensure competition in the future.

▷ **Ford TW-35**

Date 1985 **Origin** Belgium

Engine Ford 6-cylinder diesel with turbo and intercooler

Horsepower 195 hp

Transmission 16 forward, 4 reverse

In 1982, Ford's Belgian factory took over the entire TW production program, and all models were updated in 1983. This was when the TW-30 received a power boost and a new identity as the TW-35. Radar speed measurement was added as an option.

△ **Ford 7810 Silver Jubilee**

Date 1989 **Origin** UK

Engine Ford 6-cylinder diesel

Horsepower 90 hp

Transmission 8 forward, 2 reverse

Ford's factory in Basildon, Essex, marked 25 years of production with a Silver Jubilee paint finish and an extra level of specification for a limited-edition version of the 7810. Part of the Generation III range, it was introduced as an upgrade to the Series 10 tractors.

△ **John Deere 3140**

Date 1980 **Origin** Germany

Engine John Deere 6-cylinder diesel

Horsepower 97 hp

Transmission 8 forward, 2 reverse

The 40 series maintained John Deere's success story during the early 1980s. Options for the 3140 were a four-wheel drive and replacement of the standard gearbox with the Power-Syncro, change-on-the-move transmission with 16 forward speeds.

▷ **John Deere 4240S**

Date 1982 **Origin** Germany

Engine John Deere 6-cylinder turbo diesel

Horsepower 132 hp

Transmission 16 forward, 6 reverse; Quad Range

The big batch of new John Deere tractor launches during 1983 included the 40 series economy range and a 352-hp tractor with articulated steering. The new 4040S and 4240S models, featuring Quad Range transmission, also arrived in that year.

▷ **John Deere 4250**

Date 1982 **Origin** US

Engine John Deere 6-cylinder turbo diesel

Horsepower 120 hp

Transmission 16 forward, 6 reverse; Quad Range

The new 4050 and 4250 were among the first tractors exported to the UK with a full powershift transmission offering forward and reverse shifting without using the clutch. As well as the general-purpose models, there were also high-clearance, row-crop versions.

▷ **Massey Ferguson 698T**

Date 1984 **Origin** France

Engine Perkins AT4.236 4-cylinder turbo diesel

Horsepower 88 hp

Transmission 12 forward, 4 reverse

The "T" in 698T stands for turbocharger, a device for boosting the power output and improving fuel efficiency of diesel engines. The 3.9-liter Perkins engine was special as it used the latest wastegate turbo, which was more effective at slow engine speeds.

▷ **Massey Ferguson 3065HV**

Date 1992 **Origin** France

Engine Perkins 4-cylinder diesel

Horsepower 85 hp

Transmission 32 forward, 32 reverse; powershuttle

HV stands for high visibility and refers to the downward-sloping hood line that improved forward visibility from the cab. The 3065HV and other MF3000 series tractors introduced a major development in tractor electronics with Datatronic and Autotronic control and information equipment.

△ **Massey Ferguson 9240**

Date 1994 **Origin** US

Engine Cummins 6-cylinder turbo diesel with intercooling

Horsepower 226 hp

Transmission 18 forward, 9 reverse; powershift

When Massey Ferguson needed a more powerful conventional tractor, it adopted a model from the White range, which, like MF, was part of AGCO. A further complication is the fact that the MF9240's rear axle was made by David Brown.

▽ **Massey Ferguson 4270**

Date 1997 **Origin** UK

Engine Perkins 6-cylinder turbo diesel

Horsepower 110 hp

Transmission 12 forward, 3 reverse; powershuttle

The 4200 series, announced in 1997 and built at MF's Banner Lane factory in Coventry, covered the power range from 52 hp to 110 hp. Design features included a control that automatically engaged the front axle differential lock when the rear differential lock was in use.

The French-built MF825 arrived in 1960

Great Manufacturers

Massey Ferguson

Massey Ferguson can trace its history back to a small Canadian workshop in 1847, long before the first tractors were invented. Today a part of the AGCO corporation, Massey Ferguson remains a leader in the development of agricultural technology and has an excellent record of striving to increase the efficiency of its tractors.

DANIEL MASSEY AND ALANSON HARRIS were among the early pioneers of Canada's farm machinery industry. Daniel Massey started his business in 1847 with a small workshop making tools and simple machinery for local farmers in Newcastle, Ontario. Ten years later, Alanson Harris made a rather more ambitious start in Brantford, Ontario, as the owner of a small forge.

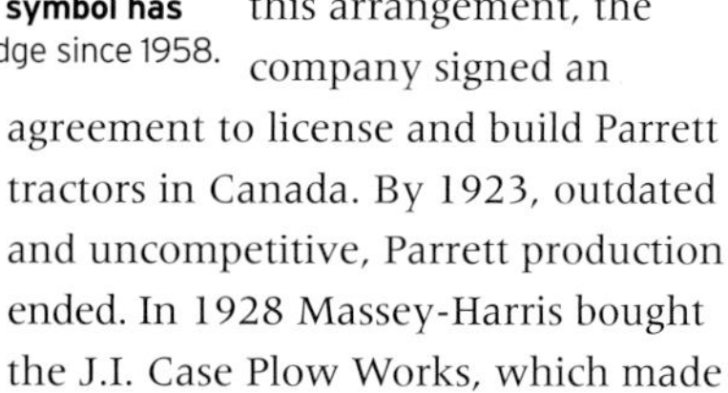

The Ferguson symbol has adorned MF's badge since 1958.

Agricultural supply was a crowded industry in the late 19th century, but Massey and Harris were more successful than most. Before they merged in 1891 to form Massey-Harris, they were the two biggest farm machinery manufacturers in Canada. The new firm was quick to develop exports and build on the companies' individual reputations. Massey machines had already appeared at an 1867 international show in France, where the Canadian firm won two gold medals and a batch of orders.

Yet Massey-Harris showed no interest in tractors until 1917, when they became the Canadian distributor for the bestselling three-wheeled US Bull tractor. When supply problems put an end to this arrangement, the company signed an agreement to license and build Parrett tractors in Canada. By 1923, outdated and uncompetitive, Parrett production ended. In 1928 Massey-Harris bought the J.I. Case Plow Works, which made the Wallis tractor, to replace it.

The well-designed Wallis established Massey-Harris as a successful tractor manufacturer.

More power
Sales leaflets for two of the models added in 1959 to offer more power to the US market. Both leaflets trumpet the tractors' ability to pull heavy implements.

In 1930 it became the first major company to build a four-wheel-drive tractor with large-diameter front and rear wheels, but the market was not yet ready for this type of tractor and sales were poor. Meanwhile, Massey-Harris continued to enjoy success in other areas of its farm machinery business, pioneering several major combine harvester developments, including the first models to use diesel power.

In 1953 Massey-Harris announced that it had bought Harry Ferguson's worldwide tractor business. The new company was initially called Massey-Harris-Ferguson, until it was shortened to Massey Ferguson in 1958.

Adding Ferguson's tractor production and marketing to the Massey-Harris machinery range placed the newly merged company at the forefront of the farm equipment industry. To strengthen its position in the marketplace, Massey Ferguson added to its arsenal by acquiring Perkins, the diesel engine specialists, in 1959.

After the Ferguson takeover, the company decided on a policy of not consolidating the Massey-Harris and Ferguson product lines or dealer networks, keeping both the gray-painted Fergusons and the bright red Massey tractors in production. This posed few problems in the UK; the smaller Fergusons did not compete against the more powerful British MH745, so it made sense to allow dealers to sell both the "gray" and "red" ranges. However, in North America the process was more difficult as product lines varied widely. New models were developed to give both sales groups access to similar tractors.

Global model
This MF165 was built at Massey Ferguson factories in the UK, France, and the US.

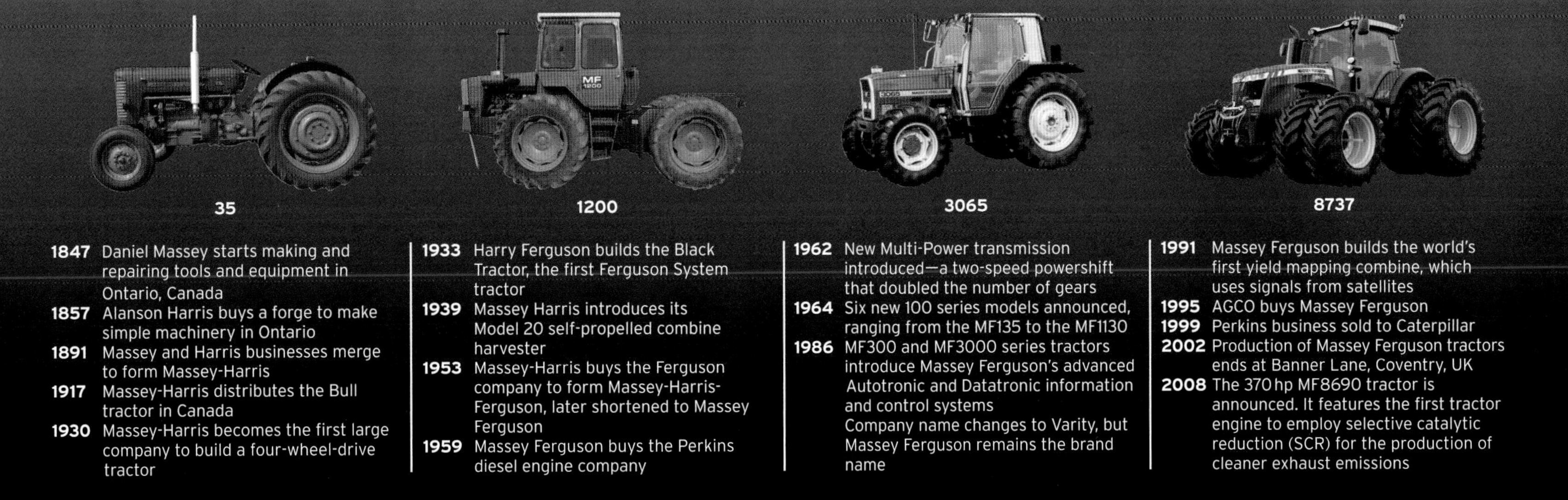

1847 Daniel Massey starts making and repairing tools and equipment in Ontario, Canada
1857 Alanson Harris buys a forge to make simple machinery in Ontario
1891 Massey and Harris businesses merge to form Massey-Harris
1917 Massey-Harris distributes the Bull tractor in Canada
1930 Massey-Harris becomes the first large company to build a four-wheel-drive tractor
1933 Harry Ferguson builds the Black Tractor, the first Ferguson System tractor
1939 Massey Harris introduces its Model 20 self-propelled combine harvester
1953 Massey-Harris buys the Ferguson company to form Massey-Harris-Ferguson, later shortened to Massey Ferguson
1959 Massey Ferguson buys the Perkins diesel engine company
1962 New Multi-Power transmission introduced—a two-speed powershift that doubled the number of gears
1964 Six new 100 series models announced, ranging from the MF135 to the MF1130
1986 MF300 and MF3000 series tractors introduce Massey Ferguson's advanced Autotronic and Datatronic information and control systems
Company name changes to Varity, but Massey Ferguson remains the brand name
1991 Massey Ferguson builds the world's first yield mapping combine, which uses signals from satellites
1995 AGCO buys Massey Ferguson
1999 Perkins business sold to Caterpillar
2002 Production of Massey Ferguson tractors ends at Banner Lane, Coventry, UK
2008 The 370 hp MF8690 tractor is announced. It features the first tractor engine to employ selective catalytic reduction (SCR) for the production of cleaner exhaust emissions

Examples of the resulting duplication included the Ferguson 40, introduced to provide "gray" dealers with a tractor in standard, row-crop, and tricycle versions. The MH50 was introduced so the "red" network had a Massey machine to match the Ferguson TO35. This expensive and complicated policy ended in 1957 when the two product lines were replaced by a uniform range of smaller and mid-sized tractors bearing red livery and the Massey Ferguson brand name.

However, Massey Ferguson still faced problems in the higher horsepower ranges, particularly in North America. Demand for more powerful machines was increasing fast and other manufacturers were responding quickly to win new customers. As a short-term solution, Massey Ferguson began buying 90 hp tractors from Minneapolis-Moline. These models were finished in Massey colors as the MF95, and a similar arrangement with Oliver produced the MF98 in 1959. The power gap soon filled with tractors developed by Massey Ferguson, including four-wheel-drive pivot-steer tractors such as 1970's popular MF1200. The 4000 series tractors continued the trend in 1978 with outputs up to 273 hp, they were equipped with electronic rear linkage control—a tractor industry first and an important step in the development of precision farming technology.

> "Designed to **actually replace horses**—under **any** soil conditions."
>
> MASSEY-HARRIS ADVERT AIMED AT SKEPTICAL FARMERS, 1930

Massey Ferguson maintained its technological leadership by introducing the electronically operated "Autotronic and Datatronic" information and control systems on the MF300 and MF3000 series in 1986. In 1991 the company helped power farming take a huge leap forward with the use of Global Positioning Systems (GPS) for combine harvester yield mapping. The technology was later rolled out to a wider range of tractor operations.

In 1995 Massey Ferguson became part of the AGCO conglomerate. From small beginnings, the Canadian firm is now AGCO's biggest brand, sitting alongside such famous names as Challenger, Fendt, Hesston, Sisu Power, and Valtra.

Operator focus
Improved "Comfort Capsule" cabs on 1970s MF500 models combined safety with reduced noise levels.

Reinventing the Plow

The increased productivity of the new breed of tractors that appeared at the end of the 20th century demanded high-capacity equipment. Changing attitudes toward arable operations brought an emphasis on soil conservation, lessening compaction, and minimal cultivation. This, combined with rising fuel costs, meant tractors had to cover as much ground as possible with as few passes of the field as necessary.

THE "PUSH-PULL" PLOW SYSTEM

A move toward minimum tillage at the end of the 1970s had seen plows go out of fashion, but problems with annual weeds led to the realization that turning the soil was essential for good husbandry. Plows enjoyed a revival during the 1980s, and new concepts in plowing included slatted, diamond, and square moldboards. Moves to make the larger multi-furrow models less unwieldy and more maneuverable led to articulated designs and the "push-pull" system. The latter consisted of a conventional plow hitched to the rear of the tractor and a second unit mounted on the front linkage to increase the number of furrows. Such systems were pioneered in the UK by two leading plow makers: Ransomes and Dowdeswell.

This six-cylinder 7810 tractor from Ford's 1989 Generation III range is equipped with a Ransomes TSR 300 Series "push-pull" plow.

North American Power

North American high-horsepower tractors were traditionally designed as two-wheel-drive machines, for working at high speeds with wide, shallow-working cultivation tools. At these power levels, very few farmers in the US and Canada operated the mounted equipment used on European farms, preferring to make use of hydraulic depth-control of trailed implements. That preference remained unchanged, but by the last two decades of the 20th century, more were specifying powered front axles. As a result, the number of their big models sold in Europe began to increase.

▷ **Case 3294**

Date 1984 **Origin** US

Engine CDC Case/Cummins 6-cylinder diesel

Horsepower 197 hp

Transmission 12 forward, 3 reverse

The 3294 was among the last tractors to be introduced by Case in its white/black livery, before parent Tenneco's purchase of International's agricultural division and subsequent adoption of the red/black colors. Production continued at the Racine, WI, factory after the merger. Unusually, the 3294 featured a full-time, four-wheel-drive system.

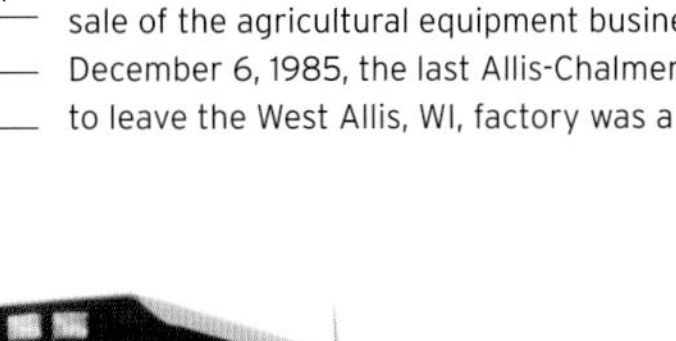

△ **Allis-Chalmers 6070**

Date 1985 **Origin** US

Engine Allis-Chalmers 4-cylinder diesel

Horsepower 80 hp

Transmission 12 forward, 3 reverse

For the 1981 sales season, Allis-Chalmers replaced its 175 and 185 tractors with the 6060 and 6080, later adding a 6070 variant. With the subsequent sale of the agricultural equipment business on December 6, 1985, the last Allis-Chalmers tractor to leave the West Allis, WI, factory was a 6070.

△ **Deutz-Allis 9150**

Date 1989 **Origin** US

Engine Deutz 6-cylinder diesel

Horsepower 155 hp

Transmission 18 forward, 9 reverse powershift

The 9100 series was the flagship North American market offering from Deutz-Allis for 1989, the last year before the buyout that formed AGCO. Comprising three models, the machines blended square US lines with hood styling that echoed the influence of German parent Klöckner-Humboldt-Deutz's tractor line.

△ **White 6195**

Date 1993 **Origin** US

Engine CDC Cummins 6-cylinder diesel

Horsepower 195 hp

Transmission 18 forward, 9 reverse powershift

White's four Workhorse 6100 series tractors shared a common platform with their "cousins" in parent firm AGCO's other major North American line of the time, AGCO Allis. The key difference was the White's use of CDC Cummins engines; the AGCO Allis equivalents featured Deutz air-cooled power.

◁ **White 125 Workhorse**

Date 1991 **Origin** US

Engine CDC Cummins 6-cylinder diesel

Horsepower 125 hp

Transmission 18 forward, 6 reverse

The 125 Workhorse was the smallest in a line of four new tractors introduced to the North American market by White in 1991. Manufactured at its plant in Coldwater, OH, the tractors featured three powershift steps in each of six gears.

◁ **White Field Boss 2-110**

Date 1982 **Origin** US

Engine Perkins 6-cylinder diesel

Horsepower 110 hp

Transmission 18 forward, 9 reverse

In 1982 White Farm Equipment (WFE) replaced its 2-105 tractor with the new Perkins-powered 2-110. A key feature of the transmission, with its six main forward gears, was a three-speed powershift, labeled Over/Under Hydraul-Shift by the maker. There was also a redesigned cab.

▽ **Case IH Magnum 7250 Pro**

Date 1997 **Origin** US

Engine CDC Case/Cummins 6-cylinder diesel

Horsepower 264 hp

Transmission 18 forward, 9 reverse powershift

In 1992 a new 7150 flagship was added to the original four-model range of Case IH 7100 series Magnum tractors. European farmers, though, had to wait until the improved 7200 series was introduced in 1994 to gain access to its successor, the 7250.

△ **International 5288**

Date 1981 **Origin** US

Engine International 6-cylinder diesel

Horsepower 180 hp

Transmission 18 forward, 6 reverse

The 5288 was the middle model in a set of three that included the 150 hp 5088 and the 205 hp 5488. The Tri-Six transmission allowed clutchless shifting between gears 1-2, 3-4, and 5-6, with other changes requiring the clutch but being synchronized. The range provided the basis for the Case IH 7100 Magnum tractors.

△ **AGCO Allis 9775**

Date 1998 **Origin** US

Engine Navistar 6-cylinder diesel

Horsepower 204 hp

Transmission 18 forward, 6 reverse powershift

With a short production run in 1998–99, the AGCO Allis 9775 was available with a choice of engines, either Cummins or Navistar. The latter was manufactured by the company formerly known, before it sold its farm equipment division, as International Harvester.

Bold Designs, Tough Times

The 1980s and early 1990s marked both the pinnacle and the beginning of the decline of British tractor manufacturing. Ford's Basildon base, Massey Ferguson's Coventry site, and Case International's Doncaster plant cemented their positions as important worldwide producers. By the late 1980s and early 1990s, though, factories such as the Case International (formerly David Brown) plant in Meltham were closed as the industry contracted. Meanwhile, most of the country's smaller domestic manufacturers failed to survive beyond the early 1990s.

◁ **International 885XL**
Date 1981 **Origin** UK
Engine International 4-cylinder diesel
Horsepower 85 hp
Transmission 8 forward, 4 reverse

When in 1981 International replaced its 84 series models with the 85 series, it was to be the last product launch under the marque for the Doncaster plant. The major development was the XL cab, which proceeded to have a long service life in subsequent products.

△ **Marshall 804**
Date 1982 **Origin** UK
Engine Leyland 4-cylinder diesel
Horsepower 82 hp
Transmission 9 forward, 3 reverse

Developed from the Leyland tractor range, which Lincolnshire farmer Charles Nickerson purchased from British Leyland in 1981, the 82 hp 804 was initially at the top end of the 02/04 series before the later introduction of the 904. The Sekura cabs were similar to those used by David Brown.

▽ **Roadless Amex Workhorse**
Date 1981 **Origin** UK
Engine Ducati 2-cylinder diesel
Horsepower 22 hp
Transmission 4 forward, 1 reverse

One of the last tractors that Roadless developed from scratch before the company's demise was the Amex Workhorse. Designed in 1981, it was intended as a low-cost machine for developing countries, where transportation made up much of the tractor's duties.

▽ **Marshall 100**
Date 1984 **Origin** UK
Engine Leyland 6-cylinder diesel
Horsepower 103 hp
Transmission 20 forward, 9 reverse

With new styling that gave the tractor's sheet steel some sharp lines, the Marshall 100, launched in 1984, looked markedly different from its predecessors. It was later joined by 115, 125, and 145 models, but these represented the company's manufacturing swan song.

△ **Roadless 120**
Date 1983 **Origin** UK
Engine Ford 6-cylinder diesel
Horsepower 130 hp
Transmission 8 forward, 4 reverse

Like fellow Ford four-wheel-drive (4WD) conversion specialists County and Muir-Hill, Roadless found the 1980s tough going. Being mainly focused on unequal-wheel 4WD, it was hit hardest by other tractor makers' entry into this sector. Production ceased in 1983, with its final tractors being purchased by British telephone company BT.

▷ County 1884

Date 1984 **Origin** UK
Engine Ford 6-cylinder diesel
Horsepower 188 hp
Transmission 16 forward, 8 reverse

Widely considered as the ultimate County tractor, the 1884 was the largest made by the Hampshire firm. Based on the Ford TW-35, it produced a rated 188 hp. While most were Q-cabbed, the last two 1884s produced were equipped with Ford's worklight-laden Super-Q cab.

▷ JWD 494 Fieldmaster

Date 1993 **Origin** UK
Engine Perkins 4-cylinder diesel
Horsepower 85 hp
Transmission 15 forward, 4 reverse

Following the demise of Marshall tractor production, the rights to produce the range were acquired by Lancashire-based former Marshall dealer John Charnley & Sons. A small number of the newly liveried Fieldmasters were made for a limited time in the late 1980s.

△ Case International 1594

Date 1988 **Origin** UK
Engine Case International 6-cylinder diesel
Horsepower 95 hp
Transmission 12 forward, 4 reverse

After the merger of the Case and International agricultural lines, parent firm Tenneco retained International's Doncaster factory and Case's Meltham plant. In 1988, though, the Meltham facility was closed and sub-100 hp production concentrated in Doncaster. Commemorative editions of the 72–108 hp 1394–1694 models marked the end of production.

◁ Case IH 4210

Date 1997 **Origin** UK
Engine Case IH 4-cylinder diesel
Horsepower 70 hp
Transmission 8 forward, 4 reverse

The influence of the 85 series tractor design launched by International in 1981 lasted right through to the Case IH 3200 and 4200 models. These were produced at the Wheatley Hall Road factory in Doncaster in the mid-1990s.

Harry Ferguson and Henry Ford with the Ford 9N

Great Manufacturers
Ford

The Ford tractor line began in 1939 with the 9N model, incorporating Harry Ferguson's hydraulic system. This little gray tractor captured the hearts and minds of American farmers as the "Blue Oval" expanded to encompass one of the largest agricultural machinery ranges in the world, with a massive global presence.

IN OCTOBER 1938 Harry Ferguson met with Henry Ford at Ford's Fair Lane residence near Dearborn, MI. The two businessmen shook hands over an agreement to collaborate on a "Ford Tractor with Ferguson System." The 9N model they built met with instant acclaim, introducing American farmers to an entirely new system of farming, and more than 300,000 were sold.

The split with Ferguson in 1946 was acrimonious, resulting in a lengthy legal battle over patent infringement that was finally resolved in 1952 after Ford settled out of court, paying Ferguson $9.25 million. Meanwhile, Ford tractor production continued in Highland Park, MI, with the 8N and subsequent NAA models, based on the 9N design but in a new red and gray livery.

> "The **largest** engineering project **ever undertaken** by any **tractor manufacturer.**"
>
> JOHN FOXWELL, CHIEF ENGINEER, FORD TRACTOR OPERATIONS, ON THE 6X, 1964-75

In 1955 Ford abandoned its one-model policy and launched a range of five new tractors in two power classes. Further options, including diesel engines, were added as the range was extended still further. A six-cylinder model, the Ford 6000, arrived in 1961, but its reputation was tarnished by reliability issues.

A new blue and gray color scheme was introduced in 1962 to herald the integration of the company's global tractor operations under a single Ford banner. The British Fordson line built in Dagenham would be swept away and replaced by a new "Worldwide" 6X range in simultaneous production in Basildon, England, in Antwerp, Belgium, and in Highland Park.

More than $90 million and one million man-hours of engineering effort were invested in the 6X program, resulting in the new Ford 2000, 3000, 4000, and 5000 tractors, unveiled in a blaze of publicity in New York on October 10, 1964. Four all-new tractors produced in largely untried facilities meant teething problems were inevitable, and the 6X range suffered from more than its fair share of failures. However, any failings were addressed by the launch of the greatly improved 6Y Ford Force range in 1968.

Ford was the rising star of the tractor industry during the 1970s as it consolidated its global presence with a worldwide sales and service network that was the envy of its competitors. New introductions during this period included the turbocharged Ford 7000, and a range of heavyweight, six-cylinder tractors that culminated in the launch of the prestigious TW series in 1979.

The top and bottom ends of the range were strengthened by manufacturing agreements with

NEW! FORD DIESEL TRACTORS

The great new FORD TRICYCLE TRACTORS for easier... better... safer... row crop farming

Ford offers more
During the 1950s, Ford's US tractor line expanded to include a range of different options: utility, all-purpose, and row-crop and diesel models were available in two different power classes.

1939 Ford 9N, named for the first year of its production, goes into production at the Rouge plant
1945 Highland Park becomes the main production center for Ford tractors
1947 Ford 8N tractor enters production
1948 Ferguson files lawsuit against Ford for patent infringement
1952 Ford settles out of court with Ferguson for $9.25 million
1953 NAA tractor introduced in year of Ford's golden jubilee

1957 Ford demonstrates its experimental gas-turbine Typhoon tractor
1958 Diesel engines offered for first time on American Ford tractors
1961 Ford Tractor Operations created to coordinate worldwide undertakings
1964 Worldwide 6X range unveiled at Radio City Music Hall in New York
1970 First Ford compact tractor supplied by Shibaura of Japan
1971 Launch of 7000, Ford's first turbocharged tractor

1974 US Ford tractor production transfers to Romeo, MI
1977 FW range, built in conjunction with Steiger, introduced
1979 TW series of high-horsepower tractors go into production in Antwerp and Romeo
1981 Romeo builds Ford's five-millionth tractor
1984 Ford produces 100,000 of the 730,000 tractors sold worldwide for a 13 percent global market share

1985 Ford acquires New Holland from Sperry Corporation
1987 Ford Tractor Operations renamed Ford New Holland and takes over Versatile; Romeo plant closes and US production is transferred to Basildon, England
1989 Basildon commemorates 25 years of production with limited-edition 7810 Silver Jubilee tractor
1991 Fiat acquires Ford New Holland and merges it with its Fiatagri division
1994 Ford name dropped from tractors

Global product, left
The Ford 6X range was designed for worldwide distribution, and more than 70 percent of the tractors built in Basildon were exported. Most were shipped by sea from the company's wharf in Dagenham.

Little and large, right
By 1979 Ford offered a tractor in almost every power class. The big articulated FW series was built for the company by Steiger in the US, while the compact tractors were sourced from Shibaura of Japan.

Steiger of North Dakota and Shibaura of Japan. Steiger supplied the FW series of articulated giants, while Shibaura built Ford's compact tractors. By 1981 the organization was offering more than 20 different tractor models in the UK alone—and that is without including all the industrial equipment. New ranges for the 1980s included the Series 10, Force ll, and Generation lll models, with greater sophistication and comfort.

By 1984 Ford was in second place in the overall tractor rankings after Massey Ferguson, and had increased its share of the North American market to take the third spot after Deere and International. It was in an unassailable position in the UK and was the market leader for the twelfth year in succession. However, a severe downturn in the world economy was affecting profits, and rumors were beginning to abound that the Ford Motor Company was preparing to divest its tractor business.

Ford felt that its tractor division would be a more attractive proposition for any prospective purchaser if it could be turned into a full-line equipment manufacturer with a range of machinery to rival that of MF, International, and Deere. In October 1985 it purchased New Holland from the Sperry Corporation. The tractor business, which now operated with a degree of autonomy from the parent company as Ford New Holland, was bolstered further by the acquisition of Versatile in 1987.

That same year, the Genesis project was inaugurated to provide the next generation of engines for an entirely new range of Ford tractors, the Series 40. By the time these new machines went into production in Basildon in October 1991, Ford New Holland was under new ownership and had been acquired by Fiat. In the years that followed, the Ford and Fiatagri agricultural machinery ranges were integrated under the New Holland flag as a new global organization was born.

Mighty power
The turbocharged, six-cylinder engine powering the TW-35 tractor from Ford's 1983 TW series developed 195 hp. Air-to-air intercooling, an exclusive feature, cooled the air entering the cylinders to accept a greater volume of fuel and increase power.

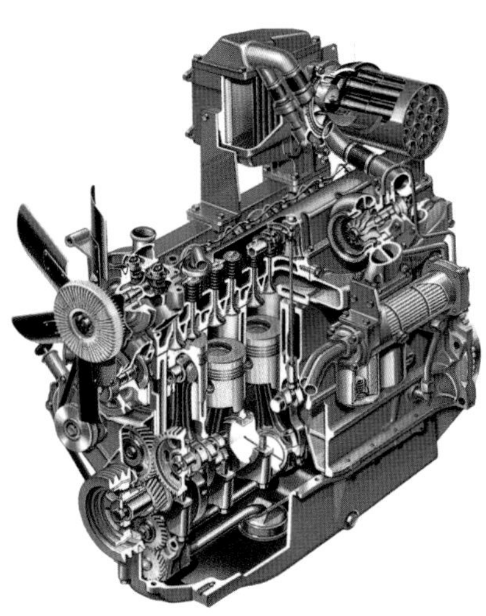

Continental Developments

As in the UK, agriculture across much of Europe rode out the recession of the early 1980s remarkably well, and so did its tractor plants. Countries such as France, Germany, Austria, Italy, and Finland continued to churn out new models that made farming faster and simpler—resulting from developments in areas such as electronics, transmissions, and operator comfort. Four-wheel-drive systems, long the preserve of after-market suppliers, were offered by the manufacturers themselves, while average horsepower increased as higher-capacity implements were developed and reversible plows became the norm.

▷ **Renault 155-54 Nectra**

Date 1991 **Origin** France
Engine MWM 6-cylinder diesel
Horsepower 145 hp
Transmission 16 forward, 16 reverse

Alongside Ford and Fiat, Renault was one of a handful of companies that had maintained involvement in tractors, cars, and commercial vehicles throughout most of the 20th century. One of its key developments was in the field of operator comfort, with the creation of its hydrostable cab suspension system.

△ **Fendt Farmer 310LS Turbomatik**

Date 1985 **Origin** Germany
Engine Fendt 4-cylinder diesel
Horsepower 94 hp
Transmission 15 forward, 4 reverse

Fendt's Farmer models featured the company's Turbomatik system, which combined a pump impeller and a turbine to provide power takeup between engine and transmission. The former injected oil into the "compartments" of the latter to provide smooth power transfer as the clutch was released and the engine sped up.

▷ **Case International 1056XL**

Date 1992 **Origin** Germany
Engine Case International 6-cylinder diesel
Horsepower 105 hp
Transmission 16 forward, 8 reverse

In 1985 the German-made 856, 956, and 1056 tractors, launched by International in 1981, gained new paint and then a new nose after the merger between Case and International. Production continued until 1992, when they were superseded by the Maxxum 5100 series.

△ **Fiat Winner F110**

Date 1990 **Origin** Italy
Engine Fiat 6-cylinder diesel
Horsepower 110 hp
Transmission 16 forward, 8 reverse

Not long before it acquired Ford's agricultural equipment business, Fiat launched a completely redesigned set of 100–140 hp tractors, collectively named the Winner series. They used a completely new cab, which did away with the angular lines found on the 80/90 series.

▷ **Steyr 8130A Turbo**

Date 1992 **Origin** Austria
Engine Steyr 6-cylinder diesel
Horsepower 120 hp
Transmission 18 forward, 6 reverse

Born out of an Austrian munitions manufacturer, the Steyr tractor business established a strong following for the quality of its products. The business and its factory were purchased by Case Corp. in 1996, and today the plant still produces Steyr and Case International tractors.

△ **Deutz-Fahr AgroXtra DX6.08**

Date 1991 **Origin** Germany
Engine Deutz 6-cylinder diesel
Horsepower 107 hp
Transmission 48 forward, 12 reverse

During the late 1980s and early 1990s a number of European farmers chose to modify their tractors to improve forward vision—a move that reflected growing use of front linkages. Recognizing this trend, Deutz-Fahr led the way among manufacturers in adopting the drop-nose design, which suited its air-cooled engined tractors.

▷ **Zetor 9540**

Date 1992 **Origin** Czech Republic
Engine Zetor 4-cylinder diesel
Horsepower 95 hp
Transmission 18 forward, 6 reverse

In 1992 Zetor launched the results of the first major redesign of its mid-sized tractors, with its 40 series machines. Beginning with the 75 hp 7540, the line was later expanded to three larger models, including the 9540.

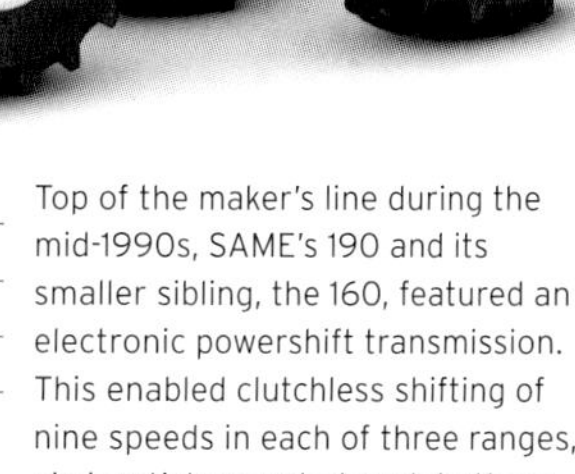

△ **SAME Titan 190**

Date 1994 **Origin** Italy
Engine SAME 6-cylinder diesel
Horsepower 189 hp
Transmission 27 forward, 27 reverse

Top of the maker's line during the mid-1990s, SAME's 190 and its smaller sibling, the 160, featured an electronic powershift transmission. This enabled clutchless shifting of nine speeds in each of three ranges, via joystick-mounted push buttons.

▽ **Valmet 8400**

Date 1995 **Origin** Finland
Engine Valmet 6-cylinder diesel
Horsepower 140 hp
Transmission 36 forward, 36 reverse

The Volvo-BM Valmet name, which had adorned the merged companies' tractors since soon after the 1979 association of their tractor businesses, had become simply "Valmet" by the later part of the decade. Despite ongoing updates, the machines still retained their distinctive, forward-nose design.

Around the World

Away from the multinational farm equipment companies, a number of regionally focused manufacturers concentrated on creating tractors primarily to suit the needs of farmers in their countries and regions. From basic, low-horsepower machines to work on small farms, to tractors built for a country's particular type of broad-acre farming, some units went on to be sold worldwide, while others remained peculiar to their country of origin. In particular, the sub-40hp compact tractors made primarily by Japanese, South Korean, Indian, and Chinese firms enjoyed a good deal of success.

◁ **Chamberlain 4480B**

Date 1982 **Origin** Australia
Engine John Deere 6-cylinder diesel
Horsepower 119hp
Transmission 12 forward, 4 reverse

Founded in 1947, Chamberlain tractor production was born out of a former munitions factory in the western Australian town of Welshpool. John Deere purchased 49 percent of Chamberlain in 1970, then bought out the company completely. The 4480B was the largest model in a 68–119hp range.

△ **Chamberlain 4490**

Date 1984 **Origin** Australia
Engine John Deere 6-cylinder diesel
Horsepower 190hp
Transmission 18 forward, 18 reverse powershift

Chamberlain tractors were given John Deere colors, while retaining Chamberlain branding, model numbers, and certain distinctive styling. The Chamberlain business was absorbed fully into John Deere, and German and US-made machines took the place of the Australian ones.

△ **CBT 8060**

Date 1989 **Origin** Brazil
Engine Perkins 4-cylinder diesel
Horsepower 60hp
Transmission 8 forward, 2 reverse

Companhia Brasileira de Tratores, based in the Brazilian city of São Carlos, began producing tractors in 1960. Its machines traditionally used a frame design in which the engine and transmission are cradled, meaning that major servicing did not require the tractor to be split.

△ **Mahindra 265 DI**

Date 1995 **Origin** India
Engine Perkins 3-cylinder diesel
Horsepower 45hp
Transmission 8 forward, 2 reverse

Mahindra is one of the world's largest tractor producers. Its chief market is its vast domestic one, but this Indian firm also exports globally. This 265 DI inherits its looks from International's B-250/275.

△ **Valmet 1780 Turbo**

Date 1989 **Origin** Brazil
Engine Valmet 6-cylinder diesel
Horsepower 170hp
Transmission 10 forward, 2 reverse

In 1960, following invitations from major manufacturers to pitch for permission to establish a domestic tractor factory, the Brazilian government selected Valmet as its preferred partner. The Finnish firm set up a plant at Mogi das Cruzes, close to São Paulo, where this 1780 Turbo was introduced in 1989.

△ **Belarus 1522**

Date 2000 **Origin** Belarus
Engine Belarus/MTZ 6-cylinder diesel
Horsepower 155hp
Transmission 16 forward, 8 reverse

Formerly the major tractor-producing factory of the USSR, the plant in Minsk, in the now-independent state of Belarus, remains the dominant tractor manufacturing facility among the former Soviet Republic states.

◁ **Kubota L3250**

Date 1989 **Origin** Japan
Engine Kubota 3-cylinder diesel
Horsepower 35hp
Transmission 8 forward, 2 reverse

Compact tractors such as this L3250 shown with tires for turf applications were the focus of Kubota's range. One of the world's major producers of this type of tractor, Kubota gradually increased the power of its larger machines to compete more fully in the agricultural sector.

◁ **Daedong D55**

Date 1996 **Origin** South Korea
Engine Daedong 3-cylinder diesel
Horsepower 55hp
Transmission 8 forward, 2 reverse

The South Korean firm Daedong initially exported its tractors under its own name, before creating the Kioti brand to appeal to a wider audience of buyers across other markets. Like many East Asian companies, it tended to focus on the sub-100hp sector.

△ **TAFE 45 DI**

Date c.1995 **Origin** India
Engine TAFE 3-cylinder diesel
Horsepower 45hp
Transmission 8 forward, 2 reverse

The Indian firm Tractors And Farm Equipment (TAFE) has had a longstanding relationship with Massey Ferguson, and has for many years used the multinational brand's components as the basis for its tractors. The internals of the 45hp 45 DI trace their heritage to the popular MF135.

Rubber Tracks Arrive

The arrival of modern rubber-tracked tractors was the culmination of many years of design and experimentation. Several unsuccessful attempts were made from the 1920s onward to combine the mobility of the rubber tire with the high tractive effort of the steel track-type tractor, mainly due to the limits of rubber technology. The breakthrough came when NASA developed rubber track for vehicles to explore the surface of the Moon—these tracks were built from continuous steel wires with rubber molded around them.

△ Waltanna 200 High Drive

Date 1990 **Origin** Australia
Engine Cummins 6-cylinder diesel
Horsepower 225 hp
Transmission Hydrostatic

James Nagorcka founded Waltanna to produce high-horsepower tractors after he had built and exhibited one at a local show, where he received requests to build further examples. He went on to create a range of conventional-looking tractors, and later modified the design for this Hi Drive model.

△ Caterpillar Challenger 65

Date 1987 **Origin** US
Engine Caterpillar 3306 6-cylinder diesel
Horsepower 256 hp
Transmission 10 forward, 2 reverse, full powershift

The Challenger 65 was the first successful rubber-track tractor. The rubber-track components were tested over a period of years using both grader tractive units and modified steel-tracked tractors. Early problems with the friction drive were solved by a rebuild program.

▽ Caterpillar Challenger 85D

Date 1997 **Origin** US
Engine Caterpillar 3196 6-cylinder diesel
Horsepower 370 hp
Transmission 10 forward, 2 reverse

Caterpillar continued to develop and widen its range of rubber-tracked tractors. The earlier Challenger 65 used an inflatable rubber tire as the front idler in the track system, which could puncture. On later models such as the 85D this was replaced by a steel idler with a bonded rubber tread. This updated feature was also necessary to cope with the very high belt tension required to transmit the ever-increasing horsepower ratings.

△ **Track Marshall TM200**

Date 1991 **Origin** Australia

Engine Cummins 6CT8.3

Horsepower 210 hp

Transmission Hydrostatic

The Track Marshall 200 was built by Waltanna in Australia with a few minor modifications for the British market. The most obvious was the color change from white to yellow—full yellow was only used on the first few machines. The two-tone yellow and black color system seen here replaced the all-yellow version.

△ **John Deere 8400T**

Date 1998 **Origin** US

Engine John Deere 6-cylinder diesel

Horsepower 235 hp

Transmission 16 speed, full powershift

The 8000T series shared their transmissions and engines with the 8000 series of wheeled-tractors, many parts being interchangeable. This was the first time for nearly 30 years that John Deere offered both wheeled and tracked versions of the same tractors.

▽ **Volgograd BT-100**

Date 1994 **Origin** Russia

Engine Volgograd 4-cylinder diesel

Horsepower 101 hp

Transmission 7 forward, 1 reverse

The rubber-track revolution was embraced by Russian agriculture as well as the West. Volgograd took the latest model of its steel-tracked tractor and replaced the metal tracks with positively driven rubber belts. Little modification to the tractor itself was required; provided the drawbar loads were kept within reason, the system worked.

◁ **Morooka MK220**

Date 1995 **Origin** Japan

Engine Cummins 6-cylinder diesel

Horsepower 220 hp

Transmission Hydrostatic

This was yet another attempt to place a rubber-tracked crawler on the market. Morooka used a positive-drive system with the sprocket driving from the front of the track for all its machines, and this tractor was no exception. The MK220 was very useful where the requirements were moderate draft and low ground pressure.

▷ **Claas Challenger 45**

Date 1997 **Origin** US

Engine Caterpillar 3116 6-cylinder diesel

Horsepower 242 hp

Transmission 16 forward, 9 reverse, full powershift

Apart from the livery, the Claas Challenger range was identical to the Caterpillar Challenger range. The Challenger 35, 45, and 55 models were the first in the Caterpillar line to use a large-diameter rear-drive wheel. Although the same tractors, the green-painted models never enjoyed the same popularity as their counterparts in Caterpillar colors.

Case IH Quadtrac

As a fully articulated, tracked machine, the Case IH Quadtrac was very unusual in its class. When other manufacturers began to offer rubber-tracked tractors, Case IH looked for a way to place one of their own on the market. To save the cost of developing a full-track machine, the company resolved to offer tracks as an option on their already-tried-and-tested articulated four-wheel-drive models.

IN JANUARY 1987, Case IH acquired the Steiger tractor manufacturing company in Fargo, ND. With this acquisition came Steiger's acclaimed line of articulated, four-wheel-drive tractors—the Midwestern company being one of the pioneers of this type of high-horsepower machine. Case IH continued to develop the product, renowned for its rugged simplicity and reliability. Its "bend-in-the-middle" type of steering eliminated the need for the complex hydraulic-over-mechanical, controlled-differential steering generally required on a full-track machine. The Quadtrac's rubber tracks were positive-drive, which allowed the tractor to operate in virtually all agricultural conditions.

Versatile power
The 9370's principal virtue was its raw strength. The Cummins N14 engine was a benchmark for well-priced, reliable power with comparatively easy maintenance. The high overall cost of the Quadtrac system was more than offset by its ability to cover acres.

FRONT VIEW

REAR VIEW

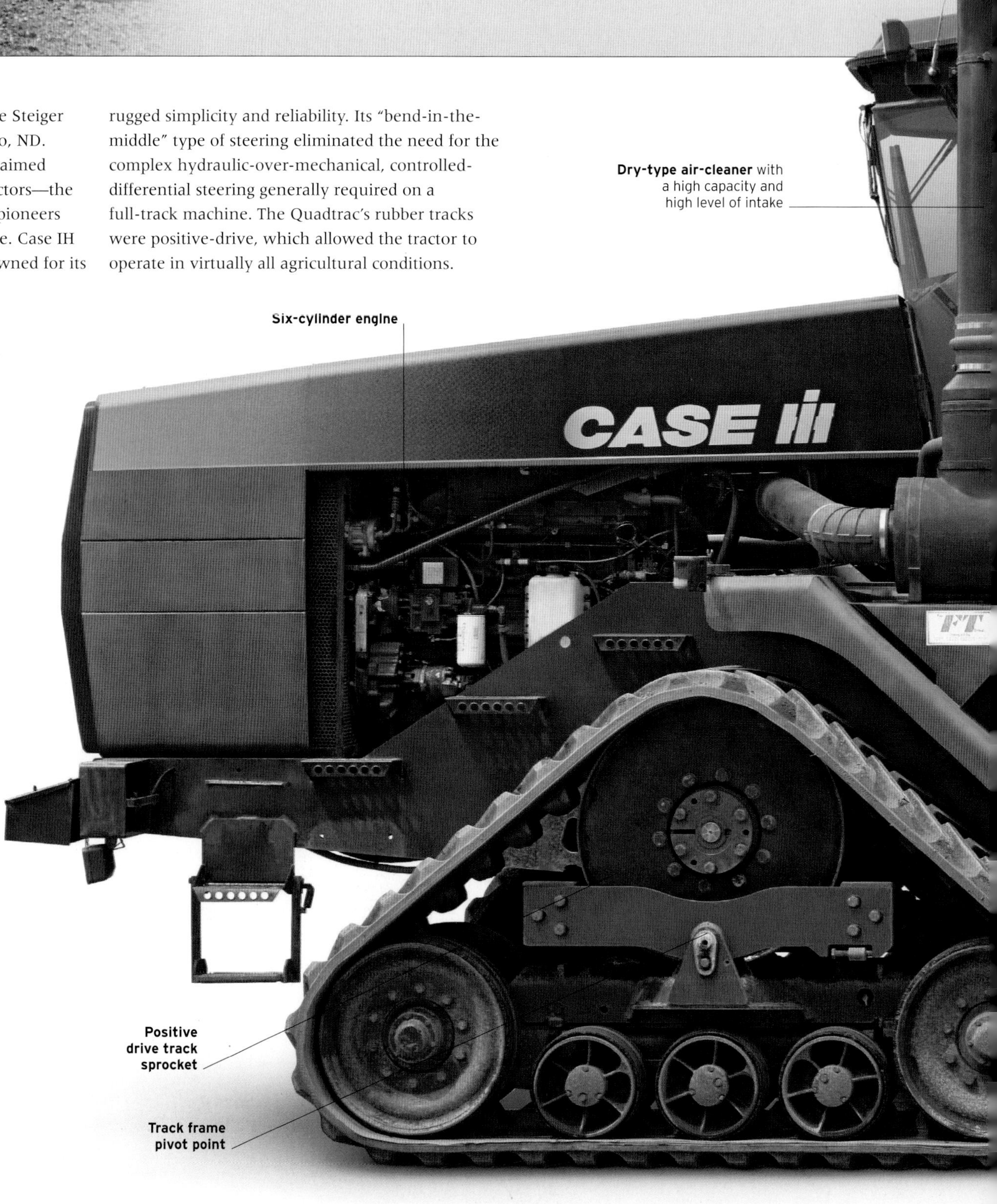

Dry-type air-cleaner with a high capacity and high level of intake

Six-cylinder engine

CASE IH

Positive drive track sprocket

Track frame pivot point

SPECIFICATIONS	
Model	Case IH Quadtrac 9370
Built	c. 1997
Origin	US
Production	Unknown
Engine	360 hp Cummins 6-cylinder turbocharged and aftercooled diesel
Capacity	855 cu in (14,000 cc)
Transmission	12 forward, 3 reverse
Top speed	18.7 mph (30.1 km/h)
Length	19 ft 6 in (5.95 m)
Weight	22 tons (20 metric tons)

THE DETAILS

1. Quadtrac motif 2. Cummins 360-hp turbocharged and aftercooled diesel engine 3. Driver's instrument panel and steering wheel 4. Tractor functions display gauges 5. Gearbox, hydraulic, and throttle controls 6. Steel-cable-reinforced rubber track

1

2

3

4

5

6

Driver's cabin is spacious, air-conditioned, and dust-free

Hydraulic, three-point linkage complete with Quic-tach attachment

Steering pivot point

Steering control hydraulic ram

Big Wheelers

The UK's four-wheel-drive tractor success started with smaller specialty companies, while the biggest manufacturers waited for the market to develop. There was a similar trend in the US and Canada when high-horsepower four-wheel drives with articulated steering were beginning to attract more customers. It was left to small companies such as Big Bud, Steiger, and Versatile to meet the initial demand—the big companies followed later. North American manufacturers dominated the high-horsepower tractor market, but large numbers were also built in Russia, with small-scale production elsewhere, including Australia and Germany.

▷ **Rome 475C**

Date 1978 **Origin** US
Engine Caterpillar V8 diesel
Horsepower 475 hp
Transmission 12 forward, 2 reverse

Rome Plow Co. is a long-established farm implement manufacturer that built high-horsepower tractors from 1978 until 1984. The 475C tractor was part of a four-model range powered by Caterpillar and Cummins engines with outputs from 375 hp to 600 hp.

△ **International 3788**

Date 1979 **Origin** US
Engine International 6-cylinder
Horsepower 170 hp
Transmission 12 forward, 6 reverse

The New 88 series models with articulated steering announced by International in 1979 earned the nickname "Snoopy" because of their unusual front-end appearance. Placing the cab near the rear, behind the hinge point, created an extra-long front section, giving a powerful appearance.

△ **Versatile 935**

Date 1980 **Origin** Canada
Engine Cummins V8 diesel
Horsepower 330 hp
Transmission 12 forward, 4 reverse

The 935 was one of a batch of new models announced in 1978, all powered by Cummins engines. The range was known as the Labor Force and the 935 was the most powerful of the five new models. Versatile was bought by the Ford New Holland organization in 1987.

△ **Baldwin DM525**

Date 1979 **Origin** Australia
Engine Cummins KTA 6-cylinder diesel
Horsepower 525 hp
Transmission 12 forward, 4 reverse

The DM525 built in 1979 with a mechanical gearbox was the Baldwin family's first production tractor. A powershift transmission version called the DP525 was available as well. Baldwin also built a 600 hp model, said to be Australia's most powerful tractor.

◁ **Big Bud 500**

Date 1985 **Origin** US
Engine Komatsu 6-cylinder diesel
Horsepower 500 hp
Transmission 12 forward, 2 reverse

Big Bud's reputation was established at the top end of the high-horsepower tractor market with models to suit the biggest farms. The standard specification for the Big 500 model included a Komatsu engine, but a 19-liter Deutz power unit was listed as an option.

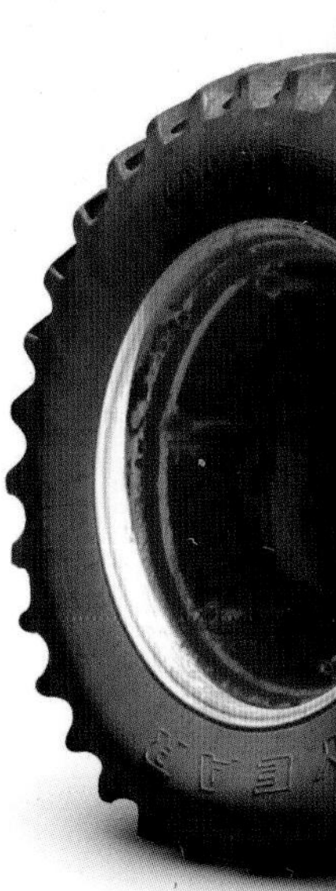

◁ **ACO 600**

Date 1989 **Origin** South Africa
Engine ADE V12 diesel
Horsepower 820 hp
Transmission 12 forward, 4 reverse

ACO was a privately owned company established in 1986 to build mainly high-horsepower tractors, including the ACO 600 model. Sales peaked in the early 1990s, but falling demand brought a change of ownership in 1999 and production ended a few years later.

△ **Case IH Quadtrac 9370**

Date 1997 **Origin** US
Engine Cummins 6-cylinder turbocharged and aftercooled diesel
Horsepower 360 hp
Transmission 12 forward, 3 reverse

Big articulated tractors in the Case range were built by Steiger and included the 9370, available with four-wheel drive or as a Quadtrac version mounted on independently suspended rubber tracks. The tracks helped spread the tractor's weight to reduce soil compaction.

▽ **Ford Versatile 9680**

Date 1993 **Origin** Canada
Engine Cummins 6-cylinder
Horsepower 325 hp
Transmission 12 forward, 4 reverse

New Holland's new Ford Versatile 80 series tractors were launched in late 1993. The 9680 model was the replacement for the earlier 946 tractor, which was powered by a 350 hp Cummins NTA-855-A engine.

△ **John Deere 8770**

Date 1993 **Origin** US
Engine John Deere 6-cylinder diesel
Horsepower 300 hp engine and 259 hp
Transmission 12 forward, 3 reverse

John Deere is one of the big success stories in the high-horsepower sector, and the four 70 series models contributed to this. The 8770 offered 300 hp but the other models in the series provided 250, 350, and 400 hp engine outputs.

JCB Fastrac

The JCB Fastrac was built in response to changing agricultural trends that saw farm tractors spending more than half their working time on transportation tasks. In 1986, JCB started developing a machine that could operate safely at high speeds on the road and carry out traditional field tasks. Launched in 1991, the line continues today.

THE FASTRAC 185 was extremely versatile. Its conventional front and rear linkages allowed the use of standard tractor implements, and tools weighing up to 6,600 lb (3,000 kg) could be mounted on the load platform behind the cab—ideal for heavier items such as agricultural crop sprayers. Self-leveling rear suspension maintained weight distribution between the front and rear axles, allowing improved traction and handling. With a powerful engine and equal-sized traction wheels, the 185 was equally comfortable carrying out heavy work in the field or rapidly transporting heavy loads.

Most sales came from large farms and agricultural contractors, but the Fastrac was also popular for municipal work as well as forestry and airport maintenance.

Fast and flexible
The Fastrac 185 was designed for speed and maneuverability. Its 170 hp engine, six-gear all-synchromesh gearbox, and two-speed splitter meant the machine could pull away with a plow in the ground or towing a heavy trailer. Its standard tires were Michelin 495/70R24, rated for a full 47 mph (75 km/h).

185 hp Cummins engine provides power and speed

Front hitch lifts up to 6,600 lb (3,000 kg)

Tires carry heavy loads over uneven ground

FRONT VIEW

REAR VIEW

THE DETAILS

1. Coil spring front suspension with telescopic shock absorbers 2. External disc brakes on all four wheels ensured powerful stopping power 3. View of dashboard 4. Hydraulic spool valve controls 5. Electronic controls for two- or four-wheel drive and differential locks, power takeoff, and hydraulic services

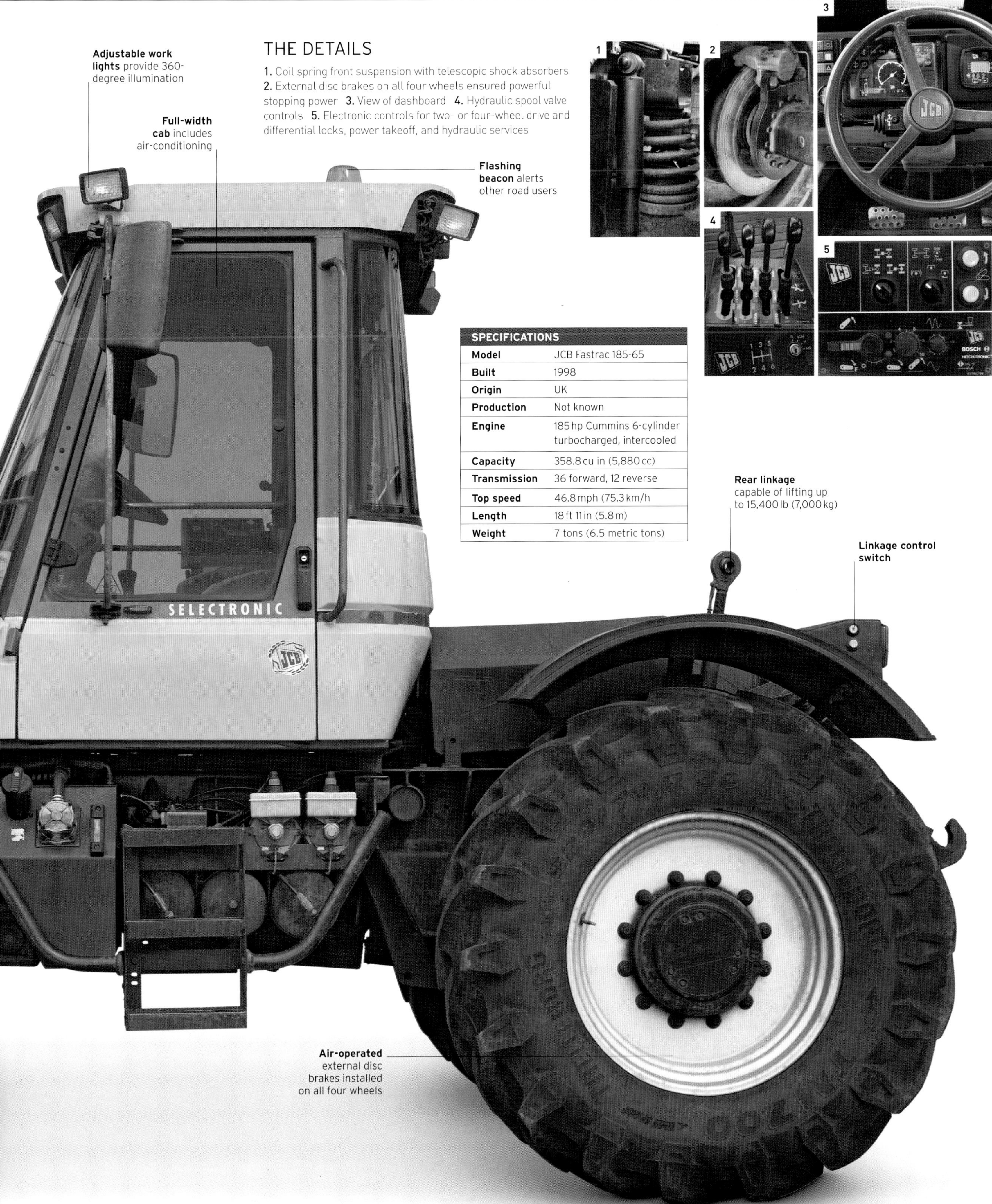

SPECIFICATIONS	
Model	JCB Fastrac 185-65
Built	1998
Origin	UK
Production	Not known
Engine	185 hp Cummins 6-cylinder turbocharged, intercooled
Capacity	358.8 cu in (5,880 cc)
Transmission	36 forward, 12 reverse
Top speed	46.8 mph (75.3 km/h)
Length	18 ft 11 in (5.8 m)
Weight	7 tons (6.5 metric tons)

Multipurpose Machines

Engineers have long sought to design multipurpose tractors or "systems" machines that can carry out two or three field operations in a single pass, enabling farmers to save time, fuel, and labor. These often combine the ability to hitch implements to both the front and rear of the machine, with a load platform on the tractor itself. Other features often include high road speeds and the ability to operate in reverse with the driver facing forward, for easier operation on some tasks.

▷ Steyr 8300

Date 1982 **Origin** Austria
Engine Steyr 6-cylinder diesel
Horsepower 245 hp
Transmission Hydrostatic

Produced from 1982 to 1987, the Steyr 8300 had an offset cab with reversible driving position, but was primarily designed to be operated in reverse drive with mounted equipment such as forage harvesters. It was replaced by an improved 8320 model before production ended in 1993.

▷ Trantor Mk II

Date 1985 **Origin** UK
Engine Leyland 4-cylinder diesel
Horsepower 80 hp
Transmission 10 forward, 2 reverse

The first Trantor was designed in response to a 1970s survey of farmers' tractor use, which showed that almost three-quarters of their time was spent on hauling and transportation tasks. Later Mark II models featured engines of up to 92 hp and a revised cab design.

△ BIMA 360

Date 1988 **Origin** France
Engine Caterpillar 6-cylinder diesel
Horsepower 360 hp
Transmission Hydrostatic

Introduced in 1983, the French-built BIMA tractor differed from most other articulated high-horsepower machines in the way it steered, with its articulation point underneath the cab and the engine to the rear. Three-point linkages and power takeoffs were installed at both ends.

◁ Mercedes-Benz MB-trac 1000

Date 1985 **Origin** Germany
Engine Mercedes-Benz 6-cylinder diesel
Horsepower 95 hp
Transmission 16 forward, 8 reverse

Mercedes-Benz introduced its MB-trac model in 1972, with an unusual and distinctive combination of equal-sized wheels and mid-mounted cab, plus a load platform to the rear for equipment such as sprayers. Later it launched larger models with a reverse drive feature; manufacturing ended in 1991.

△ **Clayton C4105**

Date 1994 **Origin** UK
Engine John Deere 4-cylinder diesel
Horsepower 110 hp
Transmission 10 forward, 2 reverse or hydrostatic

Launched in 1992, the C4105 Buggi was created as a load-carrying platform vehicle, designed primarily for use with demountable sprayers and spreaders. Optional rear linkage, hydraulic outlets, power takeoff, and drawbar increased its versatility.

△ **JCB Fastrac 185-65**

Date 1998 **Origin** UK
Engine Cummins 6-cylinder diesel
Horsepower 185 hp
Transmission 36 forward, 12 reverse

Three years after the Fastrac became fully commercially available, manufacturer JCB launched its biggest version to date, with the release of the 185-65. Powered by a Cummins 5.9-liter engine, the tractor was otherwise similar in format to the smaller, existing 135-65 and 155-65 models.

▷ **Fendt Xylon 524**

Date 1990 **Origin** Germany
Engine MAN 4-cylinder diesel
Horsepower 140 hp
Transmission 44 forward, 44 reverse

For many years Fendt had been producing tool-carrier tractors—those with engines sited underneath the cab and a clear platform out front for mounting hoes or sprayer tanks. In 1990 it developed the concept and created the mid-cabbed, equal-wheeled Xylon. Production was relatively short-lived.

△ **Moffett MFT 7840**

Date 1995 **Origin** Ireland
Engine New Holland 6-cylinder diesel
Horsepower 100 hp
Transmission 16 forward, 16 reverse

The Moffett 7840 was a special conversion of a standard Ford New Holland 7840 that allowed the driver to rotate the tractor's controls and operate it in reverse. Key benefits included unimpeded vision and rear-wheel steering when used with the firm's own rear loader.

▷ **Claas Xerion 2500**

Date 1996 **Origin** Germany
Engine Caterpillar 6-cylinder diesel
Horsepower 250 hp
Transmission Continuously variable transmission

The Claas Xerion was originally designed as a multipurpose vehicle capable of accommodating implements front and rear and on its load platform. Front and rear linkages and a movable cab allowed both operator and implement to be sited in the ideal position.

After 2000
21ST CENTURY

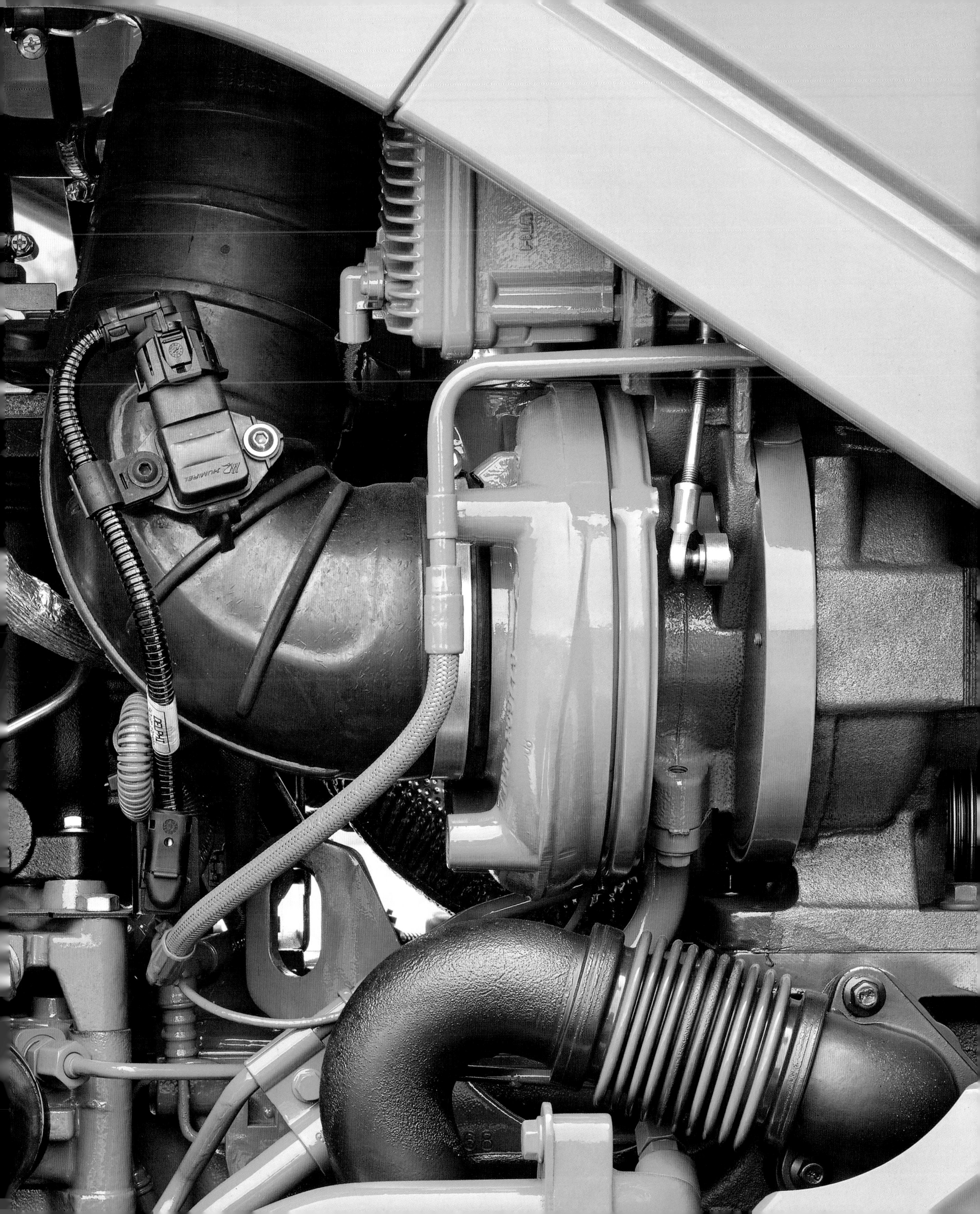

KONGSKILDE
DEMETER MULTISEED 4000
VALTRA
VALTRA
56

21ST CENTURY

Tractor development since 2000 has concentrated on improving what has gone before with the emphasis on enhanced performance, lower emissions, and advanced telematics. Average horsepower has continued to rise; in the UK it has increased from 124 hp to 150 hp since 2003. Some of the larger articulated and rubber-track machines now offer close to 600 hp. The latest developments include touch-screen displays, active suspension, and infinitely variable transmissions with different operational modes.

Farming has become a business rather than an occupation, with large companies displacing many of the family farms in intensive arable areas. Users demand high-specification machines with increased levels of reliability and durability from their components. Precision farming means many of the operations are automated by satellite guidance for accuracy, making even the driver almost a passenger.

△ **Russian tracklayers**
The Volgograd Tractor Plant, Russia's largest manufacturer of agricultural crawler tractors, became part of the Agromash Group in March 2003.

The tractor industry has had to continue changing to survive. Many of the traditional names have disappeared and factories have closed. Assembly is now carried out on a global basis with engines made in one country and transmissions in another. The three big players with a worldwide presence are John Deere, CNH (formed from the merger of New Holland with the Case Corporation), and AGCO.

Despite the growth in technology, there is still a place for a basic, low-horsepower tractor, especially in the emerging economies. This has become the fastest-growing segment of the market, served by manufacturers in India, China, and South Korea, among others. The Indian manufacturer Mahindra is now probably the largest tractor manufacturer in the world by volume, with annual sales of more than 200,000 units during 2012.

> "The **tractor** and **farm machinery** industry is **central** to the way farmers use those **resources** that are scarce."
>
> GRAHAM EDWARDS, CEO OF TRANTOR INTERNATIONAL

◁ **The latest concept from Valtra** combines articulated steering with a rotating front axle to reduce soil compaction.

Key events

- ▷ **2000** The Italian ARGO group acquires Case IH's Doncaster plant, UK, and rebrands the tractors under the McCormick banner.
- ▷ **2002** AGCO acquires the design, assembly, and marketing rights to the rubber-track Challenger.
- ▷ **2003** German manufacturer Claas takes over Renault Agriculture.
- ▷ **2004** Finnish tractor manufacturer Valtra becomes part of AGCO.
- ▷ **2006** A New Holland TM190 tractor clocks up more than 500 hours of continuous operation on biodiesel.
- ▷ **2007** An AGCO Challenger MT875B cultivates 1,591 acres (644 hectares) in 24 hours.
- ▷ **2011** SAME Deutz-Fahr builds its one-millionth tractor. Tier IVa emissions regulations become a legal requirement for engines above 174 hp.
- ▷ **2012** John Deere celebrates its 175th anniversary with record net sales and revenues of $36.2 billion.
- ▷ **2013** The Case IH Quadtrac 620 and the New Holland T8 Auto Command tractors win "Machine of the Year 2014" awards. India becomes the world's largest tractor producer.
- ▷ **2014** New Holland launches its Golden Jubilee T7 and T6 models, celebrating 50 years of tractor building in Basildon.

△ **Plowing record**
In 2005, a Case IH Quadtrac STX500 set a world record for plowing 792 acres (321 hectares) in less than 24 hours with 20-furrow plow.

Universal Workhorses

As available horsepower levels have increased over recent decades, tractors under 100 hp have been aimed at the livestock farming sector: for those raising animals and for grass-based operations. Machines above this level tend to be focused on farms with arable businesses. The two categories have distinct needs; in the under 100 hp bracket, this means a compact design for entry into low buildings and maneuverability through narrow spaces, plus ease of cab entry/exit for some tasks.

▷ **Massey Ferguson 5410**

Date 2014 **Origin** France
Engine AGCO Power 3-cylinder diesel
Horsepower 75 hp
Transmission 16 forward, 16 reverse

With the closure of the AGCO-Massey Ferguson factory in Coventry, UK in 2003, the MF4300 range was replaced with the new 5400 series. These were made in the Beauvais facility in France, and featured greater rear lift capacity, an improved cab, powershuttle, and a multiplate wet clutch.

▽ **Case IH JXU 85**

Date 2007 **Origin** Italy
Engine Iveco 4-cylinder diesel
Horsepower 86 hp
Transmission 24 forward, 24 reverse

The Case IH JXU range shared a platform with the New Holland TL-A line, with both being manufactured at the former Fiatagri tractor plant in Jesi, Italy. Model numbers were gradually expanded so that the range ultimately covered five models spanning 76–113 hp.

△ **John Deere 5090G**

Date 2014 **Origin** Italy
Engine John Deere 4-cylinder diesel
Horsepower 90 hp
Transmission 12 forward, 12 reverse

Comprising two models of 80 hp and 90 hp, powered by 4.5-liter diesel engines, John Deere's 5G tractors are available in either cab or rollbar formats, and either two- or four-wheel drive. Gears can be selected in either forward or reverse via a shuttle lever on the left of the steering column.

▷ **New Holland T4.85**

Date 2014 **Origin** Italy
Engine FPT 4-cylinder diesel
Horsepower 85 hp
Transmission 12 forward, 12 reverse

With five models from 75 hp to 115 hp, New Holland's T4 series is one of the key product lines to come out of the CNH Industrial factory in Jesi, Italy, a plant first founded in 1986 to produce Fiatagri tractors.

◁ **Zetor Major 80**

Date 2012 **Origin** Czech Republic
Engine Zetor 4-cylinder diesel
Horsepower 77 hp
Transmission 12 forward, 12 reverse

After steadily upgrading and increasing the specification of its range for a decade and expanding its power band, in 2012 Zetor introduced a new no-frills tractor model, the Major 80. The machine featured a Carraro gearbox with 12 forward and 12 reverse speeds.

△ **Lindner Lintrac**

Date 2013 **Origin** Austria
Engine Perkins 4-cylinder diesel
Horsepower 102 hp
Transmission CVT

First shown at the 2013 Agritechnica farm equipment exhibition in Germany, the key attribute of the Lintrac was that both front and rear axles steered, to give a tighter turning circle and rear wheel tracks that followed in the path of the fronts.

△ **Valtra A63**

Date 2012 **Origin** Finland
Engine AGCO Power 3-cylinder diesel
Horsepower 68 hp
Transmission 12 forward, 12 reverse

The A series was Valtra's entry-level line of tractors. Comprising five models from 50 hp to 101 hp, the three smaller machines were available as compact or orchard versions, and in the latter format had a narrower and lower design to ease movement around trees.

△ **Kubota M9960**

Date 2014 **Origin** Japan
Engine Kubota 4-cylinder diesel
Horsepower 100 hp
Transmission 36 forward, 36 reverse

Slotting in below the 110–135 hp MGX tractors, the M9960 tops Kubota's midrange line, which spans four models from 60 hp to 100 hp. Until 2014, the firm built all of its own tractors in Japan, but in 2015 three new 130–170 hp models will be made at a new factory in France.

◁ **Steyr 4065S Kompakt**

Date 2014 **Origin** Italy
Engine FPT 4-cylinder diesel
Horsepower 65 hp
Transmission 12 forward, 12 reverse

While higher-horsepower Steyr tractors are made in the marque's traditional home in the Austrian town of St. Valentin, the brand's smallest Kompakt S models—the 58 hp 4055S and 65 hp 4065S—come from the former Fiat plant in Ankara, Turkey.

John Deere 6210R

Designed to be highly versatile, the 6210R features numerous fuel and structural innovations aimed at reducing emissions and increasing fuel efficiency. With infinitely variable DirectDrive transmission based on Formula 1 technology, the 6210R has a high power-to-weight ratio, allowing it to put more power to the ground. It is equally efficient when transporting at high speeds, hauling heavy loads, or pulling large implements.

MEETING MODERN emission standards is a key challenge facing today's tractor engineers. The 6210R was John Deere's answer to balancing fuel efficiency with power and speed. A system of Intelligent Power Management was developed, carefully regulating the power and fuel used by the 210 hp engine. Extra fuel is burned only when necessary, increasing power by up to an additional 30 hp when operating demanding implements or when the tractor is traveling above 13 mph (21 km/h). In other situations, fueling is tightly controlled for maximum efficiency.

SPECIFICATIONS	
Model	John Deere 6210R
Built	2011-14
Origin	Germany
Production	Not known
Engine	210 hp John Deere 6-cylinder turbocharged, intercooled diesel
Transmission	20 forward, 20 reverse PowerQuad; 24 forward, 24 reverse DirectDrive; or AutoPowr CVT
Top speed	31 mph (50 km/h)
Length	16 ft 6 in (5.05 m)
Weight	8 tons (7.4 metric tons)

THE DETAILS

1. Fuel-efficient, diesel-only engine **2.** John Deere's "leaping deer" logo on radiator **3.** Triple link suspension on front axle provides extra traction and operator comfort **4.** Versatile lighting system **5.** Full-color, computer-control display offers optional touch screen and video capability **6.** Tractor control console moves with the seat

1

2

3

4

5

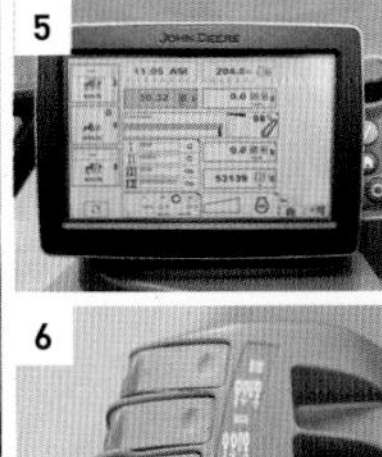

6

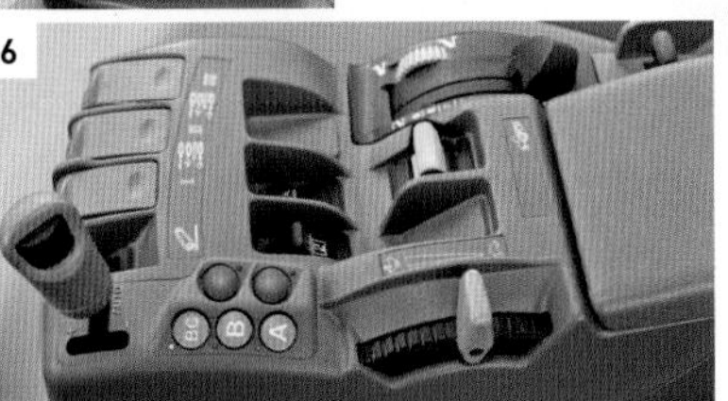

Advanced technology
All 6R series models include a CESAR Datatag security system, and a unique immobilizer key to counter theft. The 6210R could be equipped with satellite-guided AutoTrac automatic steering and hydraulic cab suspension.

Integrated front hitch and heavy-duty front power takeoff

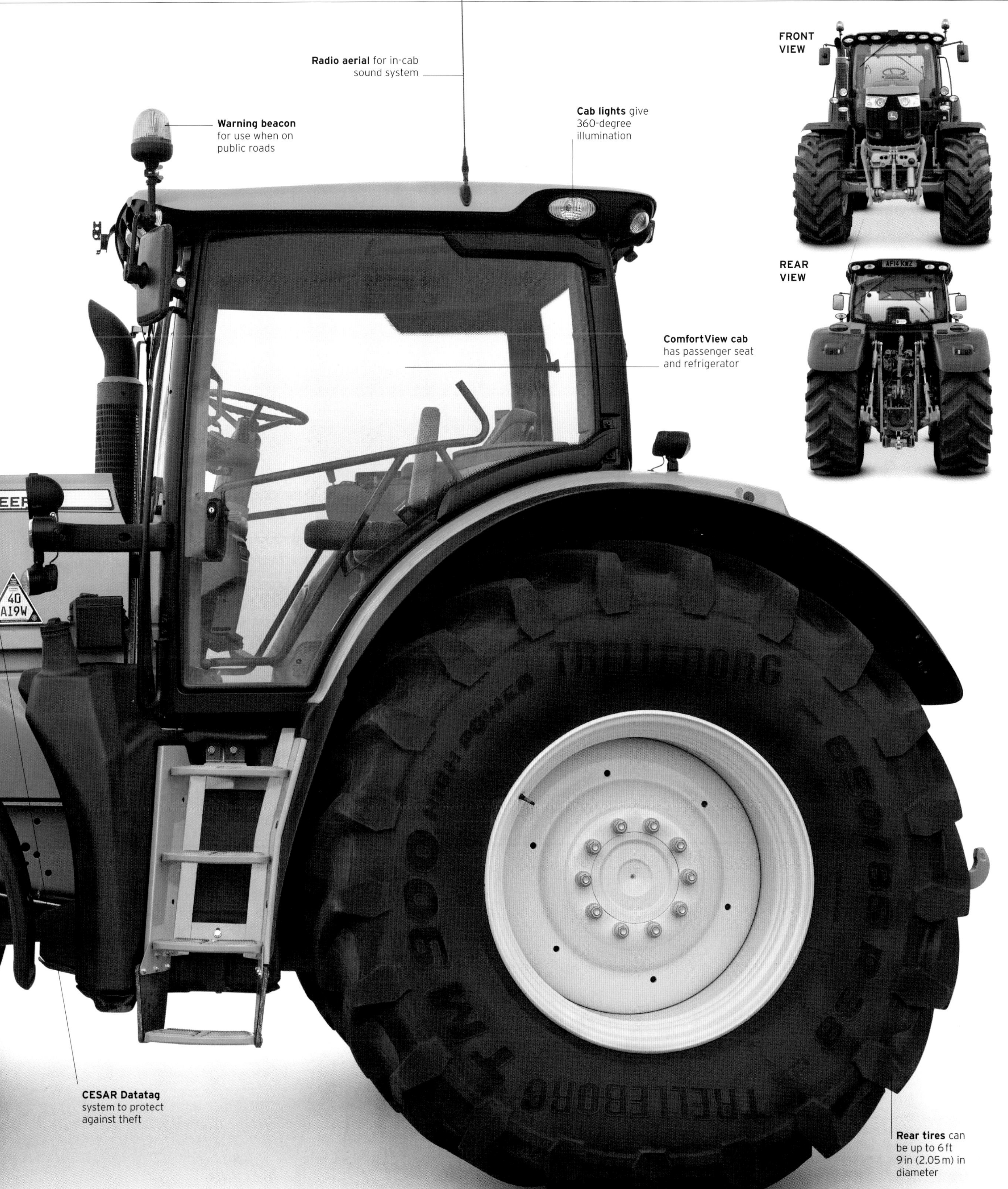
FRONT VIEW
REAR VIEW
Radio aerial for in-cab sound system
Warning beacon for use when on public roads
Cab lights give 360-degree illumination
ComfortView cab has passenger seat and refrigerator
CESAR Datatag system to protect against theft
Rear tires can be up to 6 ft 9 in (2.05 m) in diameter
TRELLEBORG
HIGH POWER
TM 900

Power and Precision

As a rule, tractors over 100 hp tend to be used for fieldwork operations, and are therefore designed with features such as powershift and continuously variable transmission (CVT), which eliminate any need for gear changing. In recent years, a considerable level of computer-aided automation has been integrated into designs to help reduce the burden on the driver, increase productivity, and reduce fuel use, which is of both economic and environmental benefit. GPS-guided auto-steering is also becoming increasingly common.

△ New Holland T8040

Date 2008 **Origin** US

Engine FPT/Iveco 6-cylinder diesel

Horsepower 255 hp

Transmission 16 forward, 4 reverse; powershift

The T8000 tractors, like their TG forebears, shared a platform with the equivalent Case IH Magnum models, but were distinct in design. They had a different engine-to-transmission connection, placing the latter below the former, rather than in line with it, resulting in a higher and shorter appearance.

◁ Renault Ares 710RZ

Date 1997 **Origin** France

Engine John Deere 6-cylinder diesel

Horsepower 145 hp

Transmission 32 forward, 32 reverse

Launched in the mid-1990s, the Renault Ares tractors were among the last of the French firm's clean sheet designs before it sold its tractor division to German harvesting specialists Claas. The new owners re-liveried and rebranded the models, but continued with them for a time before upgrading and ultimately replacing them.

△ Claas Atles 946RZ

Date 2003 **Origin** France

Engine Deutz 6-cylinder diesel

Horsepower 282 hp

Transmission 18 forward, 18 reverse

The high-horsepower Atles tractor range was launched in 2000 under the orange colors of Renault, before acquiring a new coat of green and red paint when Claas purchased the French firm's agriculture division. The range had a relatively short run under its new parent, though, before being dropped and eventually replaced by the Axion 900 line.

◁ McCormick X7.460

Date 2013 **Origin** Italy

Engine FPT 6-cylinder diesel

Horsepower 166 hp

Transmission 24 forward, 24 reverse

Launched at the biennial Agritechnica farm machinery show in Hanover, Germany, in 2013, the McCormick X7 range marked a complete departure from ARGO's former models in the 150–200 hp bracket. Among the innovations was a completely new cabin.

▷ JCB Fastrac 8310

Date 2011 **Origin** UK

Engine AGCO Power 6-cylinder diesel

Horsepower 306 hp

Transmission CVT

The first JCB tractor to breach the 300 hp mark, the 8310 and its smaller 8280 brother followed on from the redesigned 8250 in adopting an unequal-wheeled design with smaller front wheels than the rears, in a departure from the original Fastrac format.

▽ Case IH Puma 230 CVX

Date 2013 **Origin** Austria/UK

Engine FPT 6-cylinder diesel

Horsepower 228 hp

Transmission CVT

Made at the CNH facility in the Austrian town of St. Valentin, the 230 is the largest in the line of Puma tractors offered under the colors of Case IH. In CVX format, equipped with the CNH CVT, the tractors feature double-clutch technology to ensure smooth changes between the gear ranges.

△ Massey Ferguson 7618

Date 2014 **Origin** France

Engine AGCO Power 6-cylinder diesel

Horsepower 165 hp

Transmission 24 forward, 24 reverse

Built in AGCO's Beauvais factory in France, a long-serving Massey Ferguson manufacturing plant, the 7618 is the second-smallest tractor in a line of six models that spans a rated power band from 140 hp to 240 hp. All are available with either a six-step powershift transmission or a stepless CVT.

△ John Deere 6210R

Date 2011-14 **Origin** Germany

Engine John Deere 6-cylinder diesel

Horsepower 210 hp

Transmission AutoPowr CVT

This was the largest tractor ever made at John Deere's Mannheim factory in Germany, until it was superseded in 2014 by the 6215R. The 6210R was for some time the flagship of Deere's 6R range. It offered a high-horsepower package in a relatively low-weight tractor that could be ballasted up as required for heavy draft work.

◁ Fendt 936 Vario

Date 2012 **Origin** Germany

Engine Deutz 6-cylinder diesel

Horsepower 360 hp

Transmission CVT

Fendt first introduced a CVT in 1995 on its 926 tractor, which blended hydrostatic and mechanical power transfer to allow stepless travel without gear changes. It now equips all of its tractors as standard with its Vario CVT, including this 936.

SAME DA52 was the world's first diesel four-wheel-drive tractor in 1952

Great Manufacturers
SAME Deutz-Fahr

SAME Deutz-Fahr (SDF) was established in 1995 when the Italian SAME-Lamborghini-Hürlimann company purchased Deutz-Fahr from Klöcker-Humboldt-Deutz (KHD) in Germany. Today this multinational, multi-brand corporation is one of Europe's foremost tractor and diesel engine manufacturers, with production facilities in Europe and Asia.

THE ENDURING SUCCESS of SAME Deutz-Fahr can be attributed to the ingenuity and creativity of one man: the Italian company's founder, Francesco Cassani. A gifted engineer and a brilliant draftsman, Francesco Cassani was a man with extraordinary talent and vision, who, with his brother Eugenio, had spent 20 years developing diesel engines and designing tractors under his own name before World War II, introducing the Cassani 40, the world's first diesel-powered tractor, in 1927. He was just 20 years old at the time.

The SDF story begins with the establishment of the SAME (Società Accomandita Motori Endotermici) company at the height of World War II. In 1942, Francesco and Eugenio Cassani set up to manufacture tractors in an empty factory in Treviglio, Italy. Shortly after World War II, in 1948, SAME introduced the Trattorino Universale three-wheeled mini-tractor. Utilitarian in design and ungainly in appearance, this affordable and versatile, 10 hp kerosene-powered tractor featured a simple hydraulic lifting mechanism and was the first in the world to offer a reversible driver's seat.

Francesco Cassani
(1906–1973)

Impressed by the ability of the US Jeeps he had seen during the war, Francesco became determined to design a four-wheel-drive tractor. SAME accomplished this in 1952 with the launch of the world's first diesel-powered, four-wheel-drive tractor, the 25 hp twin-cylinder, air-cooled DA25.

Engineering prodigy
Francesco Cassani was just 20 when he built the world's first diesel tractor, the Cassani 40, in 1927. The tractor's engine had been designed by Francesco and his brother Eugenio.

By this time, Francesco had developed a system of building tractor engines with standardized components such as pistons, valves, and cylinder blocks. These parts could then be used universally in different engine configurations, throughout entire ranges of SAME tractors.

In 1958 the company introduced the Automatic Linkage Control Unit, which was designed according to the same principles as the Ferguson system. Francesco both admired and revered Harry Ferguson, considering the Irishman to be his only truly equal competitor.

The death of Eugenio Cassani in 1959 was a tragic loss. Yet Francesco continued to push the company forward, introducing the new Centauro range and expanding into the Netherlands, Belgium, Greece, Spain, Portugal, Switzerland, the UK, and even into Africa by the late 1960s.

In 1971 SAME acquired the tractor division of the supercar manufacturer Lamborghini, then owned by GEPI, a state-owned financial holding company. Lamborghini had its own established dealer network, an existing product line, a reputable name, and, importantly, spare manufacturing capacity. The takeover, finalized in 1973, was a success. SAME continued to grow throughout the 1970s, developing new tractor ranges—including crawlers—using technologies that had been inherited from Lamborghini.

Poster campaign
An advert for Lamborghini tractors. The supercar manufacturer's agricultural division was purchased by SAME in 1973.

In 1979 the Swiss tractor manufacturer Hürlimann was acquired, and the company name was officially changed to SAME-Lamborghini-Hürlimann (SLH). The Hürlimann factory near Zurich had an unprecedented reputation for quality, producing just a few hundred hand-assembled tractors each year. Its acquisition meant the SLH Group now had one of the widest and most comprehensive product ranges in the world market, with models spanning the 25 hp to 260 hp range. It also established the company as Italy's second largest tractor manufacturer.

A difficult period of rationalization, spawned by the world recession in the early 1980s, was turned around by

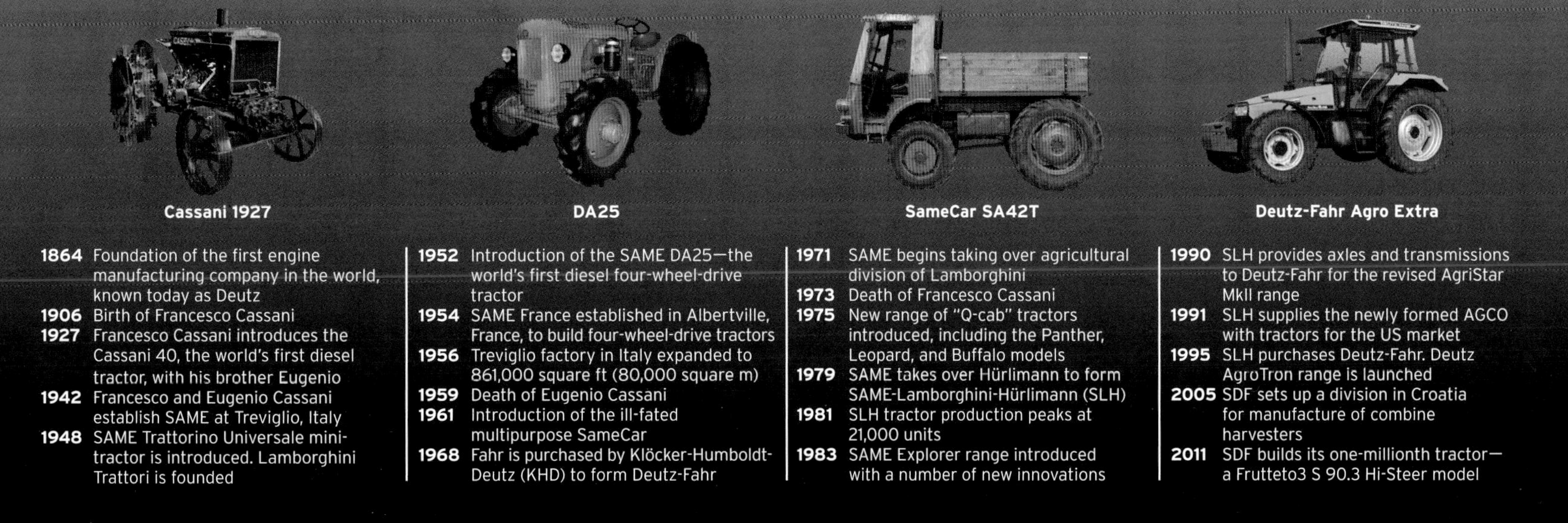

Cassani 1927

DA25

SameCar SA42T

Deutz-Fahr Agro Extra

1864 Foundation of the first engine manufacturing company in the world, known today as Deutz
1906 Birth of Francesco Cassani
1927 Francesco Cassani introduces the Cassani 40, the world's first diesel tractor, with his brother Eugenio
1942 Francesco and Eugenio Cassani establish SAME at Treviglio, Italy
1948 SAME Trattorino Universale mini-tractor is introduced. Lamborghini Trattori is founded

1952 Introduction of the SAME DA25—the world's first diesel four-wheel-drive tractor
1954 SAME France established in Albertville, France, to build four-wheel-drive tractors
1956 Treviglio factory in Italy expanded to 861,000 square ft (80,000 square m)
1959 Death of Eugenio Cassani
1961 Introduction of the ill-fated multipurpose SameCar
1968 Fahr is purchased by Klöcker-Humboldt-Deutz (KHD) to form Deutz-Fahr

1971 SAME begins taking over agricultural division of Lamborghini
1973 Death of Francesco Cassani
1975 New range of "Q-cab" tractors introduced, including the Panther, Leopard, and Buffalo models
1979 SAME takes over Hürlimann to form SAME-Lamborghini-Hürlimann (SLH)
1981 SLH tractor production peaks at 21,000 units
1983 SAME Explorer range introduced with a number of new innovations

1990 SLH provides axles and transmissions to Deutz-Fahr for the revised AgriStar MkII range
1991 SLH supplies the newly formed AGCO with tractors for the US market
1995 SLH purchases Deutz-Fahr. Deutz AgroTron range is launched
2005 SDF sets up a division in Croatia for manufacture of combine harvesters
2011 SDF builds its one-millionth tractor—a Frutteto3 S 90.3 Hi-Steer model

Agricultural engines
SDF has recently launched its own range of diesel engines. Branded as FARMotion, they are specifically designed for tractors.

the introduction of the groundbreaking SAME Explorer tractor range in 1983. The new models featured a string of innovations, including forced lubrication transmissions, front and rear power takeoff, and electro-hydraulic controls, while retaining the traditional SAME air-cooled engines and four-wheel-drive system. SAME-Lamborghini-Hürlimann was back on track, heading for what would be its most important merger to date.

In February 1995 SLH acquired the German manufacturer Deutz-Fahr, a subsidiary of the industrial conglomerate Klöcker-Humbolt-Deutz (KHD). Deutsche Bank, the principal shareholder in KHD, had announced that it wished to relinquish its share of the company, leaving KHD with no choice but to liquidate its poorly performing tractor and equipment manufacturing interests. SAME stepped in to add the respected German firm's tractor division to its arsenal. The new company would be known as SAME Deutz-Fahr (SDF).

Deutz, whose first tractor had appeared in 1919, had joined KHD in the 1930s. During the postwar period, Deutz had flourished, establishing a reputation as Germany's leading tractor producer. In 1968 the farm equipment manufacturer Fahr was purchased by KHD and the Deutz-Fahr name brand established.

The purchase of Deutz-Fahr was a colossal undertaking for SDF, and one that coincided with the launch of the underdeveloped Deutz AgroTron range. Despite initial setbacks with Deutz tractor production, the acquisition of Deutz-Fahr provided SDF with the opportunity to establish itself in the combine harvester market.

Today SDF has established manufacturing facilities across Europe and Asia, and its dealerships sell its products in more than 140 countries. In keeping with its strong heritage of diesel engine design, SAME Deutz-Fahr launched its own new FARMotion range of engines in 2014. These Tier IVa final, three- and four- cylinder engines are produced at the corporation's production facility in India.

> "SAME... was **created** not to make a **profit** but to give Italy a **prestigious** industry."
>
> FROM THE "SPIRITUAL WILL" OF FRANCESCO CASSANI

Universal components
Like other manufacturers producing multiple tractor brands, SDF shares components across its SAME, Deutz, Lamborghini, and Hürlimann ranges. SAME tractors retain their traditional orange livery.

For Fruits and Vines

Tree fruit and grape growers require specialty tractors that can travel between tree rows or vines spaced at the optimum distance for growth without damaging the tractors or the plants. This means machines designed for these environments are narrower and lower than normal. For many years, standard tractors were converted for this work, but in recent decades manufacturers have begun to offer specific machines, many of which use technology as sophisticated as their larger, field-focused counterparts.

◁ **Massey Ferguson 3350C**

Date 2001 **Origin** Italy

Engine Perkins 3-cylinder diesel

Horsepower 93 hp

Transmission 8 forward, 2 reverse

Since its relationship with Landini ended, Massey Ferguson's crawler range has been built for the company by SAME Deutz-Fahr at its Treviglio factory in Italy. Spanning the 50–100 hp sector, the line complements the firm's extensive offering in vineyard, fruit, and specialty wheeled tractors.

△ **New Holland T4.105**

Date 2014 **Origin** Italy

Engine FPT 4-cylinder diesel

Horsepower 105 hp

Transmission 24 forward, 24 reverse

This specially modified tractor originated in the CNH factory in Jesi, Italy. It has been adapted for tree work with a sleek, low-profile cab to reduce the likelihood of anything snagging on the tractor as it drives through wooded areas. The bars part low-hanging branches.

△ **SAME Frutteto 3 80S**

Date 2014 **Origin** Italy

Engine SAME 3-cylinder diesel

Horsepower 82 hp

Transmission 30 forward, 15 reverse

One of Italy's largest tractor producers, the SAME Deutz-Fahr group manufactures its own range of specialty fruit and vineyard tractors alongside its agricultural machines at its Treviglio factory in northern Italy. The Frutteto range is powered by SAME's own three-cylinder engines.

▷ **Fendt 211V**

Date 2010 **Origin** Germany

Engine AGCO Power 3-cylinder diesel

Horsepower 110 hp

Transmission Stepless CVT

Fendt's offerings in the vineyard and fruit sector are based on the 200 series tractors made in house at its Marktoberdorf plant in Germany. A unique feature of these tractors is the inclusion of a continuously variable transmission (CVT) as standard.

◁ **Ferrari Thor 85**

Date 2013 **Origin** Italy
Engine Lombardini 3-cylinder diesel
Horsepower 34 hp
Transmission 6 forward, 3 reverse

The name Ferrari is a relatively common one in Italy, and there is no known link between the tractor maker and the car company of the same name. Ferrari specializes in producing small, sub-100 hp tractors with equal-sized wheels and articulated steering.

TECHNOLOGY

Working the Vineyards

Among the typical tasks required of orchard and vineyard tractors are the pruning of tree branches, mowing grass strips between tree rows, and application of crop protection products to the plants with special sprayers, as shown below. With a considerable proportion of its farmed land area devoted to vineyards and orchards, Italy is also a high-volume producer of tractors for special applications.

Landini Rex 110F Introduced in 2010, the 110F was the largest in Landini's range of Rex specialty tractors, with a 110 hp Perkins engine.

▷ **Claas Nexos 240**

Date 2014 **Origin** France
Engine FPT 4-cylinder diesel
Horsepower 90 hp
Transmission 18 forward, 18 reverse

Claas purchased the Renault tractor business after the French firm decided to focus on car and truck production. It inherited not only a full agricultural tractor line, but also a well-established fruit and vineyard product range, going on to develop this into the new Nexos line.

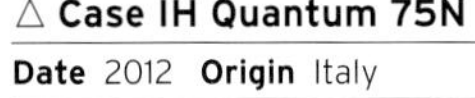

△ **Case IH Quantum 75N**

Date 2012 **Origin** Italy
Engine FPT 4-cylinder diesel
Horsepower 75 hp
Transmission 16 forward, 16 reverse

With Case IH and New Holland owned by the same parent company, its tractors are made using common platforms with different features and operator controls. Built alongside their New Holland cousins in Italy, these tractors are offered only in certain worldwide markets.

▽ **New Holland T3.55F**

Date 2014 **Origin** Italy
Engine FPT 3-cylinder diesel
Horsepower 55 hp
Transmission 16 forward, 16 reverse

Tracing their lineage back through to the Fiat fruit and vineyard ranges manufactured before the company merged its tractor interests with that of Ford, the New Holland T3.F tractors are made in house by parent company CNH.

AGCO Challenger

Rubber-belt technology arose as part of a NASA investigation into making unmanned Moon rover vehicles more stable on the lunar surface. After early developmental problems, rubber belts have all but replaced steel tracks. AGCO's Challenger combined reduced soil compaction with the high working speeds and mobility needed to move rapidly around today's large farms.

THE AGCO CHALLENGER series of rubber-belted tractors continued the revolution that began with Caterpillar's Challenger 65 in 1988. AGCO's MT series can be considered the second generation of this type of tractor. The large rear-drive wheels allowed more underbelly clearance, which, along with the readily adjustable track gauge, made the tractor eminently suitable for row-crop work. A range of different belts was available to suit various conditions. These ranged in four sizes, from 18 in (45.7 cm) for row-crop work, to 36 in (91.4 cm) for low-ground-pressure applications. Challengers featured a highly sophisticated control system consisting of six onboard computer systems for engine and transmission management, rear hydraulic functions, the GPS program, tractor output data, and maintenance and service periods.

Computerized control
The MT865C was the largest and most powerful of the MT800 series. Powered by the Caterpillar C18 diesel engine coupled to a 16-speed, full powershift transmission and with complete computer control of all mechanical and field functions, it represented the agricultural tractor at its most highly developed.

Driver's cab is fully suspended and environmentally controlled

Drive wheel with large diameter

Track rollers oscillate for smooth ride

FRONT VIEW

REAR VIEW

SPECIFICATIONS	
Model	AGCO Challenger MT865C
Built	c. 2009
Origin	US
Production	Unknown
Engine	583 hp Caterpillar C16 6-cylinder, turbocharged and aftercooled diesel
Capacity	1,105 cu in (18,100 cc)
Transmission	16 forward, 4 reverse
Top speed	25 mph (40 km/h)
Length	22 ft 6 in (6.85 m)
Weight	21 tons (19.1 metric tons)

THE DETAILS

1. Detachable front weights counterbalance the weight of raised, rear-mounted implements
2. Large-diameter front idlers reduce rolling resistance
3. Dry-type air-cleaner element ensures that clean air is delivered to the engine at all times
4. Combined three-point hitch and swinging drawbar featuring Hillside Trim Mode
5. Driver's-eye view of the tractor's steering controls and dashboard
6. Computer systems setup keyboard

Hood tilts to allow easy maintenance

Rubber belts reinforced with steel cable

Detachable centreline weights

Ever Larger Machines

The latest generation of tractors reflects the ever-growing need for more power in the hands of fewer operators. Tractors are equipped with satellite navigation, full powershift transmissions, and many have the latest Tier IVa diesel engines with reduced exhaust emissions that comply with recent legislation. With these high-powered tractors, and implements that can be more than 50 ft (15.2 m) wide, satellite navigation prevents any overlap or missed land during cultivation, which reduces the maximum output per horsepower, manpower, and fuel consumption.

△ **Kirovets K745**

Date 2002 **Origin** Russia

Engine Mercedes-Benz or Deutz V8 diesel

Horsepower 450 hp (Mercedes-Benz) to 495 hp (Deutz)

Transmission 12 forward, 2 reverse

The K series has been produced in large numbers over the years. The K745 was a robust and reliable machine equipped with autopilot satellite navigation and two rear-facing cameras. Some of its production was exported, made possible by the use of Deutz and Mercedes engines, which met emission standards.

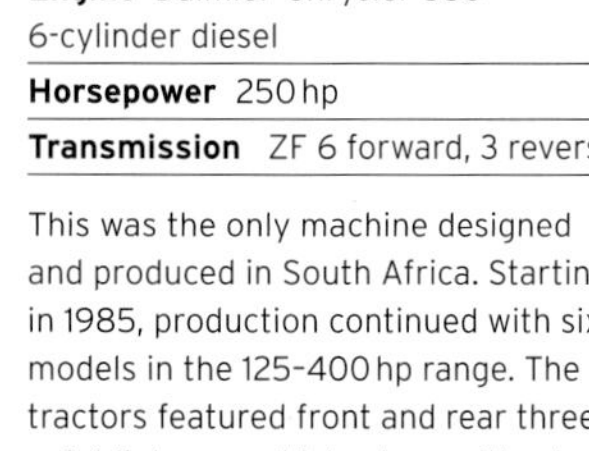

▷ **AGRICO 4+250**

Date 2003 **Origin** South Africa

Engine Daimler-Chrysler S60 6-cylinder diesel

Horsepower 250 hp

Transmission ZF 6 forward, 3 reverse

This was the only machine designed and produced in South Africa. Starting in 1985, production continued with six models in the 125–400 hp range. The tractors featured front and rear three-point linkages, which, along with a low height profile and articulated steering, gave them flexibility within each of the horsepower ranges.

△ **Case IH Steiger 535 Gold Signature Edition**

Date 2007 **Origin** US

Engine Cummins QSX15 6-cylinder diesel

Horsepower 535 hp

Transmission 16 forward, 2 reverse full powershift

Fifty Steiger 535s were produced for the 50th anniversary of the first Steiger tractor. This signature-edition machine was factory-equipped with the latest auto-guidance system, fully compliant with emission regulations, and used biodiesel.

△ **AGCO Challenger MT965C**

Date 2013 **Origin** US

Engine Caterpillar C18 6-cylinder diesel

Horsepower 510 hp

Transmission 16 forward, 4 reverse

In 2008 the MT900 series won some prestigious awards for innovation in design. With a fuel capacity of 390 gallons (1,477 liters), the MT965C is equipped for long hours of operation in the field without refueling.

Rubber Tracks

The choice of tracks or four-wheel drive is largely dependent on the type of soil, climate, and terrain in which the tractor is to be operated. The rubber-tracked machines have developed into two specific types: the full-tracked machine that is steered by its tracks, and the articulated type with four separate track assemblies. The latter is basically an adaptation of the articulated, four-wheel-drive tractor with the wheel and tracked versions sharing the same mechanical parts.

◁ **AGCO Challenger MT865C**

Date 2009 **Origin** US

Engine Caterpillar C16 6-cylinder diesel

Horsepower 583 hp

Transmission 16 forward, 4 reverse full powershift

The MT865C succeeded the MT865B with only a very slight modification. It had full electronic control for the rear linkage and incorporated satellite navigation. It had a road speed of 25 mph (40 km/h), while giving a very low noise level in the driver's cab.

▽ **New Holland T9.560**

Date 2014 **Origin** US

Engine New Holland FPT Cursor 6-cylinder diesel

Horsepower 507 hp

Transmission 16 forward, 2 reverse with transport mode

The T9.560 is a large tractor, equipped with a 24-volt starting system. The Engine Power Management (EPM) system continuously adjusts engine output to match the working conditions and automatically accounts for changes in power demand from the hydraulic, power takeoff, and transport modes.

▷ **John Deere 9560R**

Date 2014 **Origin** US

Engine John Deere PowerTech 6-cylinder diesel

Horsepower 560 hp

Transmission 18 forward, 6 reverse

The 9560R is equipped with a full powershift transmission controlled by John Deere's Efficiency Manager system. With a maximum road speed of 26 mph (42 km/h) and a 20,000-lb (9,000-kg) rear lift capacity, it is capable of massive output.

△ **John Deere 8360RT**

Date 2011 **Origin** US

Engine John Deere PowerTech 6-cylinder diesel

Horsepower 360 hp

Transmission John Deere AutoPowr/IVT forward and reverse

This track tractor had two transmission options: AutoPowr/IVT (Infinitely Variable Transmission), which offered variable speeds from 164 ft per hour (50 m/h) to 26 mph (42 km/h); Automatic PowerShift had on-the-go gear-shifting with shuttle change between forward and reverse.

▷ **Versatile Deltatrack 450DT**

Date 2014 **Origin** Canada

Engine Cummins QSX15 6-cylinder diesel

Horsepower 500 hp

Transmission Caterpillar TA22 16 forward, 4 reverse, fully programmable autoshift

A sophisticated piece of machinery, the Deltatrack has positive-drive rubber tracks, which eliminate friction in the track components. Steering is accomplished by bending the tractor around a central pivot point. It has a road speed of 22 mph (35 km/h).

SOLD & SERVICED BY FINNING
D6N

Preparing for the Coldest Journey

Antarctica had never been crossed in winter, but in March 2013, in temperatures that could fall as low as –130˚F (–90˚C) and in nearly 24 hours of total darkness, a team of five experienced explorers attempted to do just that. One hundred years after the death of Captain Scott, the team set out from Crown Bay to travel via the South Pole to Scott's base at McMurdo Sound.

EQUIPPED FOR THE CONDITIONS

The team wore heated clothes and had two special heated cabooses, one to eat and sleep in and the other for their equipment and workshop facilities. Each caboose, along with supply and fuel sleds, was towed by a Caterpillar D6N crawler tractor modified by Caterpillar dealer Finning to operate on snow and ice in extreme cold. The expedition took six months and covered more than 2,000 miles (3,200 km); including rest days and bad weather, the team was on the ice for just over 270 days, averaging 22 miles (35 km) a day. A two-man ski-unit traveled ahead of the Mobile Vehicle Landtrain (MVL) pulling a small toboggan, or pulk, equipped with ground-penetrating radar that transmitted information back to the D6Ns about the terrain and, most importantly, the presence of any crevasses.

The second of the expedition's two modified Caterpillar D6Ns is offloaded from the polar research and supply ship SA *Agulhas* at Crown Bay, Antarctica.

Power for Other Purposes

Since the development of the earliest models, tractors have found many other uses outside agriculture, from the mowing of parks and sports fields to the maintenance of roads, and from the extraction of lumber to dedicated road haulage. As more and more tasks beyond farming were developed for tractors, many mainstream makers developed dedicated models, while other specialty firms combined the features of tractors and utility vehicles.

△ Massey Ferguson 6455 Municipal

Date 2007 **Origin** France
Engine AGCO Power 4-cylinder diesel
Horsepower 105 hp
Transmission 24 forward, 24 reverse

From the maker's Beauvais factory in France, the municipal versions of Massey Ferguson's 6400 series tractors are available with a number of features to aid visibility, including a drop-nose hood and a single-piece, right-hand window for an unimpeded view of a hedgecutter head.

▷ Fendt 936 Vario Municipal

Date 2008 **Origin** Germany
Engine Deutz 6-cylinder diesel
Horsepower 360 hp
Transmission Stepless CVT

Even the largest agricultural tractors find uses in other environments. For many years the flagship in its lineup, Fendt's 360 hp 936 tractor has been equipped with the firm's Vario continuously variable transmission (CVT), first made available on the 926 model in 1995.

▷ Deutz-Fahr Agrotron K410

Date 2008 **Origin** Germany
Engine Deutz 4-cylinder diesel
Horsepower 100 hp
Transmission 24 forward, 8 reverse

Built in its Lauingen factory in southern Germany, the "Kommunal" version of Deutz-Fahr's Agrotron K410 tractor is outwardly very similar to its agricultural cousin, but the usual lime-green body panels are exchanged for highway orange livery. Models destined primarily for on-road use are often equipped with special tires.

△ Trackless MT6

Date 2010 **Origin** US
Engine Cummins 4-cylinder diesel
Horsepower 115 hp
Transmission Hydrostatic

This forward-control, articulated tractor's engine is mounted at a conventional level directly behind the operator's cab. With nothing to impede forward vision, the MT6 is well-suited to work with front-mounted implements.

▷ Reform Metrac H7X

Date 2012 **Origin** Austria
Engine VMA 4-cylinder diesel
Horsepower 70 hp
Transmission Hydrostatic

Vehicles with equal wheels, a low center of gravity, and linkage/power takeoff/hydraulic outlet packages front and rear, such as this Reform Metrac, are popular in Austria, Switzerland, and other mountainous farming areas of Europe. Their low profiles make for much safer working on slopes, where conventional tractors are at risk of overturning.

△ **Holder Sommer S990**

Date 2012 **Origin** Germany
Engine Deutz 4-cylinder diesel
Horsepower 92 hp
Transmission Hydrostatic

German manufacturer Holder has long focused on building specialty tractors, ranging from small- to mid-horsepower machines with articulated steering, to municipally targeted vehicles with forward-mounted cabs, such as this S990. Implements can be mounted on front and rear linkages—both with power takeoffs—and a load platform.

△ **Holder C270**

Date 2013 **Origin** Germany
Engine Kubota 3-cylinder diesel
Horsepower 67 hp
Transmission Hydrostatic

Holder makes a number of machines aimed at highway use with forward-control operator stations articulated just behind the cab. Brush operations like the one seen here are a key application.

△ **John Deere 4049R**

Date 2014 **Origin** US
Engine Yanmar 4-cylinder diesel
Horsepower 49 hp
Transmission Hydrostatic

Compact tractors have many uses, from work on small farms to the maintenance of grass at sporting venues. Built at John Deere's US compact tractor factory, this machine is equipped for the latter duties, with turf tires to minimize tread damage.

◁ **Steyr Profi 4130 CVT**

Date 2014 **Origin** Austria
Engine FPT 4-cylinder diesel
Horsepower 130 hp
Transmission Stepless CVT

Alongside Fendt, Austrian firm Steyr was a pioneer in the adoption of continuously variable transmissions (CVT) for tractors, first offering them in the mid-1990s. Cab protection packages, such as this one, make for easier driving and reduce the risk of damage when carrying out forestry work.

Economic and Social Progress

Tractor power reduces the workload for people and draft animals, a process that is even now at an early stage in some countries. A two-horse team will plow 1 acre (0.4 hectare) per day, but a 175-hp tractor can plow 2–3 acres (0.8–1.2 hectares) in an hour. In addition, a farmer needs about 3 acres (1.2 hectares) just to feed one working horse for a year, which reduces the farm's cash-crop production.

CHANGING FACE OF FARMING

In 1918 there were 26 million horses and mules working on North American farms, but there were fewer than eight million by 1950, providing large areas of land for additional cropping. Tractor power also releases people for other jobs. Highly mechanized farming employs under two percent of the working population, but countries that rely mainly on manual labor and animal power may have 25 percent or more working the land. Demand for food continues to grow and, although many countries do use tractors, mechanization will need to play an even greater part in the future to meet the world's needs. Tractor ranges are also being built on platforms that allow them to be tailored to serve the needs of farms of all types.

Steel cage rear wheels provide the extra grip to keep this Mahindra 475-DI working in the mud as it prepares flooded land for a rice crop in India.

TN65F 4147

World Farming

Some of the world's most popular tractor producers and best-selling models are not necessarily well-recognized in Western Europe. The biggest tractor markets include those of India, China, Russia, and the largest countries of South America, and it is in these regions, as well as in Southern and Eastern Europe and Africa, that makers focus on huge volumes of simple, straightforward machines. These include not just full-scale fieldwork tractors, but also many smaller specialty machines for specific tasks and basic applications.

◁ **TYM 1003**

Date 2014 **Origin** South Korea
Engine Perkins 4-cylinder diesel
Horsepower 100 hp
Transmission 32 forward, 32 reverse

Based in the South Korean capital, Seoul, Tong Yang Moolsan (TYM) manufactures tractors, rice-focused combine harvesters, and rice transplanters. Its tractor line focuses on the sub-100 hp sector, and the top-end 1003 is a version of the 903 understudy.

▽ **Solis 20**

Date 2014 **Origin** India
Engine Mitsubishi 3-cylinder diesel
Horsepower 20 hp
Transmission 6 forward, 2 reverse

Solis is the international brand name of Indian manufacturer Sonalika. It produces a range of tractors spanning the 20–90 hp power band, with the Solis 20 being its smallest machine. It is powered by a Mitsubishi engine driving through a constant mesh gearbox.

△ **Mahindra 6030 turbo**

Date 2012 **Origin** India
Engine Mahindra 4-cylinder diesel
Horsepower 59 hp
Transmission 8 forward, 8 reverse

One of the world's largest tractor makers, India's Mahindra tractor manufacturing business is part of the Mahindra and Mahindra industrial engineering conglomerate. It is said to sell around 85,000 tractors each year, and is the leader in its domestic market, claimed to be the largest in the world.

◁ **Valtra BT190**

Date 2014 **Origin** Brazil
Engine AGCO Power 6-cylinder diesel
Horsepower 190 hp
Transmission 24 forward, 24 reverse powershift

Produced in the Valtra plant in Mogi das Cruzes, Brazil, the BT machines are sold primarily in their South American home market, and are designed with a focus on transportation and high-speed cultivation applications. The BT190 is part of a four-model range that spans the 150–210 hp power bracket.

△ **ArmaTrac 804e**

Date 2014 **Origin** Turkey
Engine Perkins 4-cylinder diesel
Horsepower 76 hp
Transmission 16 forward, 8 reverse

Established in 1953, Turkish company Erkunt Industries began building tractors as recently as 2003. It markets its machines under the Armatrac brand, and offers an agricultural range from 50 hp to 110 hp. Key components come from long-established suppliers such as Perkins and ZF.

▷ **YTO 180**

Date 2013 **Origin** China
Engine TY295IT 2-cylinder diesel
Horsepower 18 hp
Transmission 16 forward, 4 reverse

Chinese firm YTO is one of the country's largest producers of wheeled and crawler tractors, manufacturing models of up to 180 hp. Based in Luoyang, Henan province, it also produces a number of small combine harvesters, and a wide range of grassland and tillage implements.

▽ **Kubota L3200**

Date 2014 **Origin** Japan

Engine Kubota 3-cylinder diesel

Horsepower 32 hp

Transmission 8 forward, 4 reverse or hydrostatic

The smallest model in the three-strong line, Kubota's L3200 sits squarely in the middle of compact tractor territory. As an alternative to the standard gearbox, like many compact machines, it can be equipped with hydrostatic transmission, allowing single-pedal driving without any need for changing gear.

△ **Felderman Mini 16**

Date 2014 **Origin** Czech Republic

Engine Lombardini 2-cylinder diesel

Horsepower 16 hp

Transmission Hydrostatic

Typical of the type of pivot-steer tractors with equal-sized wheels made by a number of Italian makers, the Felderman Mini 16 is in fact built in the Czech Republic. Advantages of this type of steering arrangement, where the machine articulates between the two axles, include better maneuverability in confined spaces.

▽ **Belarus 3522.5**

Date 2012 **Origin** Belarus

Engine Deutz 6-cylinder diesel

Horsepower 355 hp

Transmission 36 forward, 24 reverse powershift

This is the largest tractor produced to date at the Minsk Tractor Works in the Republic of Belarus. The main market for the 3522.5 is in Russia and the Commonwealth of Independent States (CIS) countries. It has a 10-ton rear lift capacity with electronic hitch control.

The barn where New Holland was established

Great Manufacturers
New Holland

The story of the New Holland tractor's development is entwined with the lives of inventors, pioneers, and manufacturing icons. Today, New Holland is part of the Italian multinational holding company CNH Global NV, a worldwide manufacturer of agricultural and construction equipment, with representation in 170 countries.

THE NEW HOLLAND MACHINE COMPANY was founded in 1895 by a young Mennonite blacksmith named Abe Zimmerman in a barn in New Holland, PA. Zimmerman's earliest claim to fame was his design for a "frost-proof" stationary engine, which, with its unusual bowl-shaped water jacket, resisted freezing in the harsh North American winters. Zimmerman also manufactured mills for grinding animal feed and wood saws that were powered by his engines.

New Holland's current logo was introduced in 2008.

In 1947 the flourishing New Holland Machine Company was purchased by the Sperry Corporation—later Sperry Rand. The Sperry family established their name manufacturing navigation equipment and bomb sights during World War I, and formed the Sperry Corporation in 1933.

During World War II, Sperry developed bomb sights, airborne radar, and the infamous ball turret gun used on the Boeing B-17 Flying Fortress and Consolidated B-24 Liberator bombers. Sperry's purchase of the New Holland Machine Company coincided with a major breakthrough in fodder harvesting technology: the launch of the New Holland Haybine mower-conditioner. Looking to expand, the company, now known as Sperry New Holland, acquired a controlling interest in Claeys, a Belgian farm machinery firm, in 1964.

Claeys had built its reputation manufacturing threshing machines at its factory in Zedelgum, Belgium, and at the time of the Sperry New Holland merger the company was one of Europe's largest combine harvester manufacturers. With the Belgian company on board, Sperry New Holland was in a position to introduce the world's first twin-rotor combine, the New Holland TR, in 1975.

Harvesting machinery
Sperry New Holland purchased the Belgian combine harvester manufacturer Claeys in 1964. Today, Case New Holland is a world leader in grain harvesting technology.

Full production line
Located in Basildon, Essex, the British CNH tractor manufacturing facility was built by Ford to produce its 6X tractors. Today, the factory produces the New Holland T6 and T7 in addition to Case IH Maxxum and Puma models.

The 1980s were volatile years, forcing restructuring and mergers of many major manufacturing corporations. In 1986, Ford purchased Sperry New Holland and formed Ford New Holland. At the time, Ford was the UK's leading tractor manufacturer.

In 1991, Fiat, the largest tractor manufacturer in Europe, purchased an 80 percent stake in Ford New Holland. Ford and Fiat had previously collaborated successfully with the establishment of the Ford Iveco Truck in 1986, but the Fiat merger signified the final chapter in Ford's association with tractors. New Holland Geotech, as it was then known, gradually pooled its international resources, rationalizing the components and suppliers used to build its tractors. In Europe this included the British-built Ford 40 series and the Italian Fiat Winner ranges.

In November 1995 the name New Holland replaced Ford on the 40 series tractors built in Basildon, UK, to coincide with the launch of the forthcoming New Holland tractor

There are over **1.2 miles (2 km)** of **assembly lines** at the **New Holland** plant in Basildon, UK.

8970

TS110

T8040

T9

1895 Abe Zimmerman opens his blacksmith shop in New Holland, PA
1900 Zimmerman designs his stationary "frost-proof" engine
1903 New Holland Machine Co. is founded by Zimmerman
1906 Leon Claeys establishes his threshing machine business in Belgium
1933 Sperry Corp. is founded
1947 Merger of New Holland and Sperry Corp. and introduction of New Holland Haybine

1964 Sperry New Holland purchases Claeys, the Belgian combine harvester manufacturer
1974 Sperry New Holland introduces the world's first twin-rotor combine
1986 Ford buys Sperry New Holland and forms Ford New Holland
1991 Fiat buys Ford New Holland and establishes New Holland Geotech
1992 End of production at the New Holland Machine Co. building in New Holland, PA

1995 Existing Ford and Fiat tractor ranges are re-badged as New Holland tractors
1996 New Holland launches new M/60 and L/35 series tractors in Orlando, Florida
1997 New Holland launches TS tractor range
1999 New Holland and Case Corporation merge to establish Case New Holland (CNH). New Holland introduces the Basildon-built TM series
2007 Announcement that all New Holland tractors will support 100 percent biodiesel fuel

2009 New Holland's experimental NH^2 hydrogen tractor is unveiled
2011 Introduction of SCR technology to meet 2014's Tier IVa emission requirements
2013 A prototype T6.140 methane-powered tractor unveiled at Agritechnica, Hanover, Germany
2014 New Holland's Basildon plant celebrates 50 years of continuous tractor production with limited edition Golden Jubilee tractors

ranges. At the same time the North American 70 series adopted the New Holland moniker, and terra-cotta-painted Fiat tractors were re-badged in the company's blue livery.

In January 1996, New Holland launched its all-new M/60 series in Orlando, FL. The new Basildon-built range incorporated four models, spanning 100–140 hp and featuring a combination of technologies from the 40 series, 70 series, and Fiat Winner ranges. Also launched was the L/35 series, a five-model range built in Italy, based on the Fiat 9 series. The largest tractors in this latest New Holland lineup were produced at the Versatile factory in Winnipeg, Canada. Based on the articulated Versatile 82 series and powered by Cummins engines, the range offered outputs from 260 hp to 425 hp.

Later in 1996, New Holland NV was listed on the New York Stock Exchange. By 1998 it had sold more tractors in Europe than any other manufacturer. Meanwhile, in 1997 Fiat had acquired Case IH and merged all its agricultural divisions together to form Case New Holland (CNH). As part of the deal, New Holland divested its interests in the Versatile tractor factory, which was sold to Buhler Industries Inc.

Today, although marketed and identified as separate brands, British-built New Holland and Austrian-built Case IH tractors share many of the same components and technologies, supplied by CNH factories around the world. New Holland tractors are also produced in Brazil, China, and India.

Limited edition
To commemorate 50 years of tractor production at Basildon in the UK since 1964, New Holland produced limited edition Golden Jubilee versions of its T6.160 and T7.270 tractors. The machines have a unique metallic "Profondo Blue" livery, accentuated with gold detailing on the grilles and exhaust guard.

In response to the imminent arrival of Tier IVa emission regulations, CNH reformed its tractor range in 2006 to provide full compatibility with low-sulfur biodiesel fuels.

The company was also one of the first tractor manufacturers to adopt FPT Iveco engines with selective catalytic reduction (SCR) technology. SCR relies on urea-based diesel exhaust fluid to reduce the amount of harmful gases like NO_X in diesel exhaust fumes.

Canadian construction
The New Holland 70 series tractors were manufactured in Winnipeg, Canada, from 1993 to 2002. The range was the first to feature New Holland's Super-Steer front axle.

Out of the Ordinary

Specialty tractors for specific operations, high-acreage farms, and particular types of enterprise had traditionally been the preserve of smaller firms. By the early 21st century, though, the major makers were moving into this territory, some using international agricultural shows to unveil prototypes, and others fully launching and commercially producing such machines. Getting power to the ground more efficiently, reducing the impact of machines' weight on the soil, and the search for new fuel sources are key drivers of development.

▷ Multidrive Sixtrac

Date 2013 **Origin** UK	
Engine John Deere 6-cylinder diesel	
Horsepower 240 hp	
Transmission 18 forward, 6 reverse powershift	

The UK's Gloucestershire-based Multidrive, part of US firm Alamo, offers both four- and six-wheeled variants of its load-carrying tractors, designed to accommodate sprayers and fertilizer and lime spreader bodies to create self-contained, self-propelled units. The three-axle variant was built specifically for the broad-acre farms of Australia.

△ New Holland T6050 Hi-Crop

Date 2012 **Origin** UK	
Engine New Holland 6-cylinder diesel	
Horsepower 100 hp	
Transmission 16 forward, 8 reverse	

Equal-wheel, four-wheel-drive tractors fell out of favor in the UK following their 1970s heyday. But US farmers in particular continued to value the ground clearance and traction attributes of designs such as this New Holland T6050 for working in arduous or tall-crop conditions.

◁ JCB Fastrac 4220

Date 2014 **Origin** UK	
Engine AGCO Power 6-cylinder diesel	
Horsepower 220 hp	
Transmission Stepless CVT	

Unveiled as the High-Mobility Vehicle when first shown in 1990, JCB renamed its first tractor the Fastrac when it was launched the following year. By the mid-90s there were two ranges, with a new smaller, four-wheel steer version added. In 2014, this was heavily redesigned, becoming the new 4000 series.

△ Fendt Trisix Vario

Date 2011 **Origin** Germany	
Engine MAN 6-cylinder diesel	
Horsepower 540 hp	
Transmission Stepless CVT	

A surprise exhibit at the 2007 Agritechnica farm equipment show in Germany, this concept tractor was designed to combine the benefits of wheeled and tracked tractors. Both front and rear axles steered, and the machine was capable of 37 mph (60 km/h) on the road.

New Concepts

Tractor manufacturers are continuously seeking new ways around the challenges facing modern agriculture. From cleaner fuels to alternative drive systems that lessen the impact of tractors' weight on the soil, the design of many new features is driven by environmental and economic demands. Research also extends to looking at the differing needs of farmers around the world, resulting in tractor ranges built on a platform that allows them to be tailored to serve the common needs of farms of all types.

◁ Case IH Magnum 380 CVX Rowtrac

Date 2014 **Origin** US	
Engine FPT 6-cylinder diesel	
Horsepower 380 hp	
Transmission CVT	

Eighteen years after launching its Quadtrac rubber-tracked articulated tractor, Case IH introduced a tracked machine farther down its power range, with the unveiling of the 340 hp and 380 hp Magnum Rowtrac 340 and 380. The machines, which mirrored the standard Magnum models, retained a conventional front axle arrangement for steering.

△ **Claas Xerion 3800**

Date 2010 **Origin** Germany
Engine Caterpillar 6-cylinder diesel
Horsepower 380 hp
Transmission Stepless CVT

Claas committed fully to the tractor market in 2003, purchasing Renault Agriculture, but had been working on a high-horsepower tractor of its own for the previous decade and a half. A 250 hp Xerion was unveiled in 1993, later spawning versions with more than double that amount of power

◁ **Kubota Mudder M96s**

Date 2013 **Origin** Japan
Engine Kubota 4-cylinder diesel
Horsepower 95 hp
Transmission 8 forward, 4 reverse

While tractors with equal-sized wheels

largely fell out of favor in the UK, there remained a strong demand in countries such as the US for what North American farmers call "mudder" tractors, particularly for use in vegetable-growing operations.

◁ **Massey Ferguson 4708**

Date 2014
Origin China/India/Brazil/Turkey
Engine AGCO Power 4-cylinder diesel
Horsepower 82 hp
Transmission 8 forward, 8 reverse

In 2014 AGCO's Massey Ferguson brand announced two new tractor ranges designed to be built globally to meet the needs of farmers around the world. The 2700 and 4700 series machines are simple mechanical tractors made in factories in China, India, Brazil, and Turkey.

△ **New Holland T6.140 Methane Power**

Date 2013 **Origin** UK
Engine FPT 4-cylinder methane
Horsepower 135 hp
Transmission Stepless CVT

After researching the potential for hydrogen-powered tractors, New Holland switched its focus to the practicalities of methane as an alternative power source. It showed this methane-powered T6.140 tractor at the Agritechnica exhibition in Germany in 2013.

Valtra has been manufacturing tractors in Brazil since 1960

Great Manufacturers
AGCO

Established in 1990 AGCO—an acronym for Allis Gleaner Company—is the world's largest multi-brand tractor and agricultural equipment manufacturing conglomerate, incorporating four core brands: Massey Ferguson, Fendt, Valtra, and Challenger. Today, with factories around the globe, AGCO is represented by dealers in more than 140 countries.

AGCO'S ORIGINS BEGIN with a tenuous link to the fall of the Berlin Wall and a series of corporate realignments that attempted to take advantage of budding prospects in Eastern Europe. The story began when the West German engineering firm Klöckner-Humboldt-Deutz AG (KHD) resolved to sell its interests in its North American subsidiary company Deutz-Allis. The firm had sustained consistent financial losses for KHD since its establishment in 1985, so US Deutz-Allis executives were approached by KHD in 1988 to discuss a buyout that would leave the company solely in US hands. A deal was made, and on June 20, 1990, AGCO was officially founded by four former executives of the Deutz-Allis Corporation, namely Robert Ratcliff, John Schumejda, Edward Swingle, and Jim Seaver.

Prior to the establishment of AGCO, Deutz-Allis had marketed German-built Deutz tractors capable of producing 40–150 hp alongside US-built Allis-Chalmers Gleaner combine harvesters. Larger tractors with ratings up to 200 hp were produced under contract by White Farm Equipment, which had manufactured the Deutz-powered 9100 series at its factory in Ohio since 1989.

AGCO

AGCO's corporate slogan is "Your Agricultural Company".

After its foundation, AGCO continued to market the 9100 series, changing the paint scheme from Deutz green to Allis-Chalmers orange and rebranding the tractors as AGCO-Allis models. Simultaneously, a deal was established with the Italian company SAME-Lamborghini-Hürlimann to sell its air-cooled tractors under the AGCO-Allis brand name, thereby replacing the KHD Deutz lineup.

The AGCO quest for expansion was achieved through a constant stream of acquisitions, starting with the 1991 purchase of the White Farm Equipment tractor assembly works. In the same year AGCO also purchased the Hesston Corporation and 50 percent of Hay & Forage Industries in a deal with Case IH. The expansion continued apace as, in 1992, AGCO released one half of its stock for purchase on the NASDAQ stock exchange in New York.

A key factor in AGCO's initial success was its flexibility. The conglomerate allowed its dealers to become multi-brand outlets, providing customers with not only a wide choice of tractors, but also entire ranges of farming equipment. However, tractors always remained the most important part of AGCO's growth strategy.

In 1993 the rights to distribute Massey Ferguson products in the US were acquired, adding another 1,100 dealerships to the corporation's crossover network. In June 1994 AGCO completed its purchase of Massey Ferguson, took control of the company in its entirety, and secured its own position as a global corporation. At the time, Massey Ferguson had a significant share of the sub-100 hp global market and tractors made up 85 percent of its total sales. Further AGCO acquisitions included McConnell Tractors—later to become AGCOSTAR—and South America's leading brand, Deutz Argentina SA, which ironically had been liquidated by KHD when it sold Deutz-Allis.

In 1997 AGCO purchased Fendt GmbH, a well-established German manufacturer with a reputation for cutting-edge engineering technology. Fendt was instantly recognized as one of AGCO's premium assets, and under the conglomerate's management, the high-tech brand has developed extensively. The Vario tractor series has been extended and improved, and new combine and forage harvesters have been added to the product line. The purchase of Fendt allowed AGCO to utilize the German manufacturer's pioneering technologies, such as the Vario Infinitely Variable Transmission (IVT), first employed in the Fendt Vario 900 series. IVT allows for extremely accurate speed changes

Transmission innovators
Fendt was a pioneer of IVT, introduced in 1995's 926 Vario tractor. Today, IVT is available in tractors ranging from 70 hp to the 500 hp 1050 Vario of 2015.

AGCO products are sold by more than **3,100 independent dealerships** around the **world.**

8775 AGCO-Allis

White 6144

Challenger MT 865C

Fendt 936

1938 Klöckner-Humboldt-Deutz AG (KHD) formed after a restructuring of German mechanical engineering firms
1985 KHD establishes the Deutz-Allis Corp. in North America
1988 Negotiations begin to transfer Deutz-Allis from German to US ownership
1990 AGCO established by four former senior executives of Deutz-Allis
1991 Acquisition of White tractor assembly works in Ohio and the Hesston company
1992 AGCO is listed on the NASDAQ stock exchange
1993 North American distribution rights for Massey Ferguson tractors acquired
1994 Acquisition of Massey Ferguson completed. McConnell Tractors purchased
1995 AGCO buys the assets of Tye Co., the implement and tillage equipment manufacturers
1996 Deutz Argentina bought from KHD
1997 Fendt GmbH purchased
1998 AGCO starts a joint venture with Deutz AG to manufacture engines in Argentina
2000 Self-propelled agricultural equipment manufacturer, AG-Chem Equipment Co., acquired
2001 AGCO purchases manufacturing rights to the CAT Challenger range
2002 Massey Ferguson factory at Banner Lane, Coventry, UK closed by AGCO; tractor production is transferred to Beauvais, France
2004 AGCO acquires Finnish manufacturer Valtra, and its Sisu engine plant
2007 50 percent of Italy's leading harvester manufacturer, Laverda SpA, acquired
2009 AGCO announces plans to open two manufacturing sites in China
2010 Sparex Holdings, the tractor accessory and replacement parts distributor, is acquired by AGCO. Remainder of Laverda purchased
2012 AGCO establishes the Algerian Tractor Co., retaining a 49 percent share

Tracked technology
The 2001 acquisition of CAT Challenger enabled AGCO to compete in the high-horsepower rubber-tracked sector. Challenger is AGCO's most successful brand in North America.

and the technology has now been developed for use in the conglomerate's other brands.

AGCO secured the manufacturing rights to Caterpillar's rubber-tracked CAT Challenger range in 2001. The agreement enabled AGCO to compete directly with its competitors in the high-horsepower market sector and provided the additional benefit of giving AGCO access to the CAT Mobil-Trac system, which had been pioneered by Caterpillar in 1986. AGCO also expanded into self-propelled implements with the purchase of the AG-Chem Equipment company, the manufacturer of the RoGator and TerraGator sprayers and applicators.

In 2002 AGCO's bid to buy Renault Agriculture, owned by the famous French car manufacturer, was boycotted by French labor unions. The Renault tractor division was instead acquired by the German harvester manufacturer Claas. But Claas and AGCO soon found themselves partners when the former inherited Renault's share of the GIMA transmission factory in Beauvais, France. Renault had established the factory as a joint venture with Massey Ferguson in 1994.

In December 2002, AGCO ended production at the Massey Ferguson factory at Banner Lane in Coventry, UK, after 56 years of operation. Tractor manufacture was transferred to the business's facility in Beauvais.

The Nordic tractor and engine manufacturer Valtra, formerly Valmet, joined AGCO in 2004. The deal included the acquisition of SISU-Diesel, the offroad engine supplier whose engines are currently used in numerous AGCO-branded products. The first Valmet tractor, the 15 hp 15A was built in 1951. The Valtra name was introduced in 1970 when Valmet needed a separate brand identity for its implement range of lumber cranes, frontloaders, and backhoes. The first tractors to be branded as Valtras appeared in 1997.

The AGCO story continues today with a consistent flow of new acquisitions and the integration of new technologies throughout its expanding worldwide range of agricultural products.

Full-line equipment manufacturer
Massey Ferguson is AGCO's largest key brand with the widest range of products. Besides tractors, the company produces balers, harvesting, and landscaping equipment.

Robots Ready to Take Over

Automation is already well established in the farming industry: robots stack sacks of potatoes, and on many dairy farms, cows are milked automatically. Technology for automating at least some tractor work has been available for many years. Experiments with driverless tractors date back to 1956, when a UK research team demonstrated a system that allowed a tractor to follow electrical signals carried in a network of buried cables. Research has continued with technologies such as radio control and, more recently, GPS signals from space satellites.

DO TRACTORS NEED DRIVERS?

Many of the leading manufacturers have demonstrated driverless tractors that can carry out a range of routine tasks, but the response from farmers has been cautious. There are concerns about reliability and safety, and the legalities of insurance cover and road use are uncertain.

While driverless tractors are not featuring on farmers' shopping lists, some of the robotic technology is helping drivers improve output and efficiency. An example is GPS-linked guidance, which can steer a straighter line than the most experienced plow operator.

The complex steering maneuvers needed to cultivate between the olive trees on this plantation in Spain would be difficult to automate.

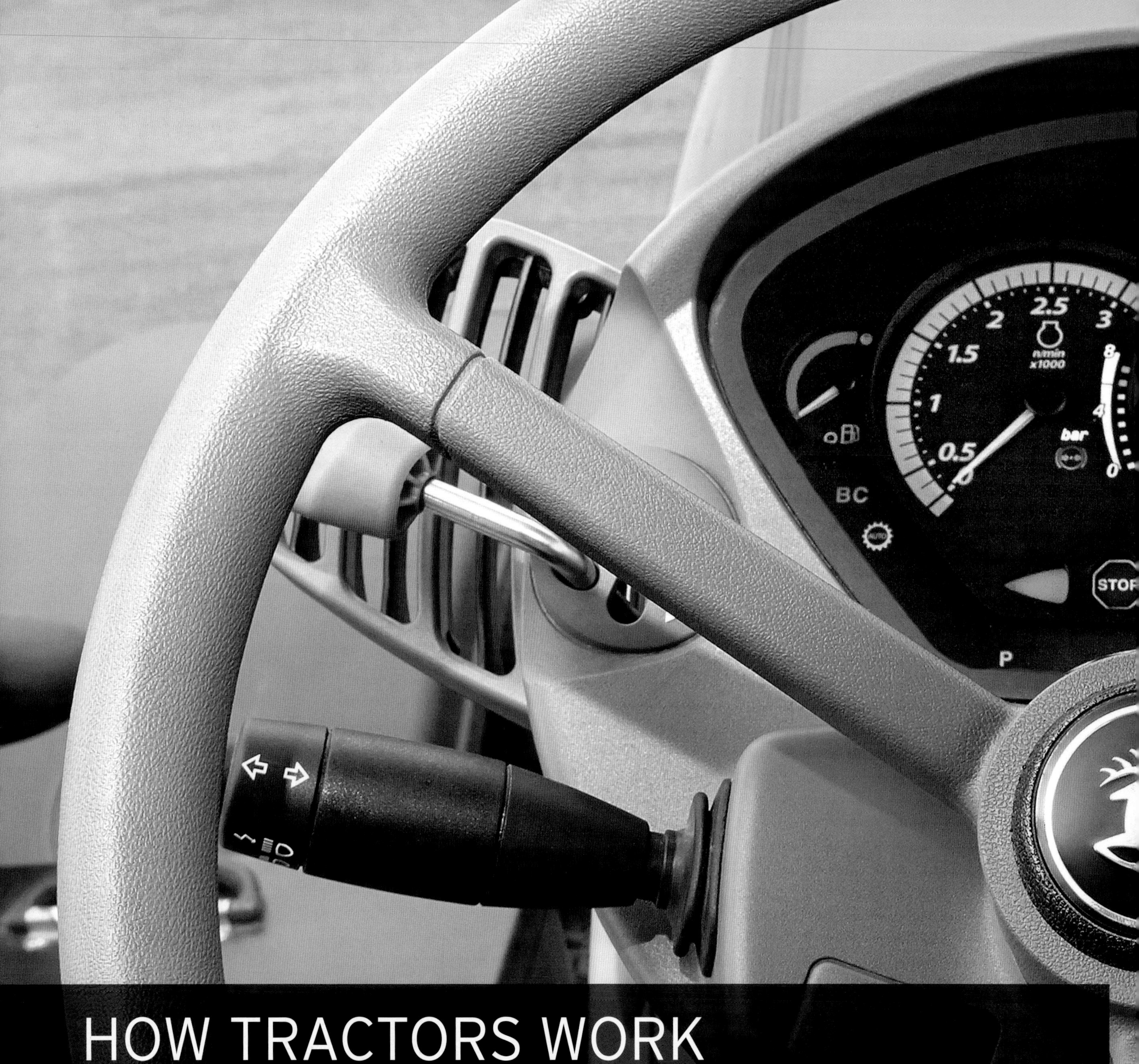

HOW TRACTORS WORK

TRACTOR TECHNOLOGY

km/h

Tractor Engines

One of the biggest problems in the early years of tractor development was engine unreliability. The primitive fuel and ignition systems designed in the first 20 to 30 years of the tractor's history meant that engines were difficult to start. Breakdowns in the field were a regular occurrence, and skilled mechanics capable of engine repairs were a rare breed. Almost all the early engines relied on gasoline to some degree, either as a fuel itself or to warm gasoline-kerosene engines to the temperature required for kerosene ignition. Despite producing less power per gallon, the low cost of kerosene made it a popular choice among farmers. Gasoline and kerosene engines grew more reliable throughout the 1920s and 1930s, until easy-starting diesels, first installed in tractors in the 1950s, became the standard. The switch to diesel has brought greater fuel economy and improved engine torque characteristics, putting more power through the wheels in difficult working conditions.

CLEANER, "GREENER" TRACTOR POWER

With rising fuel costs and growing environmental pressure to reduce exhaust emissions, today's engine designers are constantly seeking ways to use fuel more efficiently. Contemporary diesel engines electronically fine-tune fuel injection to the millisecond, minimizing waste. Turbochargers utilize exhaust airflow to improve combustion efficiency, while intercooling cools the air to the optimum temperature for efficient performance.

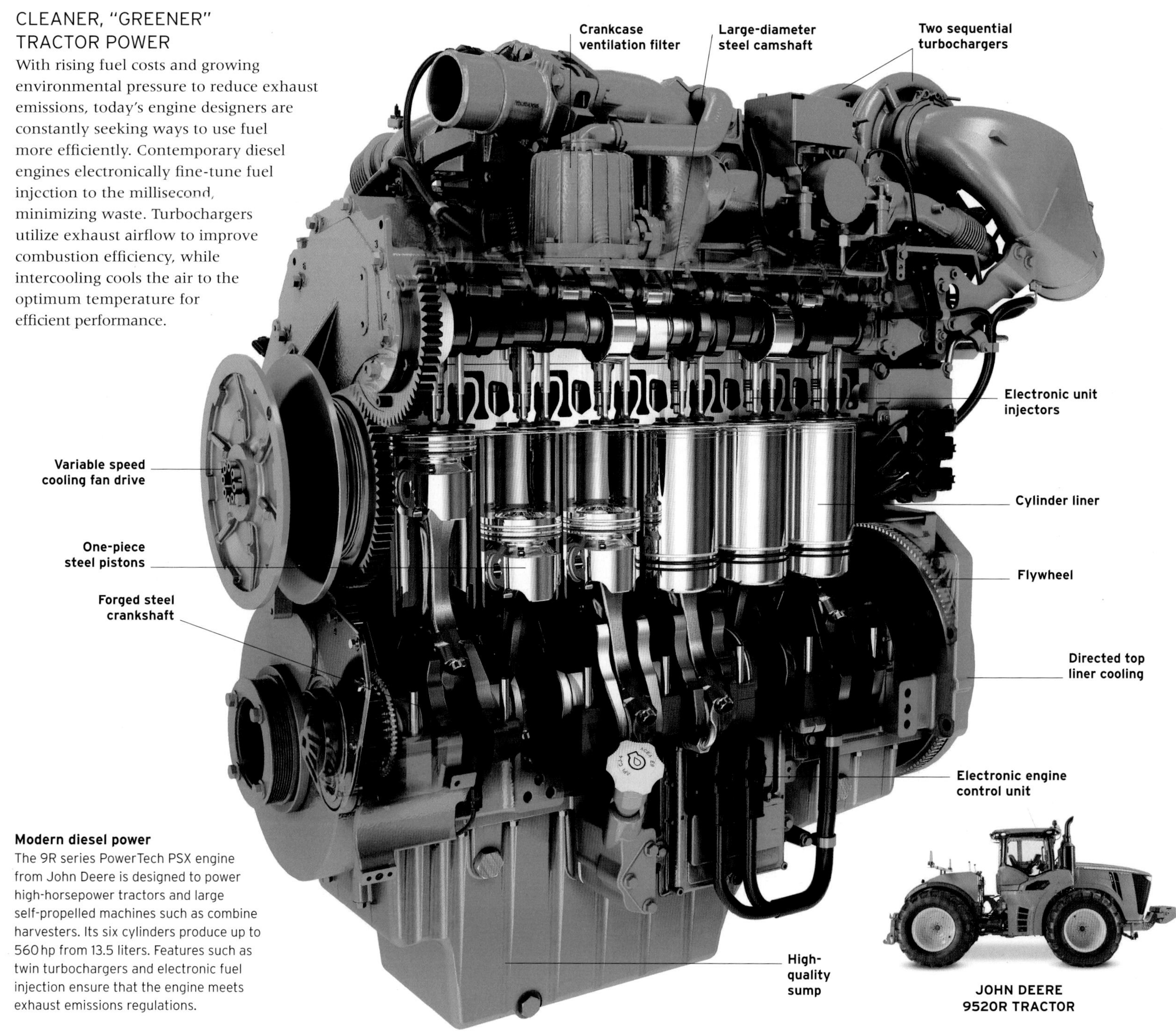

Modern diesel power
The 9R series PowerTech PSX engine from John Deere is designed to power high-horsepower tractors and large self-propelled machines such as combine harvesters. Its six cylinders produce up to 560 hp from 13.5 liters. Features such as twin turbochargers and electronic fuel injection ensure that the engine meets exhaust emissions regulations.

TRACTOR ENGINE DEVELOPMENT

The principles of tractor engines are simple: combustion of fuel in a cylinder pushes a piston whose motion is transferred via the machine's transmission to its wheels. However, the method of achieving that motion has varied widely. Single- and twin-cylinder engines employed a flywheel to maintain momentum. Hot-bulb engines used a hot metal bulb and compressed air to cause combustion. Modern diesel engines (see opposite) spray fuel into the cylinder just as the air is compressed to a great enough pressure to ignite the diesel spontaneously.

HORIZONTALLY OPPOSED TWIN-CYLINDER ENGINE

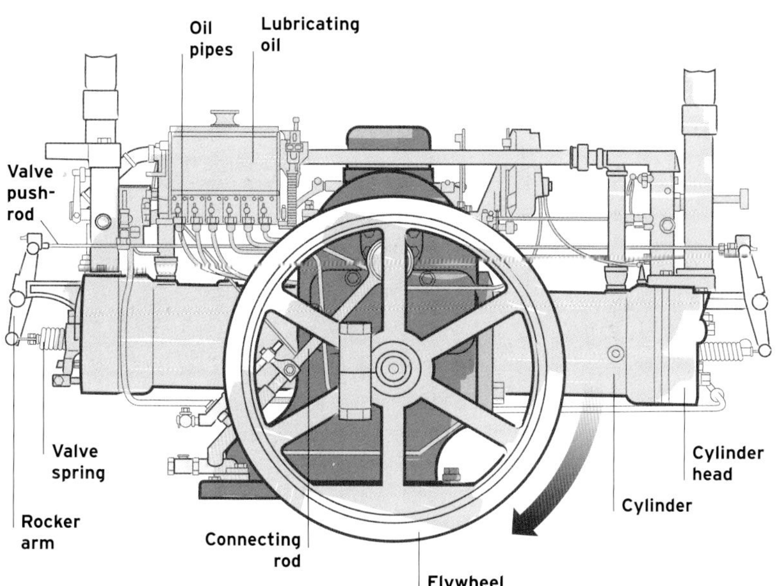

Horizontally opposed twin-cylinder engines were inherently smooth-running, the weight of each piston always in balance with its opposite. In addition, both cylinders were, in most cases, made up of identical parts. Using standard parts kept manufacturing costs down and meant dealers needed to keep far fewer parts on hand.

CASE 20-40, 1913

HORIZONTAL SINGLE-CYLINDER ENGINE

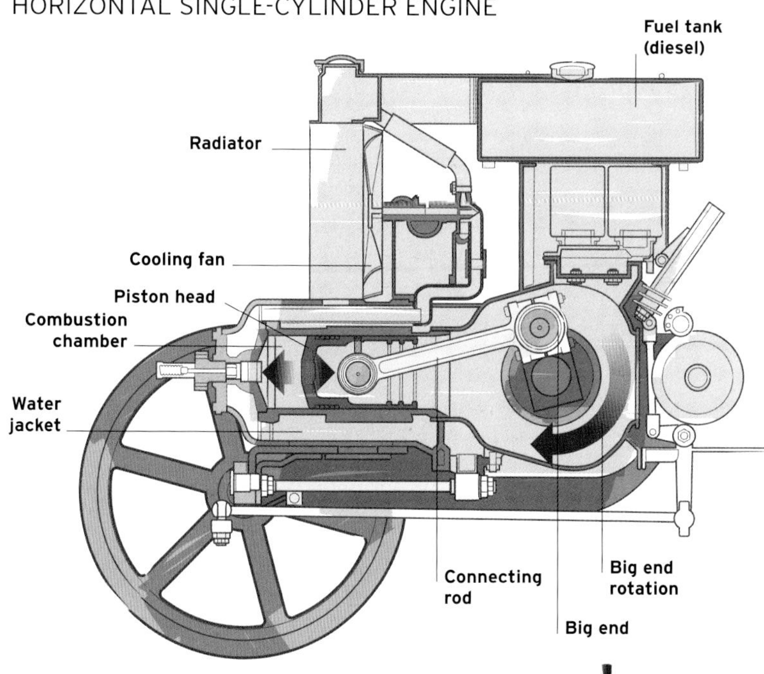

Single-cylinder engines were initially popular for their simplicity and ease of maintenance but soon lost appeal. The unbalanced cylinder configuration required a heavy flywheel and practical constraints on cylinder size limited power output to around 50 hp. Once farmers' requirements exceeded this, multiple cylinders became a necessity.

FIELD MARSHALL 3A, 1955

HOT-BULB ENGINE

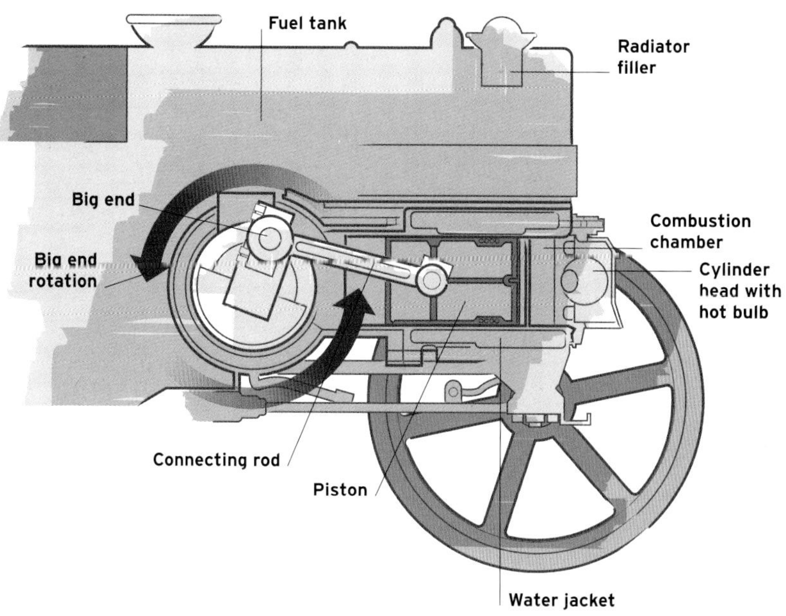

Hot-bulb or semi-diesel engines enjoyed a vast European customer base into the 1950s, largely because they were simple, reliable, and easy to maintain. The engine's ability to use low-grade fuel of varying specifications was an important asset. Their main disadvantage was the need to heat the bulb with a blowtorch to start the engine.

LANZ BULLDOG D2206, 1952

VERTICAL FOUR-CYLINDER ENGINE

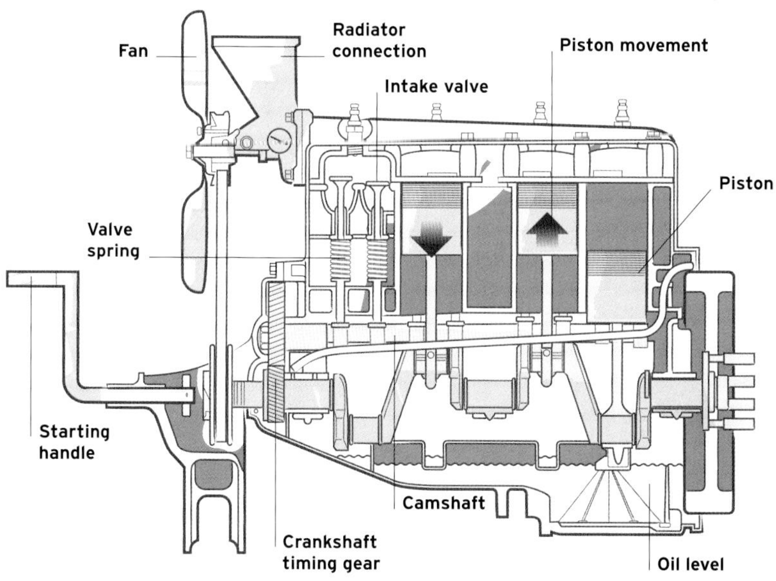

The vertical four-cylinder has become the most popular tractor engine format. The layout has advantages for users and manufacturers. It is relatively simple to build, making it a cost-effective alternative to five-, six-, and eight-cylinder engines. It runs smoothly and produces a high range of power outputs within a practical maximum engine speed.

FORDSON MODEL N, 1944

Wheels and Hydraulics

The modern tractor owes its versatility to developments dating from the early years of tractor history. The first tractors were primarily vehicles used to pull machinery and trailers, but their engines could also be equipped with a belt and pulley to power field equipment. A big step toward greater versatility was the development of the power takeoff, or PTO. A special shaft driven by the tractor engine, the PTO transmits power from the tractor to its implements. The first successful PTO was made available in 1918 on an International Harvester 8-16 "Junior" tractor. Further versatility was provided by hydraulic systems. Harry Ferguson's three-point linkage, which made its mass-market debut in 1936, allowed the use of the tractor's hydraulics to raise and lower implements. Some modern tractors generate electricity to power equipment in the field. The wide choice of tires and tracks available today gives tractors the flexibility to perform a multitude of tasks in varying ground conditions.

THREE-POINT LINKAGE

Harry Ferguson's three-point linkage was among the most important developments in tractor design. The Ferguson System allowed the use of the tractor's hydraulic power to lift and lower implements, giving drivers an unprecedented level of command over their equipment. The linkage's draft control sensed precisely the amount of force required to pull an implement and raised or lowered the linkage arms accordingly. The original linkages were rear-mounted only, but modern tractors can have a second linkage at the front, allowing the use of two implements at once.

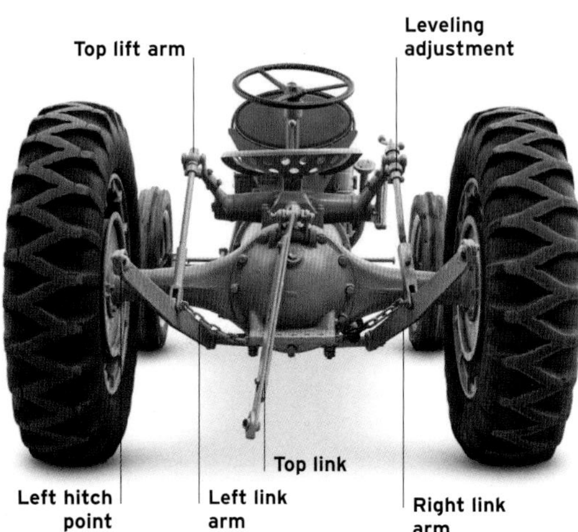

Ferguson's original linkage
The first production version of the three-point linkage is seen here on a Ferguson Type A tractor made by David Brown. This was the basis for the implement attachment system that is still standard equipment on most modern farm tractors.

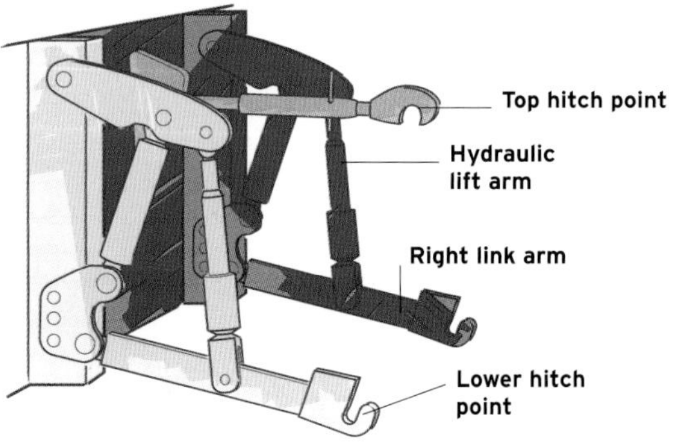

Side view of modern linkage

VERSATILE POWER SOURCE

Modern tractors like this John Deere offer plenty of engine power, which is used for much more than simply driving the tractor. Powering implements with the tractor expands the machine's role far beyond transportation and haulage, and opens up a long list of potential applications. Providing field power has been an important element of mechanization progress since the early 1900s. An early example was a tractor built in Scotland by Professor John Scott in 1903, which included a seed drill and cultivator combination that were powered by chain-and-sprocket drives connected to the tractor.

Power socket
A plug-in point to use electric power from the tractor is useful for operating lights when working on a public road. Power can be used to operate advanced equipment such as the metering system on certain seed drills.

Hydraulic power
Special spool valves that connect equipment to the tractor's hydraulic system are an important modern feature. For example, hydraulics are used to power dumping trailers and the turnover mechanism on reversible plows.

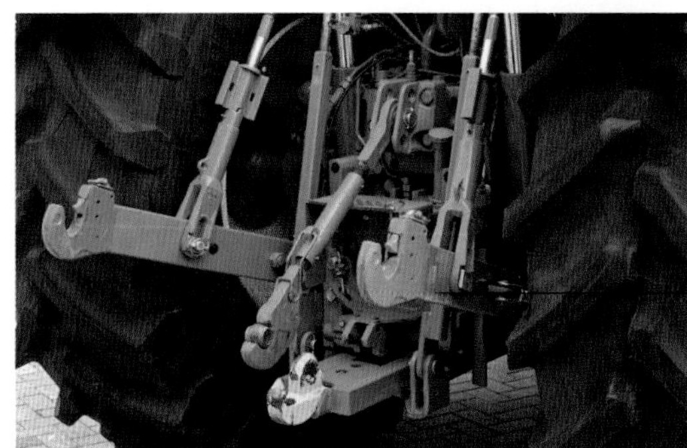

Modern linkage system
The tractor's three-point linkage is still hydraulically powered. But the refinements on modern tractors include electronically guided attachment and load sensing as well as a quick hitch design for rapidly securing lower lift arms to the implement.

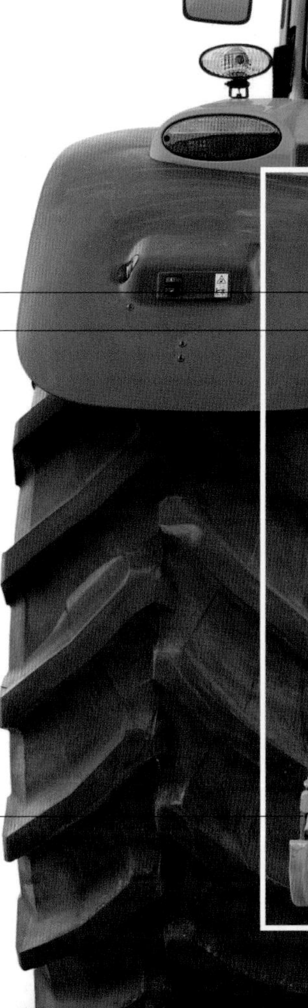

INVENTING THE WHEEL

Almost all tractors ran on steel-rimmed wheels until the mid-1930s. Plain steel rims were used for general transportation, but when extra grip was needed for field work, steel lugs were added to the rims. Tractors with lugs were usually barred from public roads because of the damage they caused, so detachable road rims were developed to fit over the lugs. Steel rims were unsuitable for fast travel, and the top speed for tractors was typically 3 mph (4.8 km/h). Speed, tractive force, and fuel economy all improved after special rubber tires were introduced by Allis-Chalmers in 1932. Within five years they were being installed on 50 percent of all new tractors.

STEEL RIMS
TWIN CITY, 1915

STEEL LUGS
ALLIS-CHALMERS, 1939

METAL ROAD RIMS
MASSEY HARRIS GP, 1932

INFLATABLE TIRES
ALLIS-CHALMERS, 1936

Remote controls
Push buttons on the rear mudguards allow tractor drivers to attach or adjust implements on the three-point linkage while keeping a safe distance from the machinery.

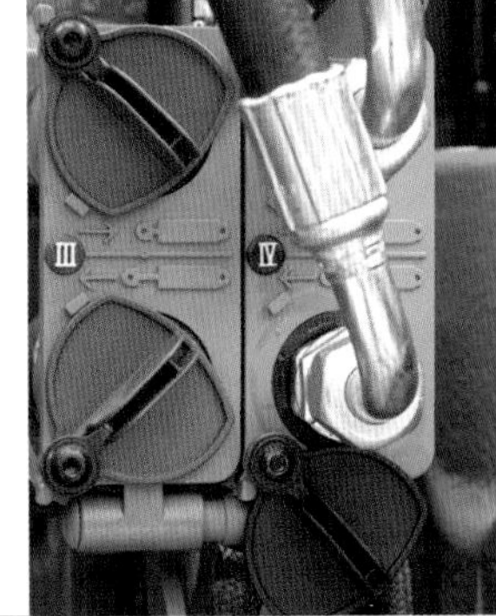

Air connection
This is used mainly for operating the air brakes of trailers and some towed machines. Using air-operated brakes provides safer, progressive stopping power and is a legal requirement for some equipment.

Power takeoff
The PTO shaft is still the most popular way to utilize engine power to operate machinery. It typically delivers up to 85 percent of the engine's maximum power to the implement.

TRACTOR TIRES

Tire tread patterns are available for special requirements. These include a ribbed pattern for non-powered front wheels, special turf tires for grass lawns, and extra-wide low-pressure tires that reduce soil compaction.

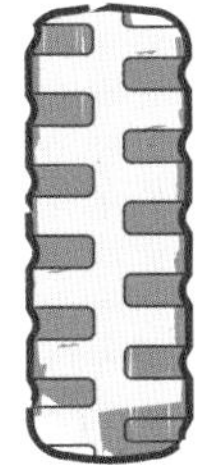

SOLID RUBBER TIRES

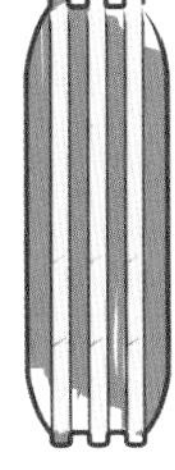

RIBBED TREAD

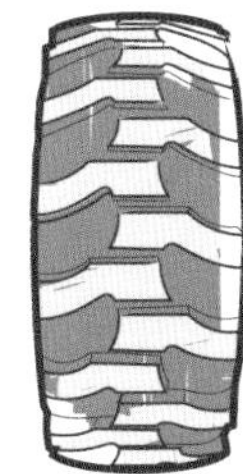

SPECIAL INDUSTRIAL TREAD

TYPICAL CHEVRON TREAD

TURF TREAD

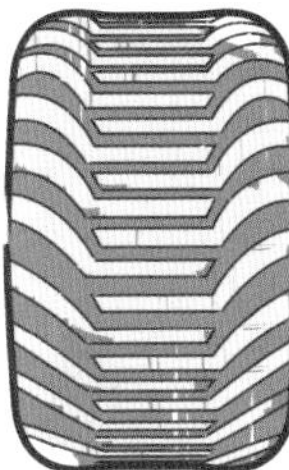

FLOTATION TIRE

TRACTOR TRACKS

Although less common than wheeled tractors, crawlers fulfill an important agricultural role. Tracklaying tractors offer a better grip and more pulling efficiency than tires, spread the tractor's weight over a bigger surface area to reduce soil compaction, and are particularly useful on rugged or mountainous terrain.

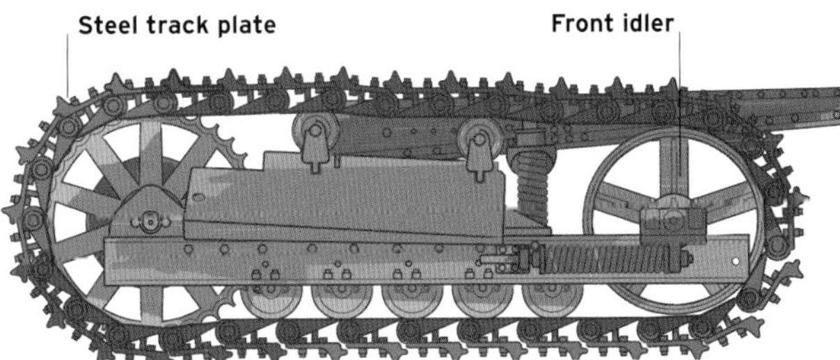

Metal tracks
For most of the 20th century, traditional steel tracks were a popular choice among farmers who wanted more pulling power, but travel speeds were slow, the machines were extremely noisy, and they were often banned from public roads.

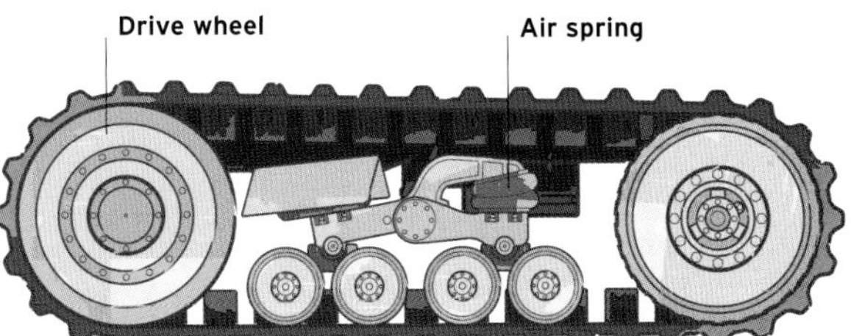

Rubber tracks
The steel-reinforced rubber tracks introduced by Caterpillar in 1987 and based on NASA technology offer all the traditional tracklayer benefits, but they are much faster and quieter than steel and, as they cause far less damage, can travel on the road.

Driving Technology

A safe, comfortable, and efficient working environment for the driver is a high priority when modern tractors are designed, but during the early years of tractor development, concern for driver welfare was almost nonexistent. Cabs offering shelter from the weather were a rare luxury during the first 60 or so years of tractor history, and it took even longer to provide safety cabs that could protect the driver if the tractor overturned. On some machines, exposed gear drives, spokes of heavy, fast-turning flywheels, and the steel lugs on drive wheels were often unguarded and could come dangerously close to the driver when the tractor lurched over rough ground. Trucks and vans offered much more evidence of concern for driver safety, taking steps to protect their operators long before similar developments took place on tractors. Both tractor manufacturers and customers who considered safety cabs a luxury must share the blame for the slow pace of improvement on farm tractors.

SITTING COMFORTABLY

Early tractor seats were shaped pieces of metal on sprung supports. Being descendants of the seats on horse-drawn farm machinery, they were not ideal for a long day's work. After World War II, padded seats quickly replaced the traditional design. Today's operators can expect seats with full adjustment and, in some cases, suspension that automatically adjusts to the driver's weight.

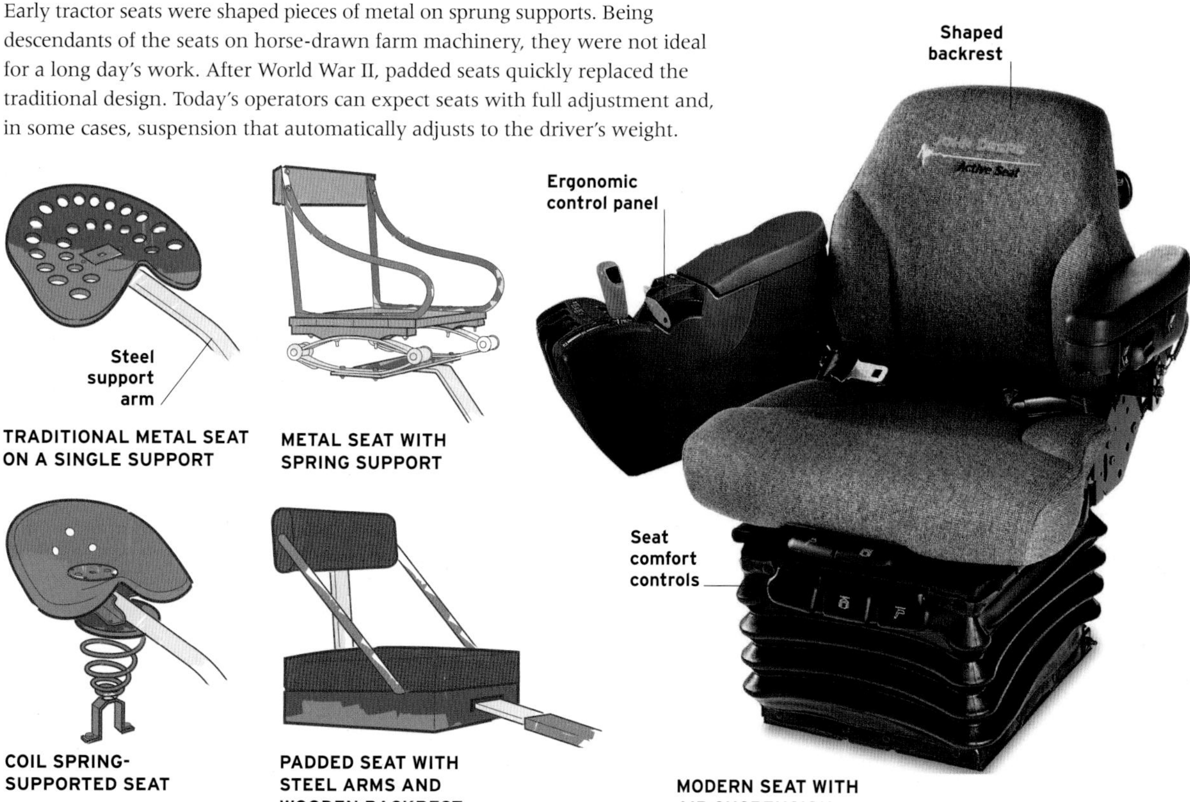

TRADITIONAL METAL SEAT ON A SINGLE SUPPORT

METAL SEAT WITH SPRING SUPPORT

COIL SPRING-SUPPORTED SEAT

PADDED SEAT WITH STEEL ARMS AND WOODEN BACKREST

MODERN SEAT WITH AIR SUSPENSION

TECHNOLOGY

Tractor Suspension

The suspension system for the axles and cab (or both) provides a smoother, more comfortable ride for the driver. For some tasks, this can mean faster working speeds and higher efficiency and productivity. For example, John Deere's front axle suspension absorbs shock loads from the wheels.

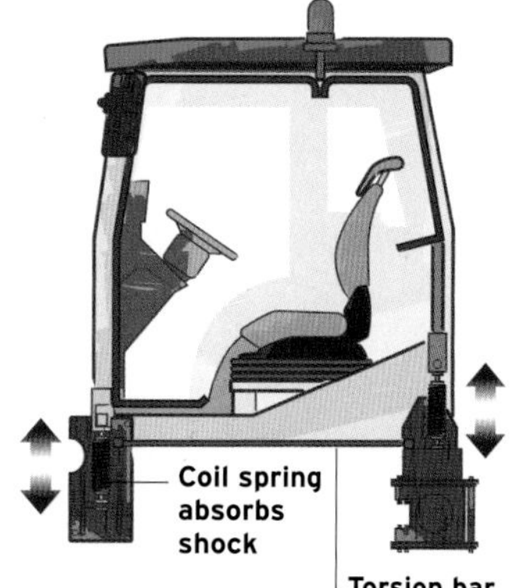

HYDROSTABLE CAB
Renault's latest suspension isolates the cab from the tractor by coiling springs around shock dampers on each corner of the cab. A torsion bar controls lateral movement.

DRIVER SAFETY

The number of drivers killed or seriously injured in overturning accidents rose as tractors grew in popularity, but effective action to resolve the problem did not arrive until 1959, when Sweden made it illegal to sell a new tractor without an approved safety cab. Other countries soon followed Sweden's example. A problem with the new cabs was that they magnified sound, and prolonged exposure to high noise levels could damage drivers' hearing. The result was a new generation of "Q-cabs" or quiet cabs.

Canopy (special IH demonstrator)
Still permitted in some countries, canopies are cheaper than cabs, but they offer limited weather and safety protection for the driver.

"Q-cab" (MF1080)
Special design features like extra sound insulation allow "Q-cabs" to achieve lower specified internal noise levels and provide a better working environment.

Modern cab (John Deere 6210R)
The latest cabs offer comforts like heated seats, refrigerators, and ergonomically designed controls, all of which allow drivers to stay in the field for longer.

FUTURISTIC TRACTORS

Modern tractors are designed to help the driver use all the machine's available power for maximum productivity. Transmission options on medium- and high-horsepower tractors are likely to include automatic gear shifting, alongside touch-sensitive controls that allow operators to exercise an extra level of precision. Many machines have eco options that automatically reduce engine speed and therefore fuel consumption when less power is required. A display screen with data from the International Organization for Standardization (ISO) keeps the operator constantly updated on the performance of equipment such as crop sprayers and seed drills.

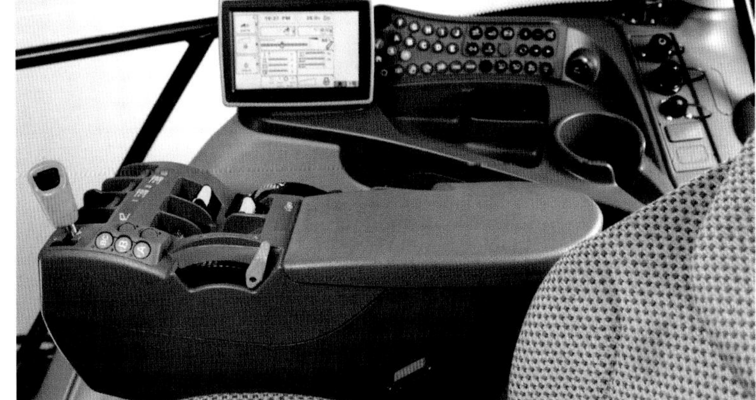

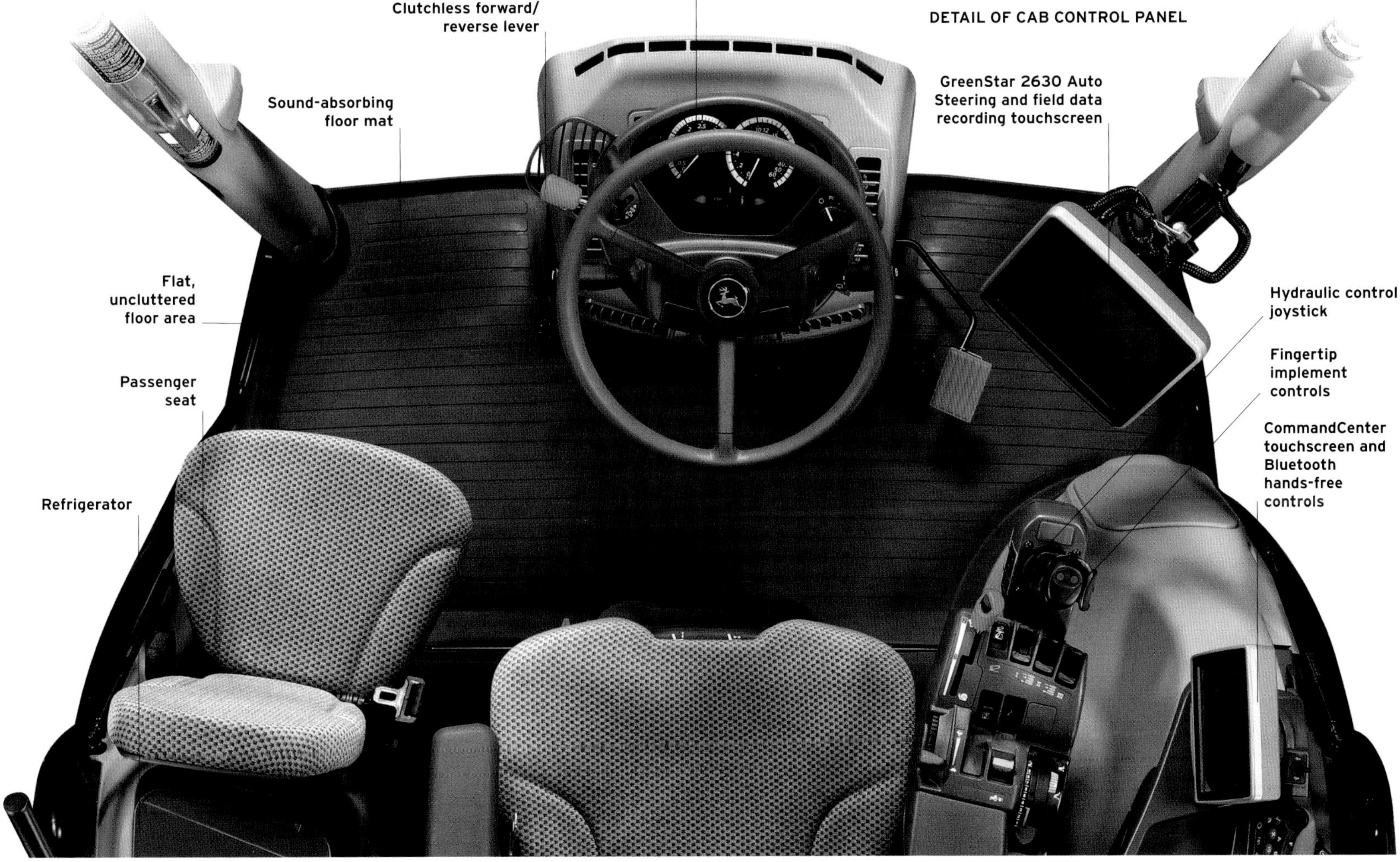

Driver-friendly cabs
The latest cabs are entirely geared toward driver efficiency. Frequently used controls for managing transmission, power takeoff, and hydraulic functions are easily accessible beside the right-hand armrest. The cab's modern features include a color display screen and even a cup holder.

DETAIL OF CAB CONTROL PANEL

Instrument panel
Clutchless forward/ reverse lever
Sound-absorbing floor mat
GreenStar 2630 Auto Steering and field data recording touchscreen
Flat, uncluttered floor area
Hydraulic control joystick
Fingertip implement controls
Passenger seat
CommandCenter touchscreen and Bluetooth hands-free controls
Refrigerator

SATELLITE-AIDED FARMING

Farm mechanization entered the space age in 1991 when Massey Ferguson unveiled a harvester linked to the Global Positioning System (GPS), a network of satellites more than 12,000 miles (19,300 km) from Earth. It was the first step in the development of precision farming, a system that uses advanced technology to achieve a greater fieldwork accuracy. GPS eases the driver's job by providing steering precision, guiding turnrow maneuvers, and automatically varying chemical application rates in different fields.

Signal receiver

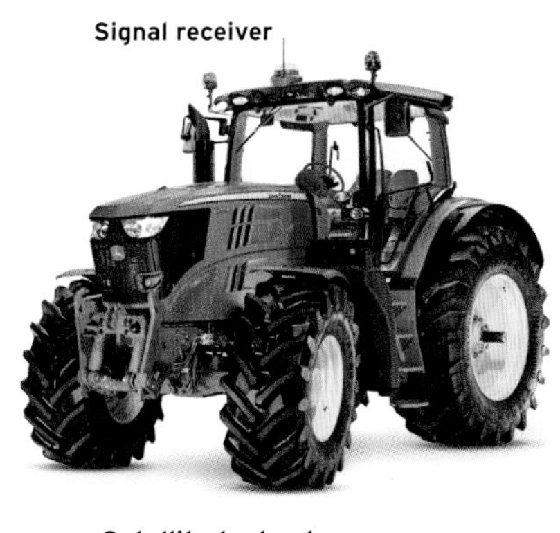

Satellite technology
This tractor's receiver picks up signals from GPS satellites and passes them on to the driver for accurate navigation.

Aerial
Electronic precision
GPS can maintain working accuracy in low-light conditions.
StarFire 3000

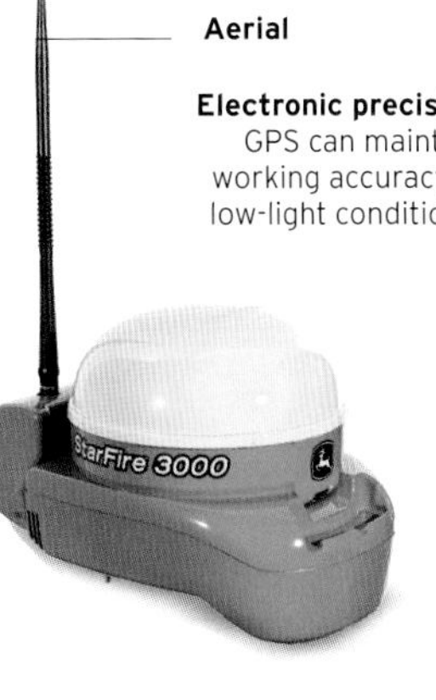

JOHN DEERE SIGNAL RECEIVER

Sky Plot
Graph
Satellite Tracking
Satellites in Solution 9
Satellites Above Elev Mask 8
Satellites Tracked 9
VDOP 1.8
HDOP 1.6
PDOP 2.4

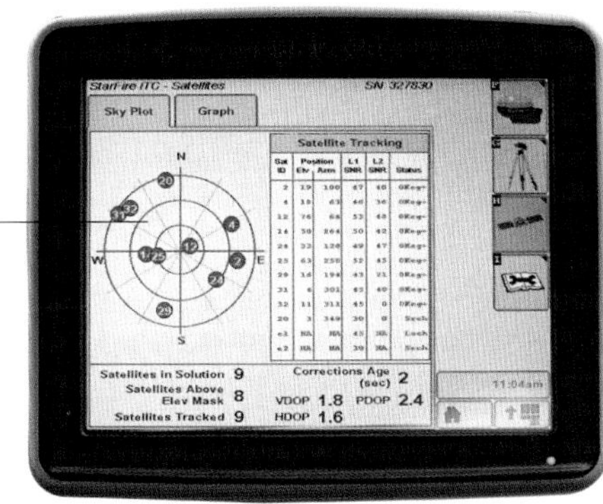

Operator interface
Information and control display screens can provide comprehensive performance data at a glance, and the touchscreen gives immediate feedback.

Glossary

Air cleaner
A device that removes dust from the air before it is drawn into the engine. Early tractors had a muslin-covered "cage." Later, a water-washer type bubbled air through a container of water. Modern machines have replaceable paper filters.

Articulation
A method of steering where the machine bends in the middle around a central pivot point. This mechanically simple method of steering is controlled by hydraulic rams actuated by the steering wheel.

Autotronic control
The electronic control of tractor brakes, differential lock, gear and power takeoff selection, and four-wheel drive. Most functions are engaged and disengaged by a push button.

Biodiesel
Fuel whose major content is diesel but has bio-additives, such as canola oil, corn oil, or soybean oil.

Bore
The diameter of each cylinder in a steam or internal combustion engine.

Cable control unit (CCU)
A unit attached to the rear transmission case of a tractor that controls, via cables, the raising, lowering, and emptying of towed scrapers. The CCU can also be arranged to control a towed ripper or a front-mounted dozer blade.

Change-on-the-move transmission
An early form of sychromesh gearbox used on a few tractors. Due to the difficulty of operating it, this gearbox was never successful. The Select-a-speed gearbox, offered at one time in Ford tractors, used a series of hydraulically controlled epicylic gear packs to give change-on-the-move gear changes.

Continuously variable transmission (CVT)
A transmission system that combines the stepless speed variability of a hydrostatic transmission with the mechanical efficiency of a traditional gear transmission.

Crab steering
A mode of steering in which all the axles on the tractor are steerable. The system can be programmed so that one axle is used for directional control. Both axles can be set to turn in opposite but synchronized angles to give a small turning circle. Both axles can be set to turn in the same direction at the same time to give the crab-steer element where the tractor moves sideways as well as moving backward or forward.

Crankshaft
The shaft that converts the oscillating motion and power of the pistons into a continuous revolving motion, which is then transferred to the drive wheels.

Cylinder
An enclosed chamber in which a piston moves to produce the power that is transmitted to wheels. In a steam engine, the piston is made to move by the force of the high-pressure steam acting against it (an external combustion engine). An internal combustion engine produces its power by burning liquid fuel introduced into the cylinder(s).

Datatronic control
A system of tractor management that allows the operator to program, coordinate, and memorize a wide range of functions such as turnrow maneuver, implement control, spool valve, and external services settings.

Diesel
A fuel distilled from crude oil comprising 75 percent saturated hydrocarbons and 25 percent aromatic hydrocarbons, with a minimum cetane number of 51(the measure of its volatility). Diesel is supplied as white diesel (derv) for road use, and red diesel (gas oil) for "off-highway" applications. Standard refined diesel, with no additives, starts to "gel" at 27.6°F (–3°C) and freezes at 17.6°F (–8°C).

Differential lock
A means of locking the axle differentials so that both or all wheels are positively driven. Useful in slippery conditions where one or more wheels lose traction and power is lost to the other wheels. Differential lock has to be disengaged to effect maneuvers or turn corners.

Direct-drive transmission
A type of transmission where the drive is transmitted mechanically by a series of gears. The number of gear ratios is dictated by the number and arrangement of the gears in the gearbox. A gear is selected for an operation and the tractor has to stop for the operator to change gear.

Direct-injection engine
A diesel engine in which the fuel is injected directly into the cylinders.

Drawbar
The point where trailed implements are attached to the tractor.

Drive wheel
The wheel (or wheels) that imparts, through the transmission, the power of the engine to the ground.

Dual-power
A form of transmission that in a constant-mesh gearbox gives the option, by means of a hydraulically operated clutch and gear pack, of two change-on-the-move speeds in any selected gear. In practice this means an eight-speed gearbox has 16 speeds.

Dual wheels
Extra wheels that are attached to the outer rim of a tractor's normal wheel and tire equipment. Dual wheels reduce the ground pressure, while at the same time increasing the tire area in contact with the ground, so improving traction.

Dynamometer car
A device to measure the tractive effort of a tractor. The dynamometer is driven from the car's wheels, and by altering and measuring the resistance applied to it, the power or force of the tractor can be recorded for any given ground conditions.

Equal-sized wheels
Wheels that are the same size on all the axles of the tractor.

Flotation tire
A tire that runs at lower pressure than a standard tire and has an extra-large cross section. This lowers the ground pressure, while at the same time increasing the area of tire in contact with the ground.

Flywheel
A heavy wheel attached to the end of the crankshaft of a steam or internal combustion engine. It stores energy from the power strokes of the engine and returns the energy to the crankshaft during non-power strokes. The flywheel also carries the transmission clutch and the ring gear to engage the starter motor during starting.

Four-stroke engine
An engine that gives a power stroke for every two revolutions of the crankshaft.

Full-line manufacturer
A manufacturer who produces a complete range of farm equipment. A full-line manufacturer can supply a farmer with all his machinery requirements.

Gas tractor
This is a term used to describe a tractor that has spark ignition and runs wholly on gasoline.

Gasoline/kerosene engine
An engine that is started on gasoline and changed over to kerosene once it has warmed up. The engine manifold is arranged so that the exhaust gases heat the incoming air/fuel mixture to aid the vaporization of the kerosene fuel.

General purpose
A tractor that can perform most of the operations required on a "mixed farm," such as plowing, cultivating, harvesting, and driving implements with the power takeoff. Stationary machinery can be driven from the belt pulley.

Hot-bulb engine
A type of engine, also known as a semi-diesel engine, that requires an external heat source for starting. This can be supplied by a blowtorch or, later, an electric spark. Once running, the blowtorch or spark is no longer required. These engines are capable of using low-grade fuel oils.

Horsepower (hp)
A unit of measurement used to express the power produced by a tractor engine. One hp is equal to 550 foot pounds per second or 745.7 watts.

Hydrostatic transmission
A system that transmits the power from the engine to the wheels or tracks through a hydraulic fluid medium. It provides an infinitely variable range of speeds in forward and reverse. This system is not very efficient and a much larger percentage of engine power is lost than in other types of transmission.

Hydraulically actuated brakes
The pressure applied to the brake pedal is transferred to the brakes by hydraulic fluid within the braking system.

Hydraulics
The use of a fluid medium under pressure in a cylinder to perform lifting, lowering, and control of actions of a tractor's implements.

Infinitely variable transmission (IVT)
A system that provides completely stepless speed changes at all engine rpm settings.

Intelligent-power management
The computerized control of the interaction between a tractor's engine and its transmission-management systems to achieve the optimum fuel economy and work output.

Kerosene
A combustible hydrocarbon fuel that was used to run tractor engines because it is cheaper than gasoline. Kerosene requires an engine to have a lower compression ratio than that needed if running continuously on gasoline. Contrary to popular belief, paraffin (UK), kerosene, and tractor vaporizing oil (TVO) are not exactly the same: paraffin has an octane rating of 0, kerosene a rating of 15–20, and TVO is rated 55–65.

Linkage
The mechanism used to attach implements and equipment to the front or rear of a tractor. The linkage lifts and lowers the implements into work, and in some cases controls the depth. It also carries all or part of an implement for road travel.

Low-tension ignition
A system in early tractors used to create the spark for ignition using a low-tension current supplied by a battery or low-tension magneto. The contact breaker points that make the spark are located inside the cylinder, and the point of ignition is achieved by mechanical means.

Multi-plate clutch
A clutch that is made up of any number of drive plates and driven plates: there are always two more driven plates than drive plates. Crawler tractor steering clutches are generally of this type. In some transmission systems, multi-plate clutches are used to engage different ratios within a given gear set.

Nebraska test
An assessment to test that a tractor's power, fuel economy, and performance meet the manufacturer's claims, first instigated in 1919 by the University of Nebraska. In the US, no tractor can be sold unless it has been the subject of a Nebraska test.

Nominal horsepower (nhp)
This is a very approximate measure of the power of early steam engines, which was used by manufacturers to give customers an idea of an engine's power when compared to that of a horse. Nhp was calculated on piston area, but did not take into account the piston stroke or boiler pressure.

Pivot steer
A method of steering that allows a machine to turn, or "pivot," on the spot. It is achieved by reversing the wheels or tracks on one side while at the same time setting the other side to forward.

Power takeoff (pto)
The shaft at the rear of a tractor to which a flexible drive shaft can be attached to transmit power from the tractor's engine to an implement that requires an external power source. The output shaft sizes, with speeds of 540 and 1,000 rpm, have been standardized by manufacturers.

Powershift transmission
Transmission that provides the ability to change gear under full power, on the move, and without the use of the clutch.

Precision farming
A system of farming that uses satellite navigation, mapping, and positioning for all the field applications. Results on crop performance are downloaded into mapping software to provide information for the future use of fertilizers and chemicals on that area of land.

Rating (horsepower):
The horsepower rating of an engine can be given in many different ways. The ratings can be shown as brake horsepower (bhp); drawbar horsepower (dhp); metric horsepower (din); Society of Automotive Engineers (SAE) specification, or kilowatts (kW). Each rating has its own specific method of measuring criteria.

Revolutions per minute (rpm)
The number of times a shaft, such as a crankshaft or power takeoff shaft, turns in one minute.

Ripper
A heavily built single or multi-tined implement that is used to loosen hard soil or rock.

Row-crop tractors
Crops such as potatoes and sugar beet are grown in rows a set distance apart. The row-crop machinery used in their cultivation and harvesting has to have wheels set the correct distance apart so that the tractor can run in the spaces between these rows without damaging the growing crop. Narrow-wheeled equipment is also required for this application.

Scraper
A heavy piece of earth-moving equipment used in construction applications to dig, carry away, and relocate soil.

Selective catalytic reduction (SCR)
The means of converting, by the use of a catalyst, nitrogen oxides contained in exhaust emissions into harmless nitrogen and water.

Semi-diesel engine
See *Hot-bulb engine*

Skid unit
These are units supplied by tractor makers to outside manufacturers who want to build their own types of machines, but who require a proprietary power unit. Skid units are supplied to the customer's specification, but typically comprise the engine (complete with hood and fuel tank), gearbox, and rear axle (with or without hydraulics). The cab, wheels, and front axle are not usually part of the package. Some customers use the tractor cab as a complete component, positioning it according to the design of their particular machine.

Sprocket
The part of a track-type tractor that engages with the tracks in order to transmit the engine power to the track components. Positive-drive, rubber-tracked tractors also use a sprocket to transmit the power to the tracks.

Steam engine
An engine that uses high-pressure steam from boiling water to drive the pistons in the cylinders. The term can also refer to any self-moving, portable, semi-portable, or stationary engine.

Super-steer
A type of steering that gives a very tight turning circle. The front axle beam can be angled into the turn by a few degrees; then the wheels are turned on their king-pins, in the conventional manner, further into the turn. This feature is very useful for applications such as turning into the ends of potato or sugar beet rows.

Synchromesh
A mechanism within a gearbox that sychronizes the speed of the gears during a gear change to make the change easier and avoid the necessity to double declutch.

Tachometer
The instrumentation that indicates the rpm of an engine and the forward speed at that rpm in any given gear.

Tier IV emission standards
Government-mandated regulations that all engine manufacturers must meet when building units for tractors in order to reduce air pollution. Tier IV emission standards were phased in from 2008 onward.

Three-point linkage
The means of attaching an implement to a tractor, which consists of two lower, or draft links (arms), and a third upper, or top, link. Originally patented by Harry Ferguson, this system incorporates hydraulic lifting, lowering, and depth-control capabilities.

Threshing machine
A machine driven by a steam engine or tractor used to mechanically separate grain from chaff and straw, and at the same time clean the grain to a marketable standard.

Tire tread
This is the pattern of the rubber that is molded to the part of the tractor tire that is in contact with the ground. Tire tread is nonreversible and designed to be self-cleaning in one direction only. Tread engages with the ground surface and transfers the power of the tractor into forward motion and useful work.

Torque
The twisting force of a turning component to rotate an object around an axis.

Total-loss lubrication system
A system used in early tractors in which the oil that is supplied to the engine cylinder, or cylinders, and bearings is metered out by drip-feed or pump at the required rate to provide adequate lubrication. The oil, once used, drains away and is lost from the system.

Track width
The width across a tractor measured from the center of the wheel or track tread on one side, to the center of the wheel or track tread on the other side.

Traction engine
The generic term for a self-moving engine driven by steam, gasoline, or oil. The engine could be used for driving and moving a threshing machine, baler, or chaff-cutter, or any other stationary plant requiring power. It could also be used for general haulage duties.

Transmission
The system that transmits or turns the power of the engine, which is supplied at relatively high revolutions into a range of land speeds, both forward and reverse.

Trembler coil ignition
An ignition system used in early tractors. A low-tension electrical current is supplied from either a battery or a dynamo built into the flywheel to trembler coils, one per cylinder. A distributor timed to the engine sends the low-tension current to each coil in turn. The coil then instantly creates a high-tension current that sparks the plug and so fires that cylinder.

Turnbuckle top-link
The third part of the "three-point linkage." A turnbuckle top link has left-hand screw threads at one end and right-hand threads at the other, and a turnbuckle between the two. The length of the turnbuckle top-link can be adjusted to level the attached implement.

Two-stroke engine
An engine that gives one power stroke for every revolution of the crankshaft.

Stroke
The distance moved by the piston from one end of the cylinder to the other in an engine. Stroke can also be expressed as measured from top to bottom dead centers of the crankshaft.

Urea-based diesel exhaust fluid
A fluid made up of 32.5 percent urea and 67.5 percent deionized water. The fluid is injected as a consumable aqueous solution into the exhaust stream, where it reacts with the emissions in order to reduce the harmful nitrous oxides produced by modern diesel engines.

Variable horsepower (vhp)
An interactive system between the gearbox and the power output of the engine. In low gears, the power setting is at the standard rating for the engine. In the higher gears, the power setting is raised to a predetermined level. This increases the work rate, but at the same time protects the transmission from the torque overload that would result if the high power settings were used in the low gears.

Yield mapping
The part of "precision farming" that collects and records information on the variations in crop yield over a given area.

Index

All page references are given in *italics*

I

J

K L

M

Acknowledgments

Dorling Kindersley would like to thank Stuart Gibbard for his support throughout the making of this book.

In addition, Dorling Kindersley would like to extend thanks to the following people for their help with making the book: Sue Gibbard; Dennis Bacon; Rory Day; Ted Everett, AGCO; Steve Mitchell, ASM Public Relations Ltd; Mark James, Wade D. Ellett, and Neil Dahlstrom, John Deere; Erik de Leye, Media Relations Representative, EAME, Caterpillar Inc., for checking the text for Caterpillar; Sarah Pickett and Richard Wiley, CNH Industrial; Jason Sankey, Marketing Communications Manager, JCB Agriculture, and Jane Cornwall, Press Office Administrator, JCB World Headquarters; Graham Barnwell, SAME Deutz-Fahr; Caroline Benson, Museum of English Rural Life, University of Reading.

The publisher would like to thank the following people for their help with making the book: Steve Crozier at Butterfly Creative Solutions, Adam Brackenbury, and Tom Morse for colour retouching; Amy Orsborne, and Daine Stahr for design help; Joanna Edwards and Catherine Saunders for editorial help; Sonia Charbonnier for technical support; Sachin Singh at DK Delhi for DTP help; Joanna Chisholm for proofreading; Helen Peters for the index.

The publisher would also like to thank the following museums, companies, and individuals for their generosity in allowing Dorling Kindersley access to their tractors for photography:

Carrington Steam and Heritage Show, 2014
Main Road, Carrington, Nr. Boston
Lincolnshire, PE22 7DZ, UK
www.carringtonrally.co.uk
With special thanks to Malcolm Robinson and Alex Bell and all the tractor owners

Chandlers (Farm Equipment) Ltd.
Main Road, Belton, Grantham, Lincolnshire, NG32 2LX, UK
www.chandlersfe.co.uk
With special thanks to Gavin Pell and Mick Thrower

James Coward, Thorney, Spalding, Lincolnshire, UK

Robert H. Crawford & Son
Agricultural Engineers & Manufacturers
Frithville, Boston,
Lincolnshire, PE22 7DU, UK
www.rhcrawford.com
With special thanks to Robert Crawford

Roger and Fran Desborough, Halesworth, Suffolk, UK

Doubleday – Holbeach
Old Fendyke, Holbeach St Johns, Spalding, Lincolnshire, PE12 8SQ, UK
www.doubledaygroup.co.uk
With special thanks to Graham Collishaw

Doubleday – Swineshead
Station Road, Swineshead, Boston, Lincolnshire, PE20 3PN, UK
www.doubledaygroup.co.uk
With special thanks to Zoe Doubleday-Collishaw and Luke Spencer

Andrew Farnham, Wisbech, Cambridgeshire, UK
with special thanks to Reg Mattless, Tracy Farnham, and David Drake

Fenland Tractors
Station Yard, Postland, Crowland, Peterborough, Cambridgeshire, PE6 0JT, UK
With special thanks to Martyn Stanley for his help during the photoshoot
www.fenlandtractors.co.uk

Henry Flashman, Gunnislake, Cornwall, UK

Geldof Tractors
Vierschaar 4, B-8531, Harelbeke, Belgium
www.geldof-tractors.com
With special thanks to Michel Geldof and Steven Vanderbeke

Stuart Gibbard
Gibbard tractors, specialists in original tractor literature
www.gibbardtractors.co.uk

Peter Goddard, Diss, Norfolk, UK

Great Dorset Steam Fair
The National Heritage Show
Tarrant Hill, Blandford Forum, Dorset, DT11 8HX, UK
www.gdsf.co.uk
Special thanks to Martin Oliver and Sarah Oliver and all the tractor owners

Happy Old Iron
B-3670, Meeuwen, Belgium
www.happyoldiron.com
With special thanks to Marc Geerkens and Lennert Geerkens

Paul Holmes, Boston, Lincolnshire, UK

Tim Ingles, Moreton in Marsh, Gloucestershire, UK

Keystone Tractor Works
880 W Roslyn Road, Colonial Heights, VA 23834, USA

Richard and Valerie Mason, Swineshead, Boston, Lincolnshire, UK

Rabtrak Ltd.
The Poplars, Fulney Drive, Spalding, Lincolnshire, PE12 6BW, UK
www.rabtrak.co.uk

Paul Rackham Ltd. Tractor Collection
Camp Farm, Roudham, Norwich, Norfolk, NR16 2RL, UK
With special thanks to Paul Rackham and to Lee Martin for his help during the photoshoot

Rural Pastimes at Euston Park
Euston, Suffolk, IP24 2QH, UK
http://www.eustonruralpastimes.org.uk
With special thanks to Tim Shelley, Henry Castle, and all the tractor owners

The Shuttleworth Collection
Shuttleworth (Old Warden Aerodrome)
Nr Biggleswade,
Bedfordshire, SG18 9EP, UK
www.shuttleworth.org

Piet Verschelde Antique Tractors
Mannebeekstraat 1 B-8790, Waregem, Belgium
www.pietverschelde.com
With special thanks to Piet Verschelde

David Wakefield, Bury, Huntingdon, Cambridgeshire, UK

The Ward Collection
with additional thanks to Oliver Wright for his help during the photoshoot

Andrew Websdale, Norwich, Norfolk, UK

Lister Wilder
The Park, Portway, Crowmarsh, Wallingford, Oxfordshire, OX10 8FG, UK

A R Wilkin, Barnham, Norfolk, UK

PICTURE CREDITS
The publisher would like to thank the following for their kind permission to allow Dorling Kindersley to photograph their vehicles:
(Key: a-above; b-below/bottom; c-centre; f-far; l-left; r-right; t-top)

All DK images shot by Gary Ombler

1 DK: Courtesy of Paul Rackham Ltd (c). **2–3 DK:** Courtesy of Paul Rackham Ltd (c). **4 DK:** Courtesy of Paul Rackham Ltd (br). **5 DK:** Courtesy of Henry Flashman (bl), Courtesy of Paul Rackham Ltd. (br). **6 DK:** Courtesy of Paul Rackham Ltd. (br); Courtesy of Richard and Valerie Mason (bl). **7 DK:** Courtesy of Doubleday Swineshead (BR); Courtesy of Paul Rackham Ltd. (bl). **8 DK:** Courtesy of Andrew Websdale (br); Courtesy of Henry Flashman (bl). **9 DK:** Courtesy of Doubleday Swineshead (br); Courtesy of Richard and Valerie Mason (bl). **10–11 DK:** Courtesy of Robert Crawford (c). **14 DK:** Courtesy of Ernie Eagle (tc); Courtesy of Happy Old Iron, Marc Geerkens (cl). **15 DK:** Courtesy of Trevor Wrench and Trish Bloomfield (tl); Courtesy of Natel Taylor (cra); Courtesy of the Farwell family (cr); Courtesy of Mary and Brian Snelgar (bl). **18 DK:** Courtesy of The Ward Collection (cr); Courtesy of Happy Old Iron, Marc Geerkens (tc); Courtesy of Robert Crawford (clb). **19 DK:** Courtesy of Malcolm Robinson (tc); Courtesy of Mike Kendall (bc). **20 DK:** Courtesy of Robert Crawford (bl); Courtesy of Robert Crawford (clb). **20–21 DK:** Courtesy of Robert Crawford (c). **21 DK:** Courtesy of Robert Crawford (all). **22 DK:** Courtesy of The Ward Collection (cra), (cla), clb); Courtesy of Geldof Tractors (tr); Courtesy of Mick Patrick (br). **23 DK:** Courtesy of The Ward Collection (crb), (clb); Courtesy of Roger and Fran Desborough (ca), (bc); **26–27 DK:** Courtesy of Paul Rackham Ltd. (c). **27 DK:** Courtesy of D. West, Canterbury (tr); Courtesy of The Ward Collection (tl); Courtesy of Happy Old Iron, Marc Geerkens (bl); Courtesy of Paul Rackham Ltd. (cra). **28 DK:** Courtesy of The Ward Collection (br), (cla); Courtesy of Keystone Tractor Works (cra); Courtesy of Geldof Tractors (CLB). **29 DK:** Courtesy of Paul Rackham Ltd (tr), (bc); Courtesy of Paul Rackham Ltd. **30 DK:** Courtesy of Simon Wyeld and family (cla); Courtesy of The Ward Collection (CRB); Courtesy of R.C. Gibbons (bl). **31 DK:** Courtesy of The Ward Collection (cra), (bc); Courtesy of Richard Vincent (tl). **34 DK:** Courtesy of The Ward Collection (crb); Courtesy of Derek Mellor (cla); Courtesy of Paul Rackham Ltd. (tr), (bl). **35 DK:** Courtesy of The Ward Collection (CRB); Courtesy of Paul Rackham Ltd. (bc), (tc). **36 DK:** Courtesy of Paul Rackham Ltd. (bl, all). **36–37 DK:** Courtesy of Paul Rackham Ltd. (c). **37 DK:** Courtesy of Paul Rackham Ltd. (all). **40 DK:** Courtesy of Happy Old Iron, Marc Geerkens (tc), (cla), (bc); Courtesy of Geldof Tractors (bl). **40–41 DK:** Courtesy of The Ward Collection (c). **41 DK:** Courtesy of The Ward Collection (br); Courtesy of James Coward (CA); Courtesy of Geldof Tractors (tr); **42 DK:** Courtesy of The Ward Collection (cr), (tc); Courtesy of Piet Verschelde (bc). **45 DK:** Courtesy of Doubleday Swineshead (ftr); Courtesy of John Bowen-Jones (tr); Courtesy of Keystone Tractor Works (tl); Courtesy of Paul Rackham Ltd. (ftl). **46 DK:** Courtesy of Paul Rackham Ltd. (c). **50 DK:** Courtesy of L. Gilbert: Chellaston (tr); Courtesy of Paul Rackham Ltd. (cr); Courtesy of Peter Goddard (clb), (br). **51 DK:** Courtesy of Andrew Farnham (tr); Courtesy of Paul Rackham Ltd. (tl); Courtesy of Peter Goddard (bc). **54 DK:** Courtesy of Paul Rackham Ltd. (cla); Courtesy of Roger and Fran Desborough (br). **55 DK:** Courtesy of Paul Rackham Ltd. (bl); Courtesy of Peter Robinson (crb); Courtesy of Roger and Fran Desborough (tl), (tr). **56 DK:** Courtesy of Paul Rackham Ltd. (clb), (bl); **56–57 DK:** Courtesy of Paul Rackham Ltd. (c); **57 DK:** Courtesy of Paul Rackham Ltd. (all). **58 DK:** Courtesy of James Coward (c); Courtesy of Geldof Tractors (bl); Courtesy of Roger and Fran Desborough (tr); Courtesy of S.M. Sheppard (br). **59 DK:** Courtesy of Happy Old Iron, Marc Geerkens (clb); Courtesy of Happy Old Iron, Marc Geerkens (cra); Courtesy of Geldof Tractors (br); Courtesy of Paul Rackham Ltd. (tr), (cla). **61 DK:** Courtesy of Geldof Tractors (tl), (tr); Courtesy of Piet Verschelde (ftr). **64 DK:** Courtesy of R. Parcell (c); Courtesy of Keystone Tractor Works (bl); Courtesy of Paul Rackham Ltd. (tr). **64–65 DK:** Courtesy of Keystone Tractor Works (c). **65 DK:** Courtesy of Robin Simons (tc); Courtesy of Peter Goddard (bl); Courtesy of Keystone Tractor Works (cla), (cra), (br). **66 DK:** Courtesy of Paul Rackham Ltd. (bl, both). **66–67 DK:** Courtesy of Paul Rackham Ltd. (c). **67 DK:** Courtesy of Paul Rackham Ltd. (all). **68 DK:** Courtesy of Keystone Tractor Works (cb); Courtesy of Paul Rackham Ltd. (bl); Courtesy of Stuart Gibbard (ca). **68–69 DK:** Courtesy of Paul Rackham Ltd. (cb). **69 DK:** Courtesy of Keystone Tractor Works (br), (cb); Courtesy of Paul Rackham Ltd. (tc), (cla); Courtesy of Peter Goddard (cra). **70 DK:** Courtesy of The Ward Collection (clb), (bl). **70–71 DK:** Courtesy of The Ward Collection (c). **71 DK:** Courtesy of The Ward Collection (all). **72 DK:** Courtesy of Happy Old Iron, Marc Geerkens (tc); Courtesy of Paul Rackham Ltd. (cra); Courtesy of Roger and Fran Desborough (cb); Courtesy of The Ward Collection (tl). **73 DK:** Courtesy of The Ward Collection (tr); Courtesy of Paul Rackham Ltd. (crb); Courtesy of Paul Rackham Ltd. (cra). **76 DK:** Courtesy of Geldof Tractors (cra); Courtesy of Piet Verschelde (tc); Courtesy of The Shuttleworth Collection (cla). **76–77 DK:** Courtesy of Roger and Fran Desborough (cb). **77 DK:** Courtesy of Geldof Tractors (cla), (tr), (tl); Courtesy of Paul Rackham Ltd. (crb); Courtesy of Piet Verschelde (bc). **79 DK:** Courtesy of Paul Rackham Ltd. (tr); Courtesy of Roger and Fran Desborough (tl). **80–81 DK:** Courtesy of Paul Rackham Ltd. (C) **84 DK:** Courtesy of Henry Flashman (tr); Courtesy of Paul Rackham Ltd. (clb) (cla); (b). **84–85 DK:** Courtesy of D. Disdel (bc). **85 DK:** Courtesy of Andrew Farnham (cla); Courtesy of Keystone Tractor Works (cr); Courtesy of Paul Rackham Ltd. (tl), (tr). **86 DK:** Courtesy of Henry Flashman (clb, both); **86–87 DK:** Courtesy of Henry Flashman (c). **87 DK:** Courtesy of Henry Flashman (all). **88 DK:** Courtesy of Paul Rackham Ltd. (cla), (cl). **88–89 DK:** Courtesy of Paul Rackham Ltd. (c); **89 DK:** Courtesy of James Coward (br), (crb); **89 DK:** Courtesy of Ken Barber (cra). **90 DK:** / Courtesy of Keystone Tractor Works (cla), (crb), (bl); **90–91 DK:** Courtesy of Paul Rackham Ltd. (tc), (bc). **91 DK:** Courtesy of Keystone Tractor Works (tc), (cra), (cb), (br). **94 DK:** Courtesy of Paul Rackham Ltd. (cr), (bc). **94–95 DK:** Courtesy of Paul Rackham Ltd. (c). **95 DK:** Courtesy of Mick Osborne (cb); Courtesy of Paul Rackham Ltd. (cl), (bc), (crb). **96 DK:** Courtesy of Paul Rackham Ltd. (bl, both). **96–97 DK:** Courtesy of Paul Rackham Ltd. (c). **97 DK:** Courtesy of Paul Rackham Ltd. (all). **98 DK:** Courtesy of Geldof Tractors (cla); Courtesy of Piet Verschelde (clb), (br) (c). **99 DK:** Courtesy of Happy Old Iron, Marc Geerkens (bl), (cb), (cr); Courtesy of Piet Verschelde (tr). **100 DK:** Courtesy of Happy Old Iron, Marc Geerkens (tr), (ca), (br); Courtesy of Piet Verschelde (clb). **100–101 DK:** Courtesy of Happy Old Iron, Marc Geerkens (c); **101 DK:** Courtesy of Geldof Tractors (br), (tr); Courtesy of Piet Verschelde (bc); Courtesy of Stuart Gibbard (cla). **103 DK:** Courtesy of Peter Rudling (tr); Courtesy of Paul Rackham Ltd. (tl), (ftl). **104 DK:** Courtesy of Dave Buckle (crb); Courtesy of John Hipperson (bl); Courtesy of Paul Rackham Ltd. (cla), (BR); Courtesy of Roger and Fran Desborough (tr). **105 DK:** Courtesy of Matthew Waters (crb); Courtesy of Paul Rackham Ltd. (ca); Courtesy of Paul Rackham Ltd. (bl). **107 DK:** Courtesy of Jim Brown (ftr); Courtesy of Paul Rackham Ltd. (tr), (ftl); Courtesy of Peter Robinson (tl). **108 DK:** Courtesy of Keystone Tractor Works (bc); Courtesy of Paul Rackham Ltd. (clb). **108–109 DK:** Courtesy of Richard and Valerie Mason (c). **109 DK:** Courtesy of Tony Jones (tr); Courtesy of A. Oglesby (tc). **110 DK:** Courtesy of Martin Shemelds (bl); Courtesy of Paul Holmes (cra). **111 DK:** Courtesy of Paul Rackham Ltd. (crb); Courtesy of Richard and Valerie Mason (tl). **114–115 DK:** Courtesy of Paul Rackham Ltd. (c). **118 DK:** Courtesy of Keystone Tractor Works (tr), (crb), (cla), (bl), (br). **119 DK:** Courtesy of Keystone Tractor Works (tr), (bl), (crb). **120 DK:** Courtesy of Peter Rudling (clb); Courtesy of Andrew Farnham (tr); Courtesy of Paul Rackham Ltd. (bc); Courtesy of Roger and Fran Desborough (cla). **120–121 DK:** Courtesy of Peter Goddard (c). **121 DK:** Courtesy of David Mason (ca); Courtesy of Matthew Waters (crb), (tr); Courtesy of John Simpson (bl); Courtesy of Paul Rackham Ltd. (br), (tl). **122 DK:** Courtesy of Andrew Websdale (bl, both). **122–123 DK:** Courtesy of Andrew Websdale (cl). **123 DK:** Courtesy of Andrew Websdale (all). **124 DK:** Courtesy of Geldof Tractors (bl); Courtesy of Paul Rackham Ltd. (tr), (cla); Courtesy of Piet Verschelde (br). **125 DK:** Courtesy of Andrew Websdale (br); Courtesy of Happy Old Iron, Marc Geerkens (crb), (tl); Courtesy of Piet Verschelde (tr). **126 DK:** Courtesy of Geldof Tractors (cla); Courtesy of Paul Rackham Ltd. (tr); Courtesy of Piet Verschelde (cb). **126–127 DK:** Courtesy of Piet Verschelde (bc). **127 DK:** Courtesy of Geldof Tractors (br); Courtesy of Piet Verschelde (tr), (cr). **128 DK:** Courtesy of Paul Rackham Ltd. (bl, both). **128–129 DK:** Courtesy of Paul Rackham Ltd. (c). **129 DK:** Courtesy of Paul Rackham Ltd. (all). **132 DK:** Courtesy of James Coward (tr); Courtesy of Piet Verschelde (cla), (bl). **133 DK:** Courtesy of Happy Old Iron, Marc Geerkens (bc); Courtesy of Henry Flashman (br); Courtesy of Paul Holmes (tr); Courtesy of Paul Rackham Ltd. (cl); Courtesy of Piet Verschelde (cla). **134 DK:** Courtesy of Keystone Tractor Works (tc); Courtesy of Paul Holmes (c); Courtesy of Paul Rackham Ltd. (bl). **134–135 DK:** Courtesy of Keystone Tractor Works (tc), (cb). **135 DK:** Courtesy of Richard and Valerie Mason (tc). **136 DK:** Courtesy of Happy Old Iron, Marc Geerkens (cr), (br); Courtesy of Keystone Tractor Works (tr); Courtesy of Mrs S. Needle (bl). **137 DK:** Courtesy of Happy Old Iron, Marc Geerkens (bl), (br); Courtesy of Keystone Tractor Works (tr), (c). **139 DK:** Courtesy of Adam Rayner (tl); Courtesy of Paul Rackham Ltd. (ftl); Courtesy of Stuart Gibbard (tr). **140 DK:** Courtesy of Keystone Tractor Works (br); Courtesy of Paul Rackham Ltd. (clb); Courtesy of Stuart Gibbard (tr). **141 DK:** Courtesy of Doubleday Swineshead (bl); Courtesy of Keystone Tractor Works (c), (tl), (tr). **142–143 DK:** Courtesy of Richard and Valerie Mason (cb). **143 DK:** Courtesy of Richard and Valerie Mason (all). **144 DK:** Courtesy of Happy Old Iron, Marc Geerkens (cla); Courtesy of Richard and Valerie Mason (bl). **144–145 DK:** Courtesy of J Hardstaff (c). **145 DK:** Courtesy of John Pickard (br); Courtesy of Richard and Valerie Mason (tl), (cr); Courtesy of Stuart Gibbard (tr). **148 DK:** Courtesy of John Bowen-Jones (c). **152 DK:** Courtesy of Happy Old Iron, Marc Geerkens (cla); Courtesy of John Bowen-Jones (bl). **152–153 DK:** Courtesy of Happy Old Iron, Marc Geerkens (ca). **153 DK:** Courtesy of Happy Old Iron, Marc Geerkens (c), (crb); Courtesy of Henry Flashman (tc); Courtesy of Keystone Tractor Works (bl). **155 DK:** Courtesy of Mick Patrick (ftl); Courtesy of Paul Rackham Ltd. (tl); Courtesy of Happy Old Iron, Marc Geerkens (tr); c. DK (ftr). **156 DK:** Courtesy of B. Murduck (clb); Courtesy of Paul Holmes (tr); Courtesy of R. Oliver (ca); **156 DK:** Courtesy of Richard and Valerie Mason (bl). **156–157 DK:** Courtesy of Peter Rash (c). **157 DK:** Courtesy of Andrew Farnham (tl); Courtesy of Paul Rackham Ltd. (tr). **156–157 DK:** Courtesy of Paul Rackham Ltd. (bc). **158 DK:** Courtesy of D. Sherwin (clb); Courtesy of Paul Rackham Ltd. (tr); Courtesy of Richard and Valerie Mason (cla). **158-159 DK:** Courtesy of D. Sherwin (bc); **159 DK:** Courtesy of Piet Verschelde (tl); Courtesy of Raymond Peter Coupland (br). **162 DK:** Courtesy of Happy Old Iron, Marc Geerkens (tc), Courtesy of Geldof Tractors (clb). **162–163 DK:** Courtesy of A.M. Smith (tc); Courtesy of Happy Old Iron, Marc Geerkens (bc). **163 DK:** Courtesy of S. Oliver (cr). **166 DK:** Courtesy of Richard and Valerie Mason (bl, both). **166–167 DK:** Courtesy of Richard and Valerie Mason (bc). **167 DK:** Courtesy of Richard and Valerie Mason (all). **168 DK:** Courtesy of A. Hardesty (crb); **168 DK:** Courtesy of Richard and Valerie Mason (tr), (cla), (bl). **168–169 DK:** Courtesy of Paul Rackham Ltd. (tc); **169 DK:** Courtesy of Gavin Chapman (tc); Courtesy of Phillip Warren (br). **170 DK:** Courtesy of Peter Tack (ca). **174–175 DK:** Courtesy of Fenland Tractors (cb). **175 DK:** Courtesy of Fenland Tractors (br). **176–177 DK:** Courtesy of Richard and Valerie Mason (c). **180 DK:** Courtesy of Doubleday Holbeach (clb); Courtesy of Stuart Gibbard (cla). **180–181 DK:** Courtesy of Doubleday Holbeach (bc); Courtesy of Bryan Bowles (tc). **181 DK:** Courtesy of Doubleday Holbeach (tr). **183 DK:** Courtesy of Andrew Farnham (ftl); Courtesy of Peter Tack (tl). **186-187 (c) DK (bc). 188 DK:** Courtesy of Doubleday

Holbeach (cla). **189 DK:** Courtesy of Richard and Valerie Mason (tr). **191 DK:** Courtesy of B. Murduck (tl); Courtesy of Bryan Bowles (ftl); Courtesy of Richard and Valerie Mason (tr); Courtesy of Stuart Gibbard (ftr). **192 DK:** Courtesy of Andrew Farnham (cla). **195 DK:** Courtesy of Robert Crawford (cla). **196 DK:** Courtesy of Tim Ingles (cb). **198 (c) DK** (bl, both). **198-199** (c) DK (cb). **199 (c) DK** (all). **201 (c) DK** (cla). **202 DK:** Courtesy of David Wakefield (bl, both). **202-203 DK:** Courtesy of David Wakefield (cb). **203 DK:** Courtesy of David Wakefield (all). **204 (c) DK** (cb). **205 DK:** Courtesy of David Wakefield (tl). **206-207 DK:** Courtesy of Doubleday Swineshead (c). **210 DK:** Courtesy of Chandlers Ltd (tr); Courtesy of Doubleday Holbeach (cla). **210-211 DK:** Courtesy of Rabtrak Ltd. (ca). **212 DK:** Courtesy of Doubleday Swineshead (bl), (cb), (cb), (bl), (clb), (bc). **212-213 DK:** Courtesy of Doubleday Swineshead (cb). **213 DK:** Courtesy of Doubleday Swineshead (all). **214 DK:** Courtesy of Doubleday Holbeach (tc), (cla), (c). **214-215 DK:** Courtesy of Chandlers Ltd (cb). **215 DK:** Courtesy of Chandlers Ltd (tr); Courtesy of Doubleday Holbeach (cla); Courtesy of Doubleday Swineshead (cr). **217 DK:** / Courtesy of Happy Old Iron, Marc Geerkens (tr). **218–219 DK:** Courtesy of Lister Wilder (bc). **220 DK:** / Courtesy of Chandlers Ltd. (clb, both). **220-221 DK:** Courtesy of Chandlers Ltd (cb). **221 DK:** Courtesy of Chandlers Ltd. (all). **222 DK:** Courtesy of Chandlers Ltd. (bc). **223 DK:** Courtesy of Doubleday Swineshead (bl). **230 DK:** Courtesy of Rabtrak Ltd. (cra), (bc). **231 DK:** Courtesy of Chandlers Ltd (tl); Courtesy of Rabtrak Ltd. (tr). **233 DK:** Courtesy of Doubleday Holbeach (tr). **235 DK:** Courtesy of Doubleday Holbeach (ca). **237 DK:** Courtesy of Chandlers Ltd. (tr), (ftr). **240–241 DK:** Courtesy of Doubleday Swineshead (cb). **243 DK:** Courtesy of Peter Goddard (bc); Courtesy of Mick Patrick (ca); Courtesy of Paul Rackham Ltd. (cra), (br). **244 DK:** Courtesy of Doubleday Swineshead (cb), (c), (bc); Courtesy of Paul Rackham Ltd. (clb). **244–245 DK:** Courtesy of Doubleday Swineshead (c). **245 DK:** Courtesy of Doubleday Swineshead (ca), (bc), (c); Courtesy of Keystone Tractor Works (cla); Courtesy of Paul Rackham Ltd. (cra); Courtesy of Peter Goddard (tlc); Courtesy of Roger and Fran Desborough (trc). **246 DK:** Courtesy of D. Sherwin (bc); Courtesy of Doubleday Swineshead (br). **247 DK:** Courtesy of Doubleday Swineshead (br).

The publisher would like to thank the following for their kind permission to reproduce theor photographs:

Key: a-above; b-below/bottom; c-centre; f-far; l-left; r-right; t-top)
SGC: Stuart Gibbard collection

12 SGC: (c). **13 SGC**: (ca, br). **14 David Parfitt**: (bl). **16 SGC**: (tl, bl, cl, cr). **17 David Parfitt**: (ftr). **SGC**: (ftl, tl, tr). **18 David Parfitt**: (br). **SGC**: (cla). **19 Gunnar Österlund**: (clb). **David Peters**: (crb). **20 SGC**: (tl). **24–25 SGC**: (c). **26 SGC**: (cla, tc, bc). **27 SGC**: (cb, bc). **28 SGC**: (tr). **29 SGC**: (tl). **30 SGC**: (tr). **31 Brian Knight**: (tr). **32–33 The Library of Congress, Washington DC**: LC-H261- 24822 [P&P] (c). **35 SGC**: (clb). **36 Roy Larkin**: (tl). **38–39 SGC**: (c). **41 SGC**: (crb). **42 David Parfitt**: (cla). **SGC**: (bl, bc, tc). **43 David Parfitt**: (cr). **SGC**: (cl). **44 John Deere**: John Deere Art Collection (bl). **John Deere**: (tl, cl). **45 John Deere**: (cr, bc). **SGC**: (cl). **48 SGC**: (c). **49 SGC**: (c, br). **50 David Parfitt**: (cla). **52 Case IH**: Wisconsin Historical Society (cl). **SGC**: (tl, bl, cr). **53 FPP**: (ftr). **SGC**: (cr). **The Library of Congress, Washington DC**: 1a35288u (bc). **54 SGC**: (tc, cr). **55 David Peters**: (cl). **56 SGC**: (tl). **60 John Deere**. **SGC**: (tl, bc). **61 John Deere**: (br). **SGC**: (cl, cr). **62-63 Museum of English Rural Life**: (c). **66 SGC**: (tl). **70 Museum of English Rural Life**: (tl). **73 Cheffins**: (clb). **Brian Knight**: (tr). **David Parfitt**: (bc). **74-75 SGC**: (c). **78 SGC**: (bc, cr). **Wikipedia**: (tl). **79 FPP**: Claas (ftr). **David Parfitt**: (ftl). **SGC**: (bl, c, br). **82 Bridgeman Images**: © The Estate of Terence Cuneo (c). **83 SGC**: (ca, br). **85 Corbis**: Bettmann (br). **86 Saskatchewan Western Development Museum**: Clark Collection, WDM-1981-S-197 (tl). **88 David Parfitt**: (tc, bl). **89 SGC**: (tl). **92–93 SGC**: (c). **94 David Parfitt**: (tr, clb). **SGC**: (cla). **96 SGC**: (tl). **99 SGC**: (br). **101 David Peters**: (cr). **102 Getty Images**: J. A. Hampton / Stringer (tl). **SGC**: (bl, cr). **103 SGC**: (ftr, cr, bc). **106 Corbis**: Bettmann (tl). **SGC**: (bc). **The Library of Congress, Washington DC**: cph.3c11278 (cla). **107 SGC**: (c, cr, fcr, br). **108 David Peters**: (bc). **SGC**: (cla). **109 David Parfitt**: (cb). **SGC**: (cr, br). **110 SGC**: (cla). **111 SGC**: (tr, cr). **112–113 SGC**: (c). **116 SGC**: (c). **117 SGC**: (ca, br, bl). **122 SGC**: (tl). **126 David Peters**: (bl). **127 Brian Knight**: (tl). **128 John Deere**: (tl). **130–131 Corbis**: Bettmann (c). **133 David Peters**: (cr). **134 Mecum Auctions**: (cla). **135 SGC**: (cr). **136 Brian Knight**: (cla). **138 SGC**: (tl, cl, cr). **138–139 SGC**: (bc). **139 SGC**: (cr). **Robert Sykes**: (ftr). **140 SGC**: (cl). **142 SGC**: (tl). **144 David Peters**: (tc, clb). **146–147 SGC**: (c). **150 SGC**: (c). **151 SGC**: (br, ca). **152 John Deere**: (crb). **154 akg-images**: Universal Images Group / Unversal History Archive (bl). **Case IH**: Wisconsin Historical Society (tl, cl). **154–155 Case IH**: Wisconsin Historical Society (bc). **155 Case IH**: (br). **157 SGC**: (br, fbr/MF 175). **159 Case IH**: Wisconsin Historical Society (tr). **David Peters**: (ca). **SGC**: (cr). **160 Archivio e Centro Storico Fiat**: (tl). **SGC**: (cl, bl, cr). **161 SGC**: (bl, crb, tl/Fiat 505C, ftl, tr/Fiat 680, ftr). **162 David Peters**: (cla, bc). **163 Brian Knight**: (tr, br). **164 FPP**: Belarus (c); Kubota (cla); Valmet (cra). **SGC**: (bc). **164–165 David Peters**: (bc). **165 SGC**: (tl, tr, cr, br). **166 SGC**: (tl). **168 SGC**: (cl). **169 FPP**: Simon Henley (tr). **170 David Peters**: (cl, bl, tr, crb). **171 David Peters**: (tc, clb). **SGC**: (br). **172–173 David Peters**: (c). **174 SGC**: (cla, tr). **175 David Parfitt**: (cr). **SGC**: (tl, tr, clb). **178 Courtesy of JCB**: (c). **179 Alamy Images**: Dinodia Photos (br). **SGC**: (c). **180 John Deere**: (bc). **181 AGCO Ltd**: (cr, br, clb). **182 AGCO Ltd**: (cl). **SGC**: (cr, fcr). **Michael Williams**: (tl, bl). **183 AGCO Ltd**: (tr, ftr). **Michael Williams**: (bc). **184–185 SGC**: (c). **186 FPP**: AGCO (bl); Deutz-Allis (cra). **David Peters**: (cl). **SGC**: (tr). **187 FPP**: (cr); AGCO (br, tr). **SGC**: (tl). **188 FPP**: International (tr). **SGC**: (bl, br, cr). **189 SGC**: (bc, crb); **Robert Sykes** (clb). **190 SGC**: (tl, fbl, bl). **191–192 SGC**: (bc). **191 SGC**: (cr, br). **192 Case IH**: (br). **FPP**: Renault (tr). **SGC**: (clb). **193 FPP**: Deutz-Fahr (cl); Steyr (tr); Zetor (cr); Same (bc); Valmet (br). **194 FPP**: CBT (ca); Valmet (br); John Deere (cla). **Mahindra & Mahindra Ltd**: (bl). **SGC**: (tc). **195 FPP**: Belarus (tr). **Brian Knight**: (br). **SGC**: (bl). **196 FPP**: (cla). **David Peters**: (tr). **197 FPP**: Claas (br). **John Deere**: (tl). **SGC**: (tr, c, bl). **198 Case IH**: (tl). **200 David Peters**: (cla, clb, bc, tr, cr). **201 Mecum Auctions**: (tr). **David Peters**: (tl, bc). **202 Courtesy of JCB**: (tl). **204 FPP**: Steyr (tr). **Brian Knight**: (cra). **David Peters**: (cla). **205 FPP**: Claas (br); Clayton (tr); Ken Topham (cr). **SGC**: (cl). **208 AGCO Ltd**: (c). **209 Alamy Images**: ITAR-TASS Photo Agency (ca). **Case IH**. (br). **210 John Deere**: (bl); **AGCO Ltd**: (crb). **211 FPP**: Kubota (clb); Lindner Traktoren (tr). **New Holland Agriculture**: (br). **Steyr**: (bc). **212 John Deere**: (tl). **214 AgriArgo UK Ltd Distributor for McCormick Tractors in UK**: (clb). **FPP**: Courtesy of JCB (cb). **216 Image supplied courtesy of the Same Deutz-Fahr Group**: (tl, c, bl, cr). **217 FPP**: Deutz-Fahr (ftr). **Image supplied courtesy of the Same Deutz-Fahr Group**: (cl, ftl, tl, br). **218 AGCO Ltd**: (cla). **New Holland Agriculture: (tr). Image supplied courtesy of the Same Deutz-Fahr Group**: (cl). **219 FPP**: ARGO (cra); Claas (c). **Lamberhurst Engineering Ltd**: (tl). **SGC**: (cr, br). **220 AGCO Ltd**: (tl). **222–223 AGCO Ltd**: (c). **222 Case IH**: (clb). **David Peters**: (tr, ca). **223 John Deere**: (crb). **SGC**: (tc). **Versatile**: (br). **224–225 www.thecoldestjourney.org**: (c) **226 AGCO Ltd**: (tr). **Tristan Balint**: (cl). **FPP**: AGCO (cla); Deutz-Fahr (cr); Reform (bc). **226–227 Steyr**: (c). **227 John Deere**: (cr). **Max-Holder**: (tr). **SGC**: (tl). **228–229 Corbis**: Martin Harvey (c). **230 FPP**: AGCO (cb); TYM (tc); ArmaTrac (clb). **Wikipedia**: Natalie Maynor (cla). **231 FPP**: Belarus (bc). **232 New Holland Agriculture**: New Holland Agriculture Photo Library (tl, cl). **SGC**: (cr, bl). **233 Alamy Images**: FLPA (bc). **SGC**: (ftl, tl), **New Holland Agriculture**: (ftr, cra). **234 AGCO Ltd**: (cr). **Case IH**: (bc). **Countrytrac**: (tr). **Courtesy of JCB**: (clb). **SGC**: (cla). **235 AGCO Ltd**: (bl). **Kubota**: (cb). **New Holland Agriculture**: (br). **236 AGCO Ltd**: (tl, cl). **236–237 AGCO Ltd**: (bc). **237 AGCO Ltd**: (cl). **AGCO Ltd**: (br). **FPP**: AGCO (ftl). **SGC**: (tl). **238-239 AWL Images**: Hemis (c). **242 John Deere**: (br). **246 Mecum Auctions**: (bl). **247 John Deere**: (bl, bc)

Images on title, contents, and introduction
page 1 International Junior 8-16
pages 2–3 Lanz Bulldog
page 4 International Junior 8-16
page 5 MM UDLX (bl), Roadless Half-Track (br)
page 6 Fowler Challenger 3 (bl), Northrop 5004T (br)
page 7 Massey Ferguson 50(bl), John Deere 6210R (br)
page 8 MM UDLX(bl), Renault N73 Junior (br),
page 9 Doe Triple-D (bl), John Deere 6210R (br)

Images on chapter opener pages
pages 10-11 1900-1920 Hornsby-Akroyd
pages 46-47 1921-1938 Case Model C
pages 80-81 1939-1951 David Brown VAK1
pages 114-115 1952-1965 Ferguson FE-35
pages 148-149 1965-1980 John Deere 4020
pages 176-177 1981-2000 Ford TW-35
pages 206-207 After 2000 John Deere 6210R
pages 240-241 How Tractors Work
John Deere 6210R